DATE DUE

DEMCO

New Approaches in Biomedical Spectroscopy

ACS SYMPOSIUM SERIES **963**

New Approaches in Biomedical Spectroscopy

Katrin Kneipp, Editor
Harvard University, Medical School

Ricardo Aroca, Editor
University of Windsor

Harald Kneipp, Editor
Harvard University, Medical School

Edeline Wentrup-Byrne, Editor
Queensland University of Technology

Sponsored by the
ACS Division of Analytical Chemistry

American Chemical Society, Washington, DC

Library of Congress Cataloging-in-Publication Data

New approaches in biomedical spectroscopy / Katrin Kneipp, editor, Ricardo
  Aroca, editor, Harald Kneipp, editor, Ediline Wentrup-Byrne, editor ;
  sponsored by the ACS Division of Analytical Chemistry.

    p. cm.—(ACS symposium series ; 963)

    "Resut of a symposium entitled New Approaches in Biomedical
Spectroscopy as part of the Pacifichem 2005 Conference held in December
2005 in Honolulu, Hawaii"—Pref.

    Includes bibliographical references and index.

    ISBN 13: 978–0–8412–7437–2 (alk. paper)

    1. Biomolecules—Spectra—Congresses. 2. Spectrum analysis—
Congresses.

    I. Kneipp, Katrin. II. Aroca, Ricardo. III. Kneipp, Harald. IV. Wentrup-Byrne,
Ediline. V. American Chemical Society. Division of Analytical Chemistry. VI.
Pacifichem 2005 (2005 : Honolulu, Hawaii)

QP519.9.S6N49   2007
572'.33—dc22

                              2006052662

PRINTED IN THE UNITED STATES OF AMERICA

Foreword

The ACS Symposium Series was first published in 1974 to provide a mechanism for publishing symposia quickly in book form. The purpose of the series is to publish timely, comprehensive books developed from ACS sponsored symposia based on current scientific research. Occasionally, books are developed from symposia sponsored by other organizations when the topic is of keen interest to the chemistry audience.

Before agreeing to publish a book, the proposed table of contents is reviewed for appropriate and comprehensive coverage and for interest to the audience. Some papers may be excluded to better focus the book; others may be added to provide comprehensiveness. When appropriate, overview or introductory chapters are added. Drafts of chapters are peer-reviewed prior to final acceptance or rejection, and manuscripts are prepared in camera-ready format.

As a rule, only original research papers and original review papers are included in the volumes. Verbatim reproductions of previously published papers are not accepted.

ACS Books Department

Contents

Analysis and Environment

Indexes

Preface

This book is the result of a symposium entitled "New Approaches in Biomedical Spectroscopy" as part of the Pacifichem 2005 Conference held in December of that year in Honolulu, Hawaii. Most of the chapters are based on talks and posters presented at this symposium, but we are delighted to also include selected contributions from colleagues who were unable to attend the meeting. We cordially thank all of our colleagues for the time and effort they invested in their articles. In particular, we appreciate the generous support of the United States Air Force Office of Scientific Research, Medical Free Electron Laser Research program, grant number FA9550-05-1-0259, for this symposium.

The development and innovative application of efficient and sensitive spectroscopic and optical techniques for studying biomedically relevant molecules and materials, structures and processes both in vitro and in vivo is a field of rapidly growing interest.

Current approaches in biomedical spectroscopy profit from the detection of both linear and non-linear spectroscopic signals. New light sources such as free electron lasers and synchrotrons open up exciting new opportunities for both the one- and two-photon absorption probing of electronic and vibrational states in biological matter using specific excitation energy regions such as the terahertz frequency range. Elastic light scattering techniques can detect morphological changes in sub-cellular units, and inelastic Raman scattering provides information on chemical structures and structural changes of biomedically relevant molecules. Interferometric measurements allow the monitoring of the mechanical motions of biological structures on the nanometer scale. Correlation methods applied to spectroscopic signals can provide valuable information on dynamical processes and the chemical kinetics of single species. Recent developments exploit the synergy of modern spectroscopy and nanotechnology. Nanoparticles can be used as multifunctional units providing sensors for spectroscopic measurements along with chemical probes and vehicles for drug delivery thus illustrating the potential for combined therapeutic and diagnostic use. Moreover, spectroscopic measurements performed in enhanced local optical fields of metal nanostructures can increase the sensitivity of the measurement by orders of magnitude. Because of the strong confinement of local optical fields, these experiments permit the collection of spectroscopic data from nanometer-probed volumes and allow lateral resolution two orders of magnitude below the diffraction limit.

The opening section of the book deals with infrared and Raman spectroscopy. These complimentary vibrational spectroscopic techniques are capable of monitoring molecular structures as well as structural changes. Such studies are of interest for understanding diseases at a molecular level as well as for developing techniques for efficient early diagnosis based on molecular structural information. The chapters in this section also demonstrate applications in proteomics and the characterization of micro organisms.

The first chapter is of general interest introducing new development in Raman spectroscopy as a powerful tool in Life Science Research. Michael Blades, Georg Schulze, Stanislav Konorov, Christopher Addison, Andrew Jirasek, and Robin Turner discuss new developments based on Fiber-Optic-Linked Raman and Resonance Raman Spectroscopy.

In the second chapter, Don McNaughton and Bayden R. Wood apply Fourier transform infrared spectroscopy imaging techniques to cancer research and the study of malignant glioma rat brain sections and melanoma sections.

Tanja Maria Greve, Niels Rastrup Andersen, Kristina Birklund Andersen, Monika Gniadecka, Hans Christian Wulf , and Ole Faurskov Nielsen discuss near infrared Raman studies for the elucidation of water structure in human and animal skin.

The detection of secondary conformation in prion protein by Fourier-transformed infrared microscopy is discussed by Norio Miyoshi, Hiroyuki Okada, Masuhiro Takata, Moriichi Shinagawa, and Kenichi Akao.

Corasi Ortiz, Yong Xie, Dongmao Zhang, and Dor Ben-Amotz describe a new technique "Drop Coating Deposition Raman Spectroscopy" and its application to proteomics.

Elizabeth Anne Carter, Craig Patrick Marshall, Mohamed Ali, Ranjini Ganendren, Tania Sorrell, Lesley Wright, Yao-Chang Lee,

Ching-Iue Chen, and Peter Andrew Lay have used infrared spectroscopy to characterize, identify and differentiate micro–organisms.

Philip Heraud, Bayden R. Wood, John Beardall, and Don McNaughton study the influence of the environment on microalgae using infrared and Raman spectroscopy.

In general, Raman scattering suffers from extremely small cross sections, however, the effect literally appears in a new light, when surface enhanced Raman scattering (SERS) takes place in the local optical fields of metal nanostructures. The effect combines high structural information content and selectivity with extremely high detection sensitivity down to the single molecule level and the potential to probe nanometer scaled volumes. The second section of the book includes chapters that demonstrate the application of SERS in the biomedical field.

In the first chapter in this section, Nilam Shah, Olga Lyandres, Chanda Yonzon, Xiaoyu Zhang, and Richard Van Duyne outline the use of surface-enhanced Raman spectroscopy in the development of biological sensors for the sensitive detection of anthrax and glucose.

Karen Dehring, Gurjit Mandair, Blake Roessler, and Michael Morris describe a novel application of surface-enhanced Raman spectroscopy for in-vitro detection of Hyaluronic Acid, a potential biomarker for osteoarthritis.

The functionalization of metal nanoparticles for surface enhanced Raman and infrared spectroscopies with adequate host molecules, that allow the selective detection of some hazardous pollutants (Polyaromatic Hydrocarbons), is discussed by Concepcion Domingo, Luca Guerrini, Patricio Leyton, Marcelo Campos-Vallette, Jose Vicente Garcia-Ramos, and Santiago Sanchez-Cortes.

Paul Goulet, Nicholas Pieczonka, and Ricardo Aroca report the fabrication and characterization of layer-by-layer (LbL) films composed of the glycoprotein avidin and metallic Ag nanoparticles. These structures show strong bio-specific interactions and are employed as selective substrates for surface-enhanced Raman and resonance Raman scattering.

Measurements of SERS spectra from bacteria and the use of SERS vibrational fingerprints for the detection and identification of vegetative bacterial cells in human blood is described by Ranjith Premasiri, Donald Moir, Mark Klempner, and Lawrence Ziegler.

SERS-based optical labels and nanosensors that deliver sensitive and spatially-localised chemical information from live cells are described by Janina Kneipp, Harald Kneipp, and Katrin Kneipp.

The next chapter develops the concept of multifunctional nanosensors. Yong-Eun Lee Koo, Rodney Agayan, Martin Philbert, Alnawaz Rehemtulla, Brian Ross, and Raoul Kopelman describe new photonic explorers for biomedical applications using biologically localized embedding (PEBBLEs). This approach demonstrates significant promise for their use as a combined therapeutic and diagnostic tool for cancer.

The measurement of intrinsic optical signals from biological objects and sophisticated optical techniques that can increase signal to background ratios as well as sensitivity in such experiments are discussed in the next section of the book.

Christopher Fang-Yen and Michael S. Feld give an overview on intrinsic optical signals in nerve tissue that can be used for functional neural imaging and discuss various spectroscopic techniques for their measurement.

Coherent anti-Stokes Raman (CARS) scattering spectroscopy probes vibrational levels based on a coherent non-linear wave mixing process and is a powerful technique for the vibrational imaging of biological structures. Daniel L. Marks and Stephen A. Boppart discuss new pulse-shaping and interferometry schemes in CARS that improve its spectroscopic capability for biological applications.

Marcel Leutenegger, Kai Hassler, Rudolf Rigler, Alberto Bilenca, and Theo Lasser describe dual-color fluorescence fluctuation spectroscopy based on an evanescent field excitation scheme as an alternative concept for single molecule detection at surfaces.

The next two chapters illustrate how photons of very different energies, in the Terahertz and in the ultra violet range, can be used to retrieve molecular structural information from native biomolecules.

Karen Siegrist, Christine Bucher, Candace Pfefferkorn, Andrew Schwarzkopf, and David Plusquellic apply high resolution THz spectroscopy to the investigation of the lowest frequency vibrational modes of crystalline peptides associated with the hydrogen-bonding networks of peptide crystals.

Yayoi Aki-Jin, Yukifumi Nagai, Kiyohiro Imai, and Masako Nagai study changes of the near-UV circular dichroism spectra of human hemoglobin upon the R→T quaternary structure transition.

The electrical properties of protein molecules adsorbed onto a gold substrate are studied by Hong Huo, Larisa-Emilia Cheran, and Michael Thompson using a scanning Kelvin nanoprobe in a microarray format.

The final chapters in the book demonstrate the powerful combination of different spectroscopic techniques for the characterization of biomolecules as well as native and engineered biomaterials.

Maria Paula Marques, Fernanda Borges, António Marinho Amorim da Costa, and Luís Alberto Esteves Batista de Carvalho combine information from Raman and Inelastic Neutron Scattering, coupled to quantum mechanical calculations for a thorough analysis of several biologically-relevant molecules from anticancer agents to drugs of abuse.

Henrik Bohr and Salim Abdali describe a set of novel instruments designed and constructed for experiments to study the folding and denaturation of proteins in conjunction with various spectroscopic methods.

Two analytical techniques: optical absorbance of Safranin-O-stained cartilage sections and energy dispersive X-ray analysis have been used by Joshua Bowden, Lew Rintoul, Thor Bostrom, James Pope, and Edeline Wentrup-Byrne to construct partial least squares models from Fourier transform infra red spectral data which can then be used to predict the constituents in native, degraded or even engineered cartilage.

Kylie M. Varcoe, Idriss Blakey, Traian V. Chirila, Anita J. Hill, and Andrew K. Whittaker use three distinct probes: positron annihilation lifetime spectroscopy (PALS), ^{1}H NMR, and ^{129}Xe NMR to study the porous structure of PHEMA hydrogels and the effect of synthetic conditions on the free volume of PHEMA.

For the development of an effective therapeutic strategy against infectious diseases, the corresponding bacterial virulence factors must be identified and suitable methods for their neutralization must be developed. Klaus Brandenburg, Jörg Howe, Manfred Rössle, and Jörg Andrä present a combination of spectroscopic methods (infrared, X-ray diffraction and fluorescence resonance energy transfer) for the analysis of the inactivation mechanism of bacterial factors.

The Pacifichem symposium "New Approaches in Biomedical Spectroscopy" was a wonderful opportunity for a diverse group from the international community of spectroscopists, with interests in biomedical applications, to come together. The basic philosophy of the symposium was to bring together both fundamental and practical information on new developments in biomedical spectroscopy. This volume is testament to the vast array of applications possible using modern spectroscopic techniques and combinations thereof. The editors are united in their appreciation to all the authors whose work made this book possible. We hope that this symposium volume will prove interesting to our colleagues in the field and will stimulate further developments and exciting new experiments. In particular, we hope this book will also encourage and inspire current students and other young researchers to enter the field and push the boundaries of future research even further.

Katrin Kneipp
Ricardo Aroca
Harald Kneipp
Edeline Wentrup-Byrne

Boston, Massachusetts; Windsor, Canada; and Brisbane, Australia

New Approaches in Biomedical Spectroscopy

Chapter 1

New Tools for Life Science Research Based on Fiber-Optic-Linked Raman and Resonance Raman Spectroscopy

M. W. Blades[1], H. G. Schulze[2], S. O. Konorov[1,2], C. J. Addison[1], A. I. Jirasek[3], and R. F. B. Turner[1,2]

[1]Department of Chemistry, The University of British Columbia, 6174 University Boulevard, Vancouver, British Columbia V6T 1Z3, Canada
[2]Michael Smith Laboratories, The University of British Columbia, 2185 East Mall, Vancouver, British Columbia V6T 1Z4, Canada
[3]Department of Physics and Astronomy, University of Victoria, Victoria, British Columbia V8W 3P6, Canada

Fiber-optic probes can exploit a favorable excitation radiation distribution within the sample that allows the use of higher laser power levels which, in turn, can yield a higher signal-to-noise ratio (SNR) for a given experiment without increasing the risk of analyte photo-damage. We have developed specialized fiber-optic probes for ultraviolet resonance Raman spectroscopy (UVRRS) that offer several advantages over conventional excitation/collection methods used for UVRRS. These probes are ideally suited for UVRRS studies involving biopolymers and small bio-molecules, in both native (e.g. physiological) and non-native (e.g. anoxic) solution environments. We have also developed novel probes based on hollow-core photonic band-gap fibers that virtually eliminate the generation of silica Raman scattering within the excitation fiber which often limits the utility of fiber-optic Raman probes in turbid media or near surfaces. These probes may offer advantages for some biomedical applications.

Many spectroscopic methods based on the Raman effect, the inelastic scattering of light by molecules, have been developed and employed in various biophysical and bioanalytical applications (*1-4*). The value inherent in all of these methods lies in the rich information content of Raman spectra due to the dependences of molecular vibrational modes on the chemical and physical microenvironment of the analytes. However, exploiting this information content can be quite challenging due to fundamental limitations in the sensitivity of the technique and/or practical limitations in the available experimental systems. Much effort has therefore focused on the development of instrumental configurations that endeavor to optimize the utility of the technique for a given application or class of applications.

Fiber-optic probes offer many obvious advantages in terms of experimental flexibility and remote measurement capability, and numerous probe designs have been reported (*5,6*). In biomedical spectroscopy applications, fiber-optic probes also offer some valuable performance advantages. For example, since the excitation light is not tightly focused, as it normally is in conventional Raman spectroscopy, substantially higher excitation power fluxes can be used, which results in higher levels of Raman signal generated within the illuminated volume. This power-distribution advantage (*7*) is particularly useful for ultraviolet resonance Raman spectroscopy (UVRRS) of biomolecules where sample damage thresholds can be quite low (<50 W/cm^2) and good signal collection efficiency is critical.

Fiber-optic probes for UVRRS in aqueous environments

One of the challenges in UVRRS is to overcome the signal-degrading effects of sample self-absorption due to the fact that signals inherently fall within a strong electronic absorption band of the analyte. In conventional (non-fiber-optic based) UVRRS experiments, a variety of schemes (e.g., Figure 1) have been employed to deal with sample self-absorption by minimizing the optical path within the sample. Such measures, though, often impose unfavorable experimental design constraints and/or require greater quantities of analyte. The use of fiber-optic probes might seem to obviate the need to take any special steps to accomplish this, owing to the obvious close proximity of the excitation and collection optics. However, the geometry of most conventional fiber-optic probe configurations (e.g., Figure 2a) does not allow for sufficient overlap of the excitation and collection volumes near enough to the fibers to be effective in UVRRS.

We have developed a fiber-optic probe configuration that overcomes the effects of sample self-absorption using a unique geometry (as illustrated in Figure 2b) where the collection volume includes the entire excited sample volume near the tip of the excitation fiber (*8*). Here, the collection efficiency is optimized for resonance Raman experiments where virtually all the signal

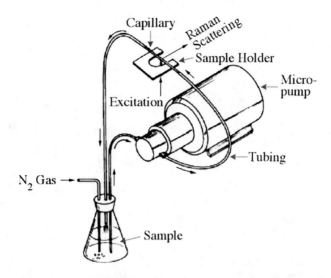

Figure 1. Example of an exposure/collection system used in non-fiber-optic based UVRRS. Here, the laser excitation is focused within the lumen of a quartz capillary, minimizing the optical path of the Raman signal within the analyte. (Adapted with permission from: G. Balakrishnan, Structure and Vibrational Spectra of Photogenerated intermediates of Quinones: A Resonance Raman Study, PhD Thesis, 1997.)

derives from a small region immediately distal to the excitation fiber tip because the excitation light is almost completely absorbed within about one probe diameter from the tip. In fact, the working curves (signal intensity *vs.* analyte concentration) are profoundly nonlinear and, for any given analyte and probe dimensions, there exists an optimal concentration above which the collected signal rapidly decreases due to absorption even over a short path between the scattering center and the collection probe. These characteristics are thoroughly modeled and explained in (*8*) along with materials and methods for fabricating such probes.

This configuration can also be used for off-resonance applications, although the collection efficiency is significantly lower than that achieved using a more conventional configuration. This comparison is illustrated in Figure 3 using ethanol and 2'-deoxyguanosine as model analytes for off-resonance and resonance conditions with 257.2 nm excitation (using a continuous-wave, frequency-doubled, argon-ion laser). Note that the probe depicted in Figure 2a is the better probe for off-resonance, but gives no measurable signal for an analyte under resonance enhancement conditions, while the probe depicted in Figure 2b gives excellent SNR spectra with a relatively short integration time (100 s). All spectra shown in Figure 2 were collected using single-collection-fiber probes.

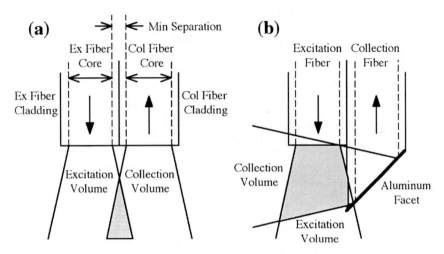

Figure 2. Alternative fiber-optic probe configurations for Raman and Resonance Raman spectroscopy. (a) represents a "conventional" geometry where the optical interfaces for both excitation and collection are the fiber endfaces; variations on this geometry where the endfaces are beveled with respect to the fiber axis can achieve improved collection efficiency, particularly for off-resonance Raman. (b) represents a more optimal geometry for resonance Raman where the optical interface for signal collection is the side of the collection fiber, through the cladding.

Multiple-collection-fiber probes based on this design have been used to help elucidate the details of the protonation state of the bound substrate in certain intradiol and extradiol dioxygenase enzymes, as well as some insights into the roles of some active site residues of the protein (9-11).

Dioxygenase enzymes are non-heme-iron containing ring-cleavage enzymes that are involved in most of the studied microbial pathways for biodegradation of aromatic compounds in the environment. These enzymes act on catecholic intermediates generated within such pathways, incorporating dioxygen into the product. In the absence of molecular oxygen, the enzymes bind, but do not cleave or release their substrates. Hence, spectroscopic studies on the stable enzyme-substrate complex are possible provided stringent anaerobic conditions are maintained, which requires a closed, tightly controlled environment. Additional constraints are imposed due the very limited availability of purified enzyme, which necessitates working with very small-volume samples (<150 µL), and photosensitivity to UV light, which necessitates short (~10 s), low-power (<4 W/cm^2) exposures. The fiber-optic probes described here are ideally suited to meet these demanding requirements.

The probe and instrumentation configurations used in this work are shown in Figure 4; the experimental details, data and analyses have been reported in (9). Here, we only wish to briefly illustrate the utility of these particular probes

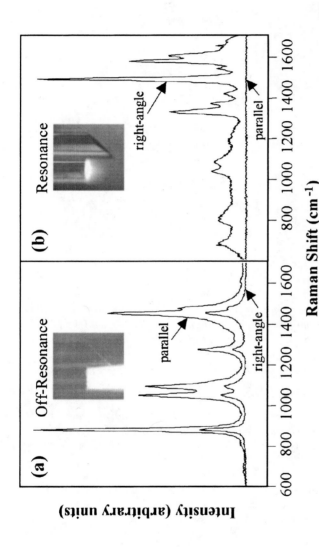

Figure 3. Comparison of parallel and right-angle collection probe geometries in both absorbing and non-absorbing media. (a) represents the non-resonance, non-absorbing case with neat ethanol. (b) spectra of 0.5 mM 2′-deoxyguanosine (dA) which is an example of an absorbing analyte under resonance conditions. The ethanol spectra were recorded with 5 s exposures with each probe while the dA spectra were recorded with 100 s exposures. All spectra were obtained with the indicated probes under otherwise identical conditions using 130 mW of 257.2 nm laser excitation. The inset photos show the penetration into the sample of laser light emerging from the excitation fiber in each case (same for both probe types).

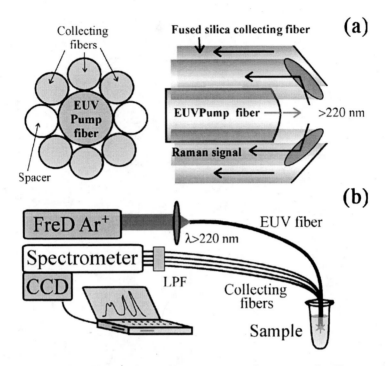

Figure 4. Multi-fiber probe and instrument configurations for fiber-optic-linked UVRRS. (a) the probe arrangement comprising six standard fused silica collection fibers (400 μm core) surrounding a single "Enhanced UV" (EUV) excitation fiber (600 μm core, FVP series, Polymicro Technologies, Phoenix, AZ); two foreshortened, unused 400 μm fibers are used as spacers to facilitate sample exchange within the excitation volume. (b) the instrumentation setup comprising an intracavity frequency-doubled argon-ion laser (FreD 90C, Coherent Inc., Santa Clara, CA) which provides several far-UV wavelengths, a long-pass filter to attenuate the laser line, a 1-m single monochrometer (model 2061, McPherson Inc, Chelmsford, MA) and a LN_2-cooled CCD detector (Princeton Instruments Spec-10 400B, Roper Scientific, Trenton, NJ).

in such research using the spectra shown in Figure 5, which is reproduced from (9) and provides a good basis for summarizing the results.

The strongest feature in the difference spectrum of Figure 5c at 1603 cm^{-1} coincides with a feature in the spectrum of monoanionic DHB in H_2O (data not shown) and is 11 cm^{-1} below that of the corresponding feature in neutral DHB. The distinctive features of dianionic DHB in the 1600 cm^{-1} region (data not shown) are not observed in the spectrum of bound DHB. Despite the apparent insensitivity of the 1603 cm^{-1} band to H/D exchange (which is discussed in 9), H/D exchange in the bound substrate is evident in several features of the difference spectrum in D_2O (Figure 5d) such as the upshift of the 1306 cm^{-1} band

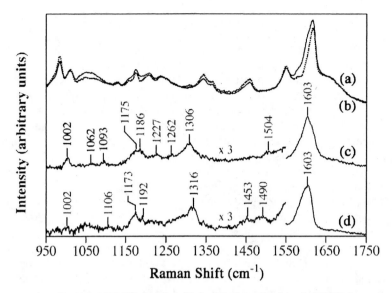

Figure 5. The FO-UVRR spectra of 2,3-dihydroxybiphenyl (DHB) bound to 2,3-dihydroxybiphenyl 1,2-dioxygenase (DHBD); the 982 cm⁻¹ peak from added Na₂SO₄ was used as an internal standard. (a) the spectrum of DHBD with DHB (where the enzyme is in molar excess, pH 8.0, 100 s). (b) the spectrum (broken line) of pure DHBD (pH 8.0, 70 s) overlaid on the same scale as (a). (c) the difference spectrum revealing bound DHB in protonated solvent (100 s). (d) the difference spectrum in deuterated solvent (pD 8.0, 130 s). The data shown are spectra averaged over a number of 10 s acquisitions with 20 mW of 248.2 nm excitation. The absence of the internal standard SO₄²⁻ signal in the difference spectra indicates proper subtraction of the parent spectra. (Adapted with permission from reference 9.)

to 1316 cm⁻¹; upshift of the 1262 cm⁻¹ band to ~1282 cm⁻¹; splitting of the 1186/1175 cm⁻¹ doublet; upshift of the weak band at 1093 cm⁻¹ to 1106 cm⁻¹; and weakening of the 1002 cm⁻¹ band. These data, together with corroborating UV-vis data, unambiguously identified the bound substrate as being monoanionic and definitively established this feature of the proposed mechanism of extradiol dioxygenases.

The details of the catalytic mechanisms and structure-function relationships of dioxygenase enzymes are of interest for the purpose of engineering variants for new bioremediation applications. There are also some related enzymes that are involved in human catabolism of aromatic amino acids and some serious metabolic disorders such as alkaptonuria are associated with mutations in homogentisate 1,2-dioxygenase (HGO) which cleaves the aromatic ring of homogentisic acid (HGA). We hope to extend our current collaborative work with dioxygenases to *in vitro* studies involving HGO.

Fiber-optic probes for RS in turbid media or near interphases

One of the disadvantages of using fiber-optic probes for RS is that the intense laser pump radiation excites silica Raman scattering throughout the excitation fiber. In some applications, like the enzyme work sited above carried out in dilute aqueous solutions, this is not problematic. However, in samples where significant Mie scattering or specular or diffuse reflections are strong, the amount of silica Raman collected by the collection fiber(s) can be substantial and difficult to appropriately correct for. We have proposed the use of hollow-core photonic crystal fibers (PCF) as excitation fibers for Raman spectroscopy (12). This follow from several recent reports describing the advantages of hollow-core PCF in a variety of non-linear optical applications (13-15).

Photonic crystal fibers guide light in a hollow core, surrounded by a microstructured cladding formed by a periodic arrangement of cylindrical air-filled channels in silica a few microns in diameter. Since only a small fraction of the excitation light propagates in silica, the intensity of silica Raman generated within the fiber is hugely reduced. The transmission properties of such fibers are determined by the periodicity and mean diameter of the microchannels and refractive index of the silica cladding. Figure 6 shows the transmission characteristics of a commercially available PCF (HC19-532-01, Crystal Fibre, Denmark) that was used to fabricate prototype excitation fibers for Raman spectroscopy using 532 nm excitation. The dispersion properties of this fiber result in high transmission centered near 532 nm with a minimum at about 540 nm. An additional feature of PCF excitation fibers for fiber-optic Raman probes is their ability to act as spectral filters due to high transmission losses at wavelengths greater than ~350 cm-1 from the central wavelength. Thus, even the small amount of silica scattering generated within the PCF is somewhat attenuated.

In the application described in (12), the hollow core of the PCF was sealed at the distal (sample) end by melting the tip using a standard fusion splicer (Ericsson AB, Stockholm, Sweden) allowing this PCF to be used in aqueous solutions. The probe was configured much as the one shown in Figure 7a, except that the (sealed) PCF protruded slightly beyond the endfaces of the three collection fibers. This probe was tested using neat ethanol as the analyte, which does not exhibit resonance enhancement at 532 nm, hence a geometry similar to that depicted in Figure 2a was used. The resulting measured spectra exhibited about a factor of 10 reduction in the silica background compared to a similarly configured probe with a solid-core silica excitation fiber. The residual silica present in the measured spectrum was shown to be due to silica Raman generated in the solid-core silica collection fibers excited by Rayleigh scattered laser light at 532 nm, plus a small amount due to the fused silica seal at the end of the PCF.

Perhaps the "worst-case" scenario for fiber-optic-linked RS, insofar as interference due to silica Raman is concerned, is measuring close to a surface.

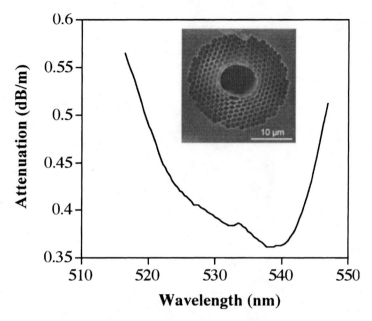

Figure 6. Attenuation of a hollow-core PCF designed for minimum attenuation centered near 532 nm. The inset shows a photomicrograph image of the PCF in cross-section.

Here, we demonstrate a PCF-based probe for use in air, where the PCF has been cleaved at the distal end, but not sealed, and configured with 3 solid-core silica collection fibers as shown in Figure 7a. This probe was tested using powdered sucrose as the analyte spread in a layer several millimeters thick on the surface of a glass microscope slide. The resulting measured spectra are shown in Figure 8a, together with results obtained using a similarly configured probe, but with a solid-core silica pump fiber, for comparison. The difference spectrum is shown in Figure 8b and, as expected, reveals only silica features. Note that, in this case, the PCF-based probe results in about a factor of 3 reduction in the silica background compared to the probe configured with a solid-core silica excitation fiber. The smaller reductive effect in this case is attributed, again, to silica Raman generated in the collection fibers. This effect is much worse here because of the large amount of diffuse reflection from the air-sample interphase, which results in substantially more 532 nm laser light being collected by the solid-core silica collection fibers, which in turn excites much more silica Raman.

In order to achieve complete elimination of silica Raman in the measured spectrum, it remains to implement some means to block the intense stray pump radiation from entering the collection fibers. One possibility, of course, is to engineer appropriate dielectric filters to deposit on the collection fibers, and this technology is certainly available, though costly. Another approach that may be

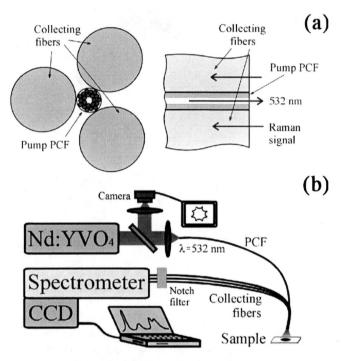

Figure 7. Multi-fiber probe and instrument configurations for fiber-optic-linked RS using a photonic crystal fiber (PCF) for excitation. (a) the probe arrangement comprising three standard fused silica collection fibers (400 μm core) surrounding a single PCF (8.6 μm x 9.5 μm core, OD 84 μm). (b) the instrumentation setup comprising a neodymium vanadate laser (Verdi V-10, Coherent Inc., Santa Clara, CA), a holographic notch filter to attenuate the laser line, a 1-m single monochrometer (model 2061, McPherson Inc, Chelmsford, MA) and a LN₂-cooled CCD detector (Princeton Instruments Spec-10 400B, Roper Scientific, Trenton, NJ).

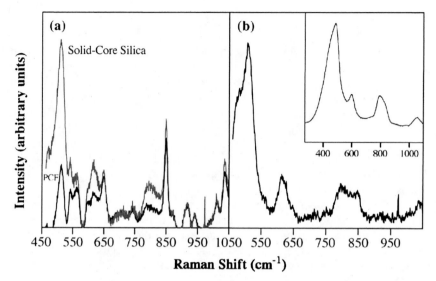

Figure 8. Comparison of sucrose Raman spectra obtained using PCF and solid-core silica pump fibers. (a) spectra obtained using about 30 mW excitation at 532 nm (10-s exposure); each spectrum was normalized to the intensity of the large sucrose peak at 850 cm^{-1}. (b) difference spectrum (solid-core − PCF) showing residual silica Raman excited by stray 532 nm light within solid-core silica collection fibers; the inset shows a (smoothed) silica Raman spectrum obtained using a different instrument with excitation at 257.2 nm.

worthwhile exploring would be the use of PCF for collection, as well as excitation. However, the PCF pass-band near 532 nm is only about 25 cm^{-1}, which is much too narrow to collect a full Raman spectrum, although this pass-band could be engineered to coincide with some Raman band of interest for a specific application. Alternatively, a collection fiber bundle of different PCFs could be used to cover a range of Stokes shifts. Regardless of the collection fiber configuration, the use of PCF for excitation should at least substantially reduce the interference due to silica Raman, which could offer significant advantages for biomedical applications of RS involving tissues or turbid biological fluids.

Finally, it is worthwhile mentioning a further advantage of hollow-core PCF fibers that may be attractive for other biomedical applications, namely their capacity to transmit very high-power laser pulses. PCF have been shown to allow transmission of near infrared (NIR) pulse trains where the laser radiation fluence coupled into the core of the fiber exceeded the breakdown threshold of fused silica by nearly an order of magnitude (*16*). This has been exploited in the PCF-guided laser ablation of dental plaques (*17*) and may also offer promise for new applications in biomedical imaging using multi-photon fluorescence or optical coherence tomography.

12

Concluding Remarks

We have described some novel technologies for implementing Raman-based spectroscopies that offer some unique advantages for fiber-optic-linked applications. These fiber-optic probe designs address problems relating to signal collection efficiency in resonance Raman applications, and interference due to silica Raman scattering in turbid media or measurements at or near surfaces in visible or NIR Raman applications. Our group's future work in this area, in terms of fiber-optic probe development will likely focus on UVRRS applications. In particular, we intend to explore the possibility of employing PCF excitation fibers in the ultraviolet, and incorporating them in UVRRS probes of the type depicted in Figure 4a, as some interesting *in vitro* applications of UVRRS also involve turbid media or surfaces. In the far UV, the Stokes shifts are not as large in terms of wavelength, so it may also be feasible to design hybrid PCF/silica collection fibers that exploit the right-angle collection geometry in a short segment of silica coupled to a PCF waveguide. We also plan to continue developing new biophysical and bioanalytical applications for fiber-optic-linked UVRRS.

Acknowledgments

The authors wish to acknowledge the financial support provided by the Natural Sciences and Engineering Research Council of Canada for the work described here. We also acknowledge the infrastructure support provided by the Canada Foundation for Innovation, the British Columbia Knowledge Development Fund, and The University of British Columbia.

Literature Cited

1. Hanlon, E.B.; Manoharan, R.; Koo, T.-W.; Shafer, K.E.; Motz, J.T.; Fitzmaurice, M.; Kramer, J.R.; Itzkan, I.; Dasari, R.R.; Feld, M.S. *Phys. Med. Biol.* **2000**, *45*, R1 - R59.
2. Schweitzer-Stenner, R. *J. Raman Spectrosc.* **2005**, *36*, 276 - 278.
3. Benevides, J.M.; Overman, S.A.; Thomas Jr., G.J. *J. Raman Spectrosc.* **2005**, *36*, 279 - 299
4. Tuma, R. *J. Raman Spectrosc.* **2005**, *36*, 307 - 319.
5. Utzinger, U.; Richards-Kortum, R. *J. Biomed. Optics* **2003**, *8*, 121 - 147.
6. Utzinger, U.; Richards-Kortum, R. In *Encyclopedia of Spectroscopy and Spectrometry;* J. C. Lindon, G. E. Tranter, and J. L. Holmes, Eds.; Academic Press: London, 2000, pp. 513–527.
7. Barbosa, C.J.; Vaillancourt, F.H.; Eltis, L.D.; Blades, M.W.; Turner, R.F.B. *J. Raman Spectrosc.* **2002**, *33*, 503 - 510.

8. Greek, L.S.; Schulze, H.G.; Blades, M.W.; Haynes, C.A.; Turner, R.F.B. *Applied Optics* **1998**, *37*, 170 - 180.
9. Vaillancourt, F.H.; Barbosa, C.J.; Spiro, T.G.; Bolin, J.T.; Blades, M.W.; Turner, R.F.B.; Eltis, L.D. *J. Am. Chem. Soc.* **2002**, *124*, 2485 - 2496.
10. Jirasek, A.I.; Vaillancourt, F.H.; Barbosa, C.J.; Eltis, L.D.; Blades, M.W.; Turner, R.F.B. *Am. Pharm. Rev.* **2004**, *7*, 49 - 53.
11. Horsman, G.P.; Jirasek, A.I.; Vaillancourt, F.H.; Barbosa, C.J.; Jarzecki, A.A.; Xu, C.; Spiro, T.G.; Lipscomb, J.D.; Blades, M.W.; Turner, R.F.B.; Eltis, L.D. *J. Am. Chem. Soc.* **2005**, *127*, 16882 - 16891.
12. Konorov, S.O.; Schulze, H.G.; Addison, C.J.; Turner, R.F.B.; Blades, M.W. *Optics Letters* **2006**, (in press).
13. Benabid, F.; Knight, J. C.; Antonopoulos, G.; Russell, P. St. J. *Science* **2002**, *298*, 399-402.
14. Konorov, S.O.; Fedotov, A.B.; Zheltikov, A.M. *Opt. Lett.* **2003**, *28*, 1448-1450.
15. Fedotov,A.B.; Konorov, S.O.; Mitrokhin, V.P.; Serebryannikov, E.E.; Zheltikov, A.M. *Phys. Rev. A* **2004**, *70*, 045802.
16. Konorov, S.O.; Fedotov, A.B.; Kolevatova, O.A.; Beloglazov, V.I.; Skibina, N.B.; Shcherbakov, A.V.; Wintner, E.; Zheltikov, A.M. *J. Phys. D* **2003**, *36*, 1375-1381.
17. Konorov, S.O.; Fedotov, A.B.; Mitrokhin, V.P.; Beloglazov, V.I.; Skibina, N.B.; Shcherbakov, A.V.; Wintner, E.; Scalora, M.; Zheltikov, A.M. *Appl. Opt.* **2004**, *43*, 2251-2256.

Chapter 2

Applications of Fourier-Transform Infrared Imaging in Cancer Research

Don McNaughton and Bayden R. Wood

Centre for Biospectroscopy and School of Chemistry, Monash University, Clayton, Victoria 3800, Australia

False color images of cervical tissue sections, malignant glioma rat brain sections and melanoma sections have been constructed from FTIR hyperspectral data. The images are directly compared with stained sections and shown to be useful in distinguishing cervical pathology, malignant glioma and class I human lymphocyte antigen (HLA) expression in melanoma. Infrared spectra extracted from the images are shown to be useful for determining the macromolecular changes that accompany disease. Techniques for building three dimensional images to determine the extent of tissue change are described.

Introduction

Over the last 10-15 years Fourier Transform Infra-Red (FTIR) spectroscopy and more particularly infrared micro-spectroscopy, has been investigated and shown to have a role in the diagnosis and understanding of a wide range of cancers, including colon, breast, liver, skin, lung and brain. A recent review details much of the progress to date and indicates the current breadth of the field (1), although a full review of the field is not yet available. Most of these studies have concentrated on single point spectroscopy, either in the laboratory with large apertures, or using synchrotron IR sources to approach diffraction limited resolution (2). For tissue samples point to point mapping has been used in an attempt to emulate and extend visible pathology. Although this methodology showed some promise, the time involved in obtaining sufficient data was prohibitive (3). However the advent of focal plane array (FPA) based imaging spectrometers (4) and other instruments based on linear arrays has provided instrumentation capable of rapidly obtaining IR hyper-spectral maps of thin tissue sections at close to diffraction limited resolution. These systems now provide the basis for the further development of infrared spectroscopy as a tool in the diagnosis of cancer and in following the effects of cancer progression and treatment. We outline here some recent work on the use of FPA imaging in cervical cancer diagnosis, brain cancer and in the investigation of HLA expression in melanoma. We also briefly describe the development of 3D imaging and its application to pathology.

In this work all spectral images have been collected with a Varian Stingray imaging system in fast scan mode, using a MCT 64 × 64 array and a 15× objective with a theoretical pixel resolution of 5.5 µm. Unless otherwise stated spectra were collected at 6cm^{-1} resolution with 16 co-added scans. A number of strategies were used to achieve images of larger areas: A 2× magnifying objective was placed in the instrument to give a theoretical pixel resolution of 11 µm; larger images were built by constructing tiled images from a number of individual tiles; for computer intensive data processing pixel aggregation was necessary to generate files that could be processed in real time. Tissue samples were formalin-fixed in paraffin blocks, sectioned at 4-5 µm, deposited on "Kevley MirrIR low e microscope slides" and then washed in xylene to remove paraffin. Spectra were recorded in absorption/reflection. As we have shown in a previous publication (5) the spatial resolution using FPA technology with this methodology approaches 10 µm rather than the diffraction limit. Spectral preprocessing and processing was carried out using mainly Cytospec (6), purpose built routines using Matlab for 2D image stitching and Scirun (7) for 3D imaging .

Once hyperspectral data blocks have been collected the data processing, using imaging programs such as Cytospec, can be divided into a number of phases. Firstly, poor quality spectra, resulting from too thin or thick sample areas, noisy pixels or gaseous water contamination, need to be removed. Secondly, a pre-processing routine consisting of baseline correction, or spectral derivativization and normalization is required to eliminate thickness effects and baseline variation. When using absorption-reflection slides at diffraction limited resolution, dispersion and scattering artifacts also become important and these must also be eliminated or minimized (8,9) in the preprocessing phase. The final phase is image construction. FTIR image data can be processed in a univariate mode where chemical maps (also called functional group maps) based on peak intensity, peak area or peak ratios can be routinely generated with the software supplied or using Cytospec. While these methods can provide information on the distribution and relative concentration of a particular functional group and hence a specific major biomolecule, they are not very useful in terms of classifying anatomical and histopathological features within the tissue matrix and multivariate image reconstruction is required. Typical methodology includes unsupervised hierarchical cluster analysis (UHCA), K-means clustering, principal components analysis, linear discriminant analysis, artificial neural networks and fuzzy C-means clustering. These methods are aimed at classifying spectra based on similarity and thus are used to discern anatomical and histopathological features based on underlying differences in the macromolecular chemistry of the different cell and tissue types that constitute the sample. We and others have found UHCA to be the most useful technique for direct pathological comparison, although this severely restricts the size of images due to the computing overheads involved in calculating the distance matrices. Using a desk top PC with a Windows^{TM} operating system the largest images that can be processed with UHCA are 128 × 128, hence the need for the pixel aggregation mentioned above.

FTIR imaging for Cervical Cancer Pathology

In the early 1990s Wong et al. (*10,11*) investigated the infrared spectra of exfoliated cervical cells and reported spectral differences between samples from patients diagnosed normal by cytology and those from patients diagnosed with dysplasia or cancer. Subsequent work by the Diem group and independently by McNaughton and co-workers indicated that the spectral changes observed may not be related to the number and molecular composition of dysplastic cells *per se* but other factors such as inflamation (*12-13*), the number of dividing versus non-dividing cells (*14*) as well as the cells overall divisional activity (*15*). The presence of biological components such as mucin, erythrocytes and leukocytes was also shown to obscure diagnostic regions of the spectra (*12,13*). Multivariate and artificial neural networks have been applied to the analysis of the spectra of exfoliated cells in an attempt to circumvent these confounding variables.(*16-20*) Some of these techniques provide information on the important variables that may distinguish normal from diseased samples, but they are limited in that they do not provide information on the cervical cell types and their stage of differentiation and maturation within the cervix. Another factor that was shown to affect the analysis was the stage of the menstrual cycle at cell collection (*21*). In summary these studies demonstrated that a detailed understanding of the spectral features of the cell types and spectral variations resulting from differentiation, maturation, and cell cycle stages is a pre-requisite for the interpretation of the spectral differences between normal and dysplastic cytological diagnosed samples.

In this study we applied FTIR FPA imaging spectroscopy and unsupervised hierarchical clustering (UHCA, D-values, Ward's algorithm) to investigate the tissue architecture of the cervical epithelium. UHCA was performed on the whole spectral region (1800 – 950 cm^{-1}), the amide I region (1700-1570 cm^{-1}) and the ν_{asym} PO$_2^-$ region (1300-1200 cm^{-1}) in order to ascertain which region or regions are appropriate for analysis. In the final step all spectra in a cluster are assigned the same color and the final image derived from a false color map. Such a map, using the whole spectral region, is presented in grayscale in figure 1 together with the mean 2nd derivative spectra for each cluster and an adjacent haematoxylin and eosin (H&E) stained section of cervical epithelium. The mean spectrum of a cluster represents all spectra in a cluster and can be used for the interpretation of the chemical or biochemical differences between clusters. The number of clusters to be used is chosen by the user and we investigated from 2 – 8 clusters to determine the optimum number for each tissue sample. At five clusters most of the cell types apparent in the stained cervical epithelium tissue section, i.e. superficial, intermediate, parabasal, and connective tissue correlate well with the color coded clusters, although the small stromal inclusion of intermediate cells is not apparent in the parabasal region. At 8 clusters the narrow basal cell region is well defined, the small stromal inclusion is apparent and a region between intermediate and parabasal is apparent as a separate cluster. Results using just the amide I region are similar to those using the whole spectrum but the basal layer and the small inclusion are apparent at lower cluster numbers. There is also further differentiation of the connective tissue apparent at high cluster numbers. Given that only 5 or 6 major cell types are expected in such a normal sample this is what one would expect. Thus by minimizing the spectral region for analysis and using fewer clusters no useful information has been lost. The inclusion of further clusters in the analysis resulted in differentiation based mainly on baseline variation and artifacts introduced by the mosaic nature of the images. Investigation of the nucleic acid spectral region in the same fashion was not as useful. Differentiation of all but the three major cell types was lost even at cluster numbers > 3. This is due to overlying absorptions from glycogen, the amount of which varies with non-specific disease and confounds analysis.

The major spectral differences observed in the mean spectra are: changes in the bands due to protein where both spectral shifts, intensity changes and broadening are apparent in amide I; nucleic acids, where the ν_{asym} PO$_2^-$ intensity shows great variation; glycogen, where the bands vary markedly, possibly due more to baseline effects than to real spectral changes. These changes are as expected and the mean spectra from the analysis using only the protein region are very similar, indicating that using a reduced spectral region and consequently less computer time, gives almost identical discrimination. By selecting the spectral window for data processing we show that the combination of FPA

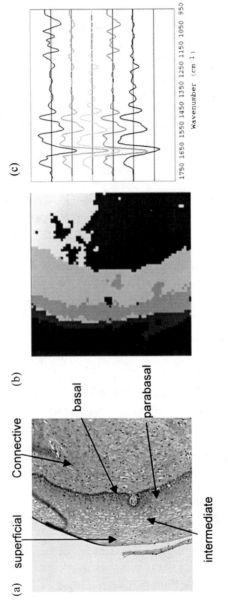

Figure 1. Cervical epithelium section (a) H&E stained section showing the five major cellular tissue types; (b) UHCA image using 5 clusters and (c) Mean spectra from each cluster in the image.

technology and UHCA is a rapid non-subjective analytical tool for the identification of anatomical features in cervical tissue.

The same type of analysis can be carried out on a diseased tissue section (*22*) and Figure 2A and B, depict a H&E stained section from a patient diagnosed with high-grade dysplasia showing a potential metastatic inclusion within the glandular endometrium, whilst 2C and D are UHCA images of the outlined areas. The spectra of this particular section were taken by point to point mapping with a 20×20 μm^2 aperture size chosen to allow sufficient S/N for further analysis within a reasonable timeframe (5.5hr). From these spectra an image of the *ca* 1200×1200 μm^2 section along with another image recorded over a 300×300 μm^2 have been constructed for comparison with the H&E stained section. Even though the images are in gray scale the correlation of the cell types and dysplastic regions within the stained section and the cluster maps is easily apparent. The mean extracted spectra for each of the clusters shown in Figure 2E display a broad range of differences across the spectrum, indicating significant molecular differences between the different cell types. One spectrum shows bands indicative of collagen at 1229, 1239 and 1250 cm^{-1} while the other spectra seem to be relatively devoid of collagen but show an increase in the symmetric and asymmetric phosphate modes indicative of nucleic acids. In particular the cells associated with the cancer cells have quite pronounced $v_{sym}(PO_2^-)$ and $v_{asym}(PO_2^-)$ at 1081 and 1240 cm^{-1} wavenumber values, respectively. With the glycogen region precluded from the analysis as described above, correlation with the major anatomical and histopathological features observed in the H&E stained section and the 5 clusters is excellent. For these samples the overlying "confounding variables" such as regions of leukocytes, erythrocytes and the underlying connective tissue appear as distinct clusters in the image.

Although the use of an FPA FTIR imaging system reduces the time taken for obtaining images to a matter of minutes, the UHCA takes considerable processing time, as mentioned above, and becomes the limiting step in analysis. To overcome this we have explored ANN imaging as a means of reducing this time. We have built a database of spectra from each cell type and this is used to train an ANN. This has greatly decreased the processing time and produces results to date that are as good as the UHCA images presented above.

FTIR imaging of glioblastoma multiforme in rat brain

Brain tissue and tumors have rarely been studied using FTIR, although Salzer *et al* (23) have recently described the use of FTIR mapping as a basis for the classification of human astrocytic gliomas. Glioblastoma multiforme (GBM) is a highly malignant human brain tumor and numerous clinical studies and

20

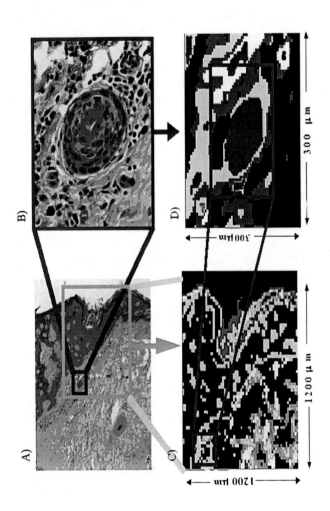

Figure 2. A) H&E stained section of a high grade dysplastic lesion. B) "Blow up" of potential micro-metastacy. C) UHCA map generated from spectra recorded over the area shown in A. D) UHCA map generated from spectra recorded over area shown in B. E) Mean extracted spectra from UHCA map D. (Reproduced from reference 22)

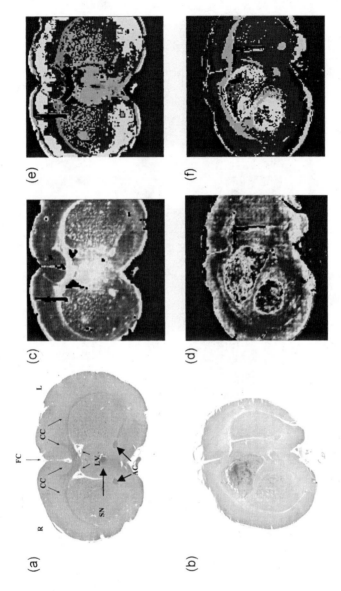

Figure 3. Coronal H&E stained sections of (a) healthy rat brain and (b) with malignant glioma, FC: site of Falx cerebri (interhemispherical fissure), AC: anterior commissure, CC: corpus callosum; LV: lateral ventricle; SN: septal nucleus, R: right hemisphere; L: left hemisphere. Chemimage (amide I) of (c) healthy and (d) malignant. UHCA (6 cluster) image of (e) healthy and (f) malignant. (adapted from reference 24)

animal experiments are under way with the goal of understanding tumor biology and developing potential therapeutic approaches. C6 cell glioma, a commercially available cell line, in the adult rat is a frequently used animal model for the malignant human glial tumor. By combining standard analytical methods such as histology and immunohistochemistry with FTIR microspectroscopy and imaging, we have explored a tumor diagnostic which allows us to obtain information about structure and composition of tumor tissues that is not easily obtained with either method alone (24). Two adult Wistar rats, one serving as a host for the brain tumor and one as an age matched control, were used.

The H&E stained sections in figure 3 show all the easily identifiable anatomical structures and these are labeled on the healthy tissue image. In the tissue of the tumor bearing rat the tumor growth is easily seen and the brain morphology is distinctly distorted. Figures 3c and d show chemimages constructed from the integrated intensity of the amide I band (1680-1620cm^{-1}). The good correlation between these chemimages and the stained sections show that the protein concentration alone is useful for delineating anatomical features. By using two clusters in the UHCA, distinction between the tumor and normal tissue was easily apparent, whilst for the healthy tissue the two clusters essentially identified white and grey matter as separate clusters. The white matter spectra showed increased intensity for protein bands and phosphodiester bands (1235 and 1078 cm^{-1}). Figures 3e and f show images constructed from six clusters. For the healthy tissue the anatomical structures are well delineated and the UHCA is able to isolate the spectra of each cell type with much of the difference attributed to the degree of myelination present in the different brain structures. At larger cluster numbers (8 clusters) the UHCA also differentiates layers in the brain cortex which correspond well with the known Brodmann anatomical layers of the neocortex. Much less structural differentiation is apparent in the tumor tissue, although there is distinct clustering of tumor/non tumor tissue.

FTIR imaging for detection of HLA class I expression in Melanoma tumors.

Human Lymphocyte Antigen (HLA) class I molecules are ubiquitously expressed on the surface of tissues and have a major role in both immunosurveillance and tumor immunity. Tumor cells decrease the expression of HLA class I molecules (termed downregulation) in order to avoid eradication by the immune system (25). This downregulation of HLA class I molecules renders T cell- based immunotherapies ineffective because they rely on HLA class I molecules and as a result, downregulation of HLA class I molecules is also found to be associated with poor prognosis (26). Normally HLA class I molecule expression on the surface of melanoma tissues is identified using immunofluorescence and appropriate HLA type I monoclonal antibodies (27). This is a time consuming and costly methodology. We have recently investigated the use of FTIR imaging as an alternative method of determining HLA class I expression on the surface of melanoma samples (28). A grey scale color image of a metastatic melanoma tumor section immunohistochemically stained with HLA class I heavy chain (HC-10) antibody is presented in figure 4a, whilst figure 4b contains a UHCA IR image of an adjacent section. Figure 4b shows there is a separation between HLA non-expressing (light cluster) and HLA expressing (dark cluster) cells which correlates well with the immunochemically stained sections. This excellent correlation was shown in all sections examined in tissue sections originating from a number of different patients.

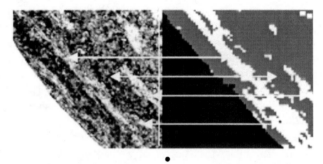

Figure 4. Digital photograph and 2-D false colour cluster map (grey scale) of consecutive tissue sections obtained from a metastatic melanoma tumour dissected from the lymph node (a) Immunohistochemistry staining of tumorous tissue section using primary antibody specific for free HLA class I heavy chain (HC-10); (b) Cluster image of adjacent tumour tissue section with the tissue edge pixels removed.(adapted from reference 28)

From the cluster average spectra representing HLA expressing and non-expressing cells there are consistent characteristic spectral differences. The amide I band of the spectra of HLA expressing cells is more intense and spectra from non-expressing cells show a small shift of this band to a lower wavenumber value, indicating differences in protein conformation. The area under the amide I peak is significantly different (by t-test) between expressing/non expressing cells and this could be used in identifying HLA positive and negative areas on tissues. Bands at *ca* 1400 and 1450cm^{-1}, attributed to methyl and methylene deformation modes associated with amino acid side chains also show lower absorbance in the spectra of expressing cells as do the asymmetrical and symmetrical phosphate stretching modes at 1244 and 1080cm^{-1} respectively. The latter difference may be attributable to higher concentrations of RNA and DNA in the expressing cells.

In summary FTIR imaging enables the distinction of HLA class I positive areas from class I negative areas in melanoma tissue in an accurate, rapid and cost effective manner without the use of antibodies. However, more work is required to determine if the observed changes between class I expressing and non-expressing cells is the direct result of the FTIR technique detecting differences in molecules on the cell surface, or alternatively changes in the underlying molecular architecture responsible for the expression of these molecules.

2D and 3D imaging of tissue

There are other areas where FTIR imaging can be extended to provide further useful tools in cancer research. Using a MATLAB® routine developed by our group, four FTIR images of adjacent sections were stitched together side by side to give a single large 2D image frame. Spectra that passed a quality test were converted to second derivative spectra using the Savitsky-Golay algorithm. UHCA (D-values, Ward's algorithm) was performed on second derivative spectra over the 1272-950 cm^{-1} spectral window. The resultant cluster map was then divided back into the 4 component 2D cluster maps, each map corresponding to one of the FTIR images.

This is particularly useful for the comparison of adjacent sections where the same cell types are present or when sections have been subjected to different treatments and a direct comparison needs to be made. Moreover, the ability to generate and visualize 3D FTIR cluster maps provides a new avenue to assess variation in multiple tissue sections and to determine the extent of penetration of histopathological structures based on the underlying macromolecular structure of the diseased tissue. The thin sections (4 μm) required for use with the Kevley slides are less than the thickness of a single cervical cell, consequently multiple sections enable the analysis of whole cells and also minimize the effects of orientation artifacts that can arise during tissue sectioning.

Figure 5a shows a UHCA image of four adjacent sections of adenocarcinoma containing cervical tissue stitched together, allowing the penetration of histological features to be easily visualized. In general terms the clusters represent red blood cells embedded in the stromal matrix, stroma, lymphocyte exudates and glandular tissue indicative of adenocarcinoma. In tissue sections 3 and 4 there is an increase in the area of connective tissue relative to glandular tissue when compared to sections 1 and 2, indicating penetration of the glandular tissue into the connective layer. Mean spectra can be

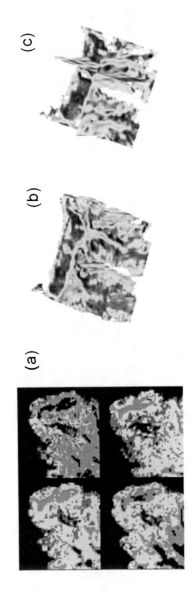

Figure 5. (a) UHCA five cluster maps of four adjacent cervical sections with adenocarcinoma; (b) Chemimage (amide I intensity) block constructed from four adjacent sections of Monkey gut; (c) Sections through the block in (b).

extracted from the 3D image and compared to provide biochemical information and hence may be used for diagnostic purposes.

Using our Matlab modules the sections can be stacked to provide a three dimensional block as shown in figure 5b for monkey gut tissue where the gut villi can easily be seen projecting into the tissue block. These blocks can then be cut vertically or horizontally to interrogate the tissue architecture as seen in figure 5c.

Software products such as ScirunII (7) are also available to allow larger 3D blocks to be constructed and these powerful programs allow the user to manipulate the colors and make the blocks transparent. The SCIRunII software suite provides a graphical user interface for rapid development of "networks" of instruction routines for the stacking and rendering of the input data. For example the 4 cluster maps of figure 5b were "stacked" into a scalar volume field of numbered clusters from which 3D cluster maps were rendered. Using the ANN maps mentioned above to speed up processing the generation of these 3D blocks and visualization of pathological features within a block is possible within a reasonable time scale.

Acknowledgements

We thank the National Health and Medical Research Council (NHMRC) of Australia for financial support and our many collaborators who made this work possible in particular Brian Tait, Sock Fern Chew, Keith Bambery, Corey Evans, Max Diem, Elizabeth Schultke and Sarah Rigley MacDonald. Dr Wood is funded by an Australian Synchrotron Research Program Fellowship Grant and a Monash Synchrotron Research Fellowship. Mr Finlay Shanks is thanked for instrumental support and Mr Clyde Riley (Royal Women's Hospital, Melbourne) for sectioning.

References

1. Sabu, R. K.; Mordechai, S. *Future Oncology* **2005**, *1*, 635-647.
2. Tobin, M. J.; Chesters, M. A.; Chalmers, J. M.; Rutten, F. J. M.; Fisher, S. E.; Symonds, I. M.; Hitchcock, A; Allibone, R.; Dias-Gunasekara, S. *Faraday Discussions* **2004**, *126*, 27-39.
3. Wood, B. R.; McNaughton, D.; Chiriboga, L.; Yee, H.; Diem; M. *Gynecol. Oncol.* **2003**, *93*, 59-68.
4. Lewis, E. N.; Treado, P. J.; Reeder, R. C.; Story, G. M.; Dowrey, A. E.; Marcott, C.; Levin, I. W. *Anal. Chem.* **1995**, *67*, 3377-3381.
5. Bambery, K. R; Wood, B. R.; Quinn' M. A.; McNaughton, D. *Aust J. Chem.* **2004**, *57*, 1139-1143.

6. Lasch, P. CytoSpec 1.05 – an Application for FT-IR Spectroscopic imaging: see http://www.cytospec.com for details
7. Weinstein, D. M.; Parker, S. G.; Simpson, J. K.; Zimmerman, K.; Jones, G. In *The Visualization Handbook,* Hansen C.D; Johnson, C. R. Ed., Elsevier, New York, NY, 2005; pp. 615-632.
8. Romeo, M.; Diem, M. *Vibrational Spectroscopy* 2005, *38*, 129-132.
9 Mohlenhoff, B.; Romeo, M.; Diem, M.; Wood, B. R. *Biophysical Journal* 2005, *88*, 3635-3640.
10. Wong, P. T. T.; Wong, R. K.; Caputo, T. A.; Godwin, T. A.; Rigas, B. *Proc. Natl. Acad. Sci. USA.* 1991, *88*, 1088-10992.
11. Wong, P. T. T.; Wong, R. K.; Fung, M. F. K. *Appl. Spectrosc.* 1993, *47*, 1058-1063.
12. Chiriboga, L.; Xie, P.; Vigorita, V.; Zarou, D.; Zakim, D.; Diem, M. *Biospectroscopy* 1997, *4*, 55-59.
13. Wood, B. R.; Quinn, M. A., Tait; B.; Hislop, T.; Romeo, M. *Biospectroscopy* 1998, *4*, 75-91.
14. Diem, M.; Boydston-White, S.; Chiriboga, L. *Appl. Spectrosc.* 1999, *53*, 148A-161A.
15. Boydston-White, S.; Gopen, T.; Houser, S.; Bargonetti, J.; Diem, M. *Biospectroscopy* 1998, *5*, 219-227.
16. Wood, B. R.; Quinn, M. A.; Burden, F. R.; McNaughton, D. *Biospectroscopy* 1996, *2*, 143-153.
17. Romeo, M.; Burden, F. R.; Wood, B. R.; Quinn, M. A.; Tait, B.; McNaughton, D. *Cell Mol. Biol.* 1998, *44*, 179-187.
18. Romeo, M.; Wood, B. R.; Quinn, M. A.; McNaughton, D. *Vibrational Spectroscopy* 2003, *72*, 69-76.
19. Cohenford, M. A.; Godwin, T. A.; Cahn, F.; Bhandare, P.; Caputo, T. A. ; Rigas, B. *Gynecol. Oncol.* 1997, *66*, 59-65.
20. Cohenford, M. A.; Rigas, B. *Proc. Natl. Acad. Sci. USA* 1998, *95*, 15327-15332.
21. Romeo, M.; Wood, B. R.; McNaughton, D. *Vibrational Spectroscopy* 2002, *28*, 167-175.
22. Wood, B. R.; Bambery, K. R; Miller, L. M.; Quinn, M.; Chiriboga, L.; Diem, M.; McNaughton, D., In *Biomedical Applications of Micro- and Nanoengineering II*, Nicolau, D. V. Ed., Swinburne Univ. of Technology, Australia, Proceedings of SPIE Vol. 5651, 2005, pp. 78-84
23. Krafft, C.; Sobottka, S. B.; Schackert, G.; Salzer, R. *Analyst* 2004, *129*, 921-925.
24. Bambery, K. R; Schultke, E; Wood, B. R.; Rigley MacDonald, S. T.; Ataelmannan, K.; Griebel, R. W.; Juurlink, B.H.J.; McNaughton, D., *BBA Biomembranes* 2006 in press available on line May 06.

25. Wood, B. R.; Tait, B. D.; McNaughton, D. *Hum. Immunol.* **2000**, *61*, 158-65.

26. van Duinen, S.G.; Ruiter, D.J.; Broecker, E.B.; van der Velde E A; Sorg C.; Welvaart K.; Ferrone S. *Cancer Res.* **1988**, *48*, 1019-1025.

27. Mendez, R.; Serrano, A.; Jager, E.; Maleno, I.; Ruiz-Cabello, F.; Knuth, A.; Garrido, F. *Tissue Antigens* **2001**, *57*, 508-519.

28. Chew, S. F.; Wood, B. R.; Kanaan, C.; Browning, J.; MacGregor, D.; Davis, I.; Cebon, J.; Tait, B. D.; McNaughton, D., **2006**, in press Tissue Antigens.

Chapter 3

Biomedical Aspects of Water Structure in Human and Animal Skin: A Near Infrared–Fourier Transform–Raman Study

T. M. Greve[1,2], N. Rastrup Andersen[2], K. Birklund Andersen[2],
M. Gniadecka[3], H. C. Wulf[3], and O. Faurskov Nielsen[1,*]

[1]Department of Chemistry, University of Copenhagen,
Universitetsparken 5, 2100 Copenhagen, Denmark
[2]Spectroscopy and Physical Chemistry, LEO Pharma, Industriparken 55,
2750 Ballerup, Denmark
[3]Department of Dermatology, Bispebjerg Hospital, Bispebjerg Bakke 23,
2400 Copenhagen, Denmark

NIR-FT-Raman spectra of skin from pig ear, guinea pig and mouse were recorded with excitation at 1064 nm and compared to spectra of human skin. The $R(\overline{v})$-representation was used to eliminate the intense Rayleigh band. The total water content in each sample was estimated from the intensities of the OH-stretching vibrations at about 3200 cm^{-1}. A low-wavenumber band around 180 cm^{-1} (~ 5.5 terahertz) was characteristic of a bulk-like liquid water structure. Water content and structure in skin from pig ear, guinea pig and human were similar and different from mouse skin. Differences in loss of bulk water were observed for skin samples after freezing and thawing. Skin biopsies of human skin with various skin tumors showed an increase of water with a bulk-like structure in skin with malignant skin tumors.

Water is the most abundant substance in human cells. Water accounts for about 70 %(w/w) of a cell's weight, and most intracellular reactions occur in an aqueous environment (*1*). We know that water is important for protein dynamics and function in living cells. However, the role of water in bio-systems is far from well understood. In a review with the title "Water: now you see it, now you

© 2007 American Chemical Society

don't" Lewitt and Park ask four questions about water molecules interacting with proteins (2). Where are they? How long do they stay there? How strongly do they interact with protein? How do they affect protein structure and stability? These questions illustrate the common point of view when looking on protein/water interactions. We look on the influence of water on the protein structure and dynamics. But what about the structure and dynamics of water itself? An excellent review on "Unsolved mysteries of water in its liquid and glassy phases" has been given by Stanley et al. (3). In protein interiors water can be bound in very different environments (4). Puppels et al. have determined the water concentration in brain tissue and in skin (5,6). The skin studies were performed by confocal Raman spectroscopy and allowed a determination of the water concentration in the stratum corneum with an impressive depth resolution of 5 μm (6). The water concentration was found from the relative intensities of the OH- and CH-stretching vibrations in the region 2500 to 3800 cm^{-1} (6). This region gives an estimate of the total water concentration, i.e. water bound to biomolecules and water with a structure like the one in bulk liquid water. Because the latter is not bound to biomolecules it is sometimes referred to as "free" water although it is still hydrogen bonded in liquid water. THz spectroscopy has been used to study the intermolecular structural relaxation times in liquid water (7). In the present contribution the high wavenumber NIR-FT-Raman spectrum from 2500 to 3500 cm^{-1} and the low wavenumber spectrum from around 50 to 400 cm^{-1} will be used in studies of animal and human skin samples. In THz the latter region is from 1.5 to 12 THz.

Instrumental

In order to avoid fluorescence, Raman spectra of the skin samples were obtained by excitation at 1064 nm (NIR-FT-Raman). The Rayleigh line was suppressed by use of a filter allowing the Raman spectrum to be recorded to a low wavenumber Raman shift of 80 cm^{-1}. A Raman spectrum of liquid water was recorded with excitation at 532 nm (VIS-Raman). The lower limit for this spectrum was 10 cm^{-1} and a better overall shape of the low-wavenumber Raman band was obtained.

VIS-Raman

A Raman spectrum of liquid water was obtained in a 90° scattering configuration on a DILOR Z-24 triple additive spectrometer with excitation at 532 nm. Laser power 400 mW and 3 cm^{-1} spectral resolution.

NIR-FT-Raman

Raman spectra were recorded on a BRUKER IFS66 spectrometer equipped with a FRA106 Raman module or a BRUKER RFS100 Raman instrument. A liquid nitrogen cooled Ge-detector was used in both instruments. Laser excitation at 1064 nm with a laser power up to 300 mW and 1500 mW, respectively in an 180° scattering configuration. In all cases spectra with a very low laser power were initially recorded in order to assure that no degradation of the samples occurred in the laser beam. Spectral resolution 4 cm⁻¹. No white light background correction was performed.

Raman Spectroscopy of Liquid Water

Walrafen assigned both the internal and external vibrations in the Raman spectrum of liquid water in 1972 (8). The spectral bands were assigned in terms of a water "five-molecule" model. In this a central molecule was hydrogen-bonded to 4 other water molecules in a tetrahedron. Agmon made a more comprehensive assignment of the vibrational Raman bands of this $(H_2O)_5$ molecule (9).

Figure 1A shows the Raman spectrum of liquid water recorded with 532 nm (VIS) and 1064 nm (NIR) excitation. The OH-stretching vibrations between 3000 and 3500 cm⁻¹ are very different in the two spectra. This is caused by the rapid drop in sensitivity of the Ge-detector in the NIR-FT-Raman spectrum towards 3500 cm⁻¹. Thus of the two stretching bands in the VIS-spectrum only the lower wavenumber Raman band is seen. Due to the often encountered problem of fluorescence in Raman spectra of skin with laser excitation in the VIS-region only NIR-FT-Raman spectra of skin were obtained. Even in the NIR-FT-Raman spectra a correlation between the broad background intensity and skin pigmentation was found (10). The NH-vibration of collagen at 3329 cm⁻¹ does not contribute significantly to the intensity of the OH-vibration in the NIR-FT-Raman spectrum (6, 11).

A serious problem in the low-wavenumber region of a Raman spectrum is the high intensity of the Rayleigh line often extending to several hundred wavenumbers. This problem can in many cases be solved by use of the $R(\overline{\nu})$-representation (12-20). This representation is easily calculated from the intensity, $I(\overline{\nu})$ in the Raman spectrum, $R(\overline{\nu}) = \overline{\nu}[1 - \exp(-hc\overline{\nu}/kT)] I(\overline{\nu})$, where $\overline{\nu}$ is the Raman shift (cm⁻¹), c is the velocity of light, T is the absolute temperature, h is Planck's constant and k is Boltzmann's constant. The $R(\overline{\nu})$-representation converts the Rayleigh line to a weakly declining plateau, thus allowing low-wavenumber Raman lines to be more easily recognized. Figure 1B shows the low-wavenumber VIS-Raman spectrum of liquid water and the calculated $R(\overline{\nu})$-representation. This representation clearly shows two bands, one with a maximum around 60 cm⁻¹ and another stronger band with a maximum at about 180 cm⁻¹. As we have previously reported isotopic substitution showed that this band was caused by vibrational motions of the oxygen atoms in tetrahedral surroundings (12,19,20). In water at room temperature 19% of the hydrogen bonds are broken (21). Thus, in bulk-liquid water the amount of water molecules with 4

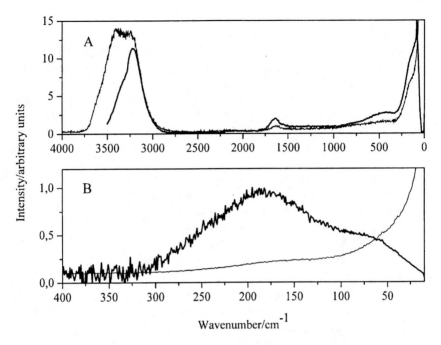

Figure 1. Raman spectra of liquid water. In Figure A the thicker curve shows the NIR-FT- spectrum and the thinner curve the VIS-spectrum. In Figure B the thinner curve shows the low-wavenumber part of the VIS-spectrum and the thicker curve the R($\overline{v}$)-representation. Adapted from Journal of Molecular Structure, 661-662, M. Gniadecka, O.F. Nielsen, H.C. Wulf, Water content and structure in malignant and benign skin tumours, 405-410 Copyright 2003, reference 26, with permission from Elsevier.

tetrahedral hydrogen bonds is $(0.81)^4 = 43\%$. The band in the R($\overline{v}$)-representation of water at 180 cm^{-1} is significant for the presence of bulk-like water.

Skin Samples from Humans, Pig Ears, Hairless Guinea Pigs and Mice

All human samples were from women. Three of the samples were from breast reduction surgery and one from an abdominal reduction surgery. After surgery the skin samples were kept on gauze moisten with a phosphate buffer solution. The samples were kept in a refrigerator (5°C) for around 12 hours before the Raman spectra were recorded. The hairless guinea pig skin samples were taken from the back of the thigh and mouse skin samples from the back. Spectra of pig ear, hairless guinea pig and mouse samples were recorded a few hours after the animals were killed.

The sizes of the skin samples were 2.5x2.5 cm². Each sample was mounted on a piece of plastic foam using small pins to maintain a natural stretching.

Figure 2 shows Raman spectra of skin from humans (4 samples), pigs (5 samples), hairless guinea pigs (6 samples) and mice (4 samples). All spectra are normalized to the same intensity value for the aliphatic CH-stretching vibrations around 2930-2940 cm⁻¹. Variations are seen in the water content from different samples within each group. Evidently the water content in mouse skin is much smaller than that of the other three groups. The water contents of humans, pigs and hairless guinea pigs are rather similar. However, the tendency for these groups is that pig skin contains more water than skin from hairless guinea pig which contains more water than skin from human.

The low-wavenumber R($\overline{v}$)-representation for the skin samples are given in Figure 3A. An average of all spectra in a group is shown. The rather sharp band with a maximum at 80 cm⁻¹ is an artifact from the laser. A very broad band with a maximum around 110-120 cm⁻¹ arises from hydrogen bonding in proteins (*17*). The band at 180 cm⁻¹ from "free" water with a bulk-like structure is observed as a high wavenumber shoulder on this band. The shapes of the R($\overline{v}$)-representations for human, pig and hairless guinea pig are very similar, whereas the R($\overline{v}$)-representation for mouse shows a lower intensity on the high wavenumber side of the protein band. Also shown in Figure 3A is the difference between the R($\overline{v}$)-representations for human and mouse. The overall slope of

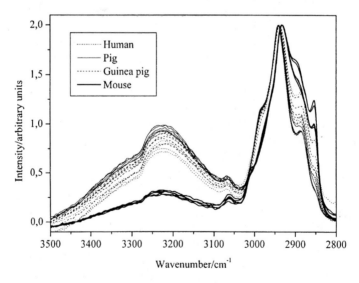

Figure 2. NIR-FT-Raman spectra in the OH/CH-stretching region of skin from humans (dotted lines), pigs (thinner full lines), hairless guinea pigs (broken lines), mice (thicker full lines).

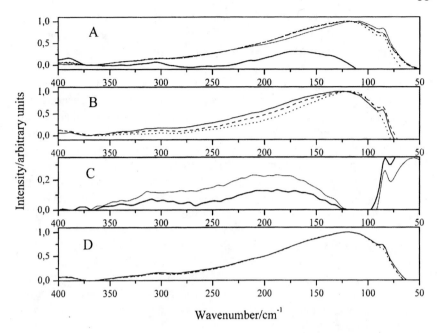

Figure 3. Low-wavenumber Raman spectra in the R($\overline{v}$)-representation of skin samples from humans and animals. A: Human (dotted line), pig (broken/dotted line), guinea pig (broken line), mouse (thinner full line). For each group is shown an average of the spectra obtained and the thicker line shows the difference between human and mouse skin. B: Fresh human skin, S0 (full line); after freezing for one week, S1 and thawing (broken line; after freezing for one week more and thawing, S2 (dotted line). C: Difference between the samples in Figure B; the thinner line shows the difference between the fresh sample, S0 and the sample frozen for one week, S1, and the thicker curve this difference for the fresh, S0, and two weeks, S2, frozen samples. D: Spectra obtained by the same procedure as in Figure B, but for another human skin sample.

this difference spectrum is very similar to the slope of water in Figure 1B, showing that a larger amount of water in the human samples has a water structure like that in bulk-liquid water.

Freezing and Thawing of Skin Samples

The influence of freezing and thawing was investigated for the human and pig ear samples. The samples were frozen at -18°C for one week. After thawing Raman spectra were recorded. The samples were once more frozen at -18°C for another week and after thawing Raman spectra were again recorded. The 4 human samples behaved differently to each other. Two of the samples showed a loss of water in the OH-stretching region while the intensities of the OH-stretching vibrations were identical before and after thawing for the other two samples. Figure 3B shows $R(\overline{v})$-representations for one of the samples that showed an altered water content after freezing. In Figure 3C the spectrum of the two times frozen sample, S2, was subtracted from the fresh sample, S0, and the sample frozen only once, S1, in Figure 3B. The water band with a maximum around 180 cm^{-1} is more intense in the fresh sample, S0, showing that there is more water with a bulk-like structure in S0 than in the sample frozen one time, S1. But also S1 shows the "free" water band. This means that there was a loss in the amount of bulk water during the freezing and thawing of this human skin sample. Figure 3D shows $R(\overline{v})$-representations of another human skin sample after freezing and thawing for one and two weeks. The shapes of the curves for these two samples are identical to that for the fresh sample showing that no change in water content occurs upon freezing and thawing for this skin sample.

Skin samples from pig ears were also frozen and thawed two times with a one week interval. No change in water content was observed for these samples.

It is difficult to explain why the loss of water differs by freezing and thawing for different skin samples. Further experiments are necessary. However, knowledge of what happens by freezing and thawing of skin samples is very important, because skin biopsies often are frozen before pathological examination. The water content of animal skin may also be important for selection of animal skin in pharmaceutical skin penetration studies. In laboratory studies animal skin with water content similar to human skin should be chosen. The low-wavenumber Raman spectrum in the $R(\overline{v})$-representation is a way to follow similarities and changes in the content of "free" water in skin samples.

Malignant and Benign Skin Tumors

Near-infrared spectroscopy and THz pulse imagining have recently shown an increasing content of water in tumor bearing tissue (22, 23). Some of us have used NIR-FT-Raman spectroscopy to investigate the structure of water in proteins and lipids in human skin hair and nail (24) and water and protein structure in photo aged and chronically aged skin (25). In a more recent paper we studied the water content and structure in malignant and benign skin tumors (26). In this paper (26) we collected biopsies from malignant melanoma, MM (5 patients); basal cell carcinoma, BCC (5 patients); pigmented nevi, NV (8 patients); seborrheic keratosis, SK (5 patients); and normal skin (5 young individuals, 21-37 years old and 3 aged individuals, 70-95 years old). The OH/CH-stretching regions of the spectra recorded in connection with our previously published work (26) are shown in Figure 4. Curve fitting was previously performed and the ratio between the intensities of the OH and CH stretching vibrations were found (26). With the exception of SK, no significant

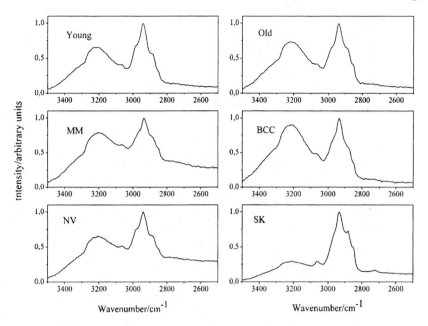

Figure 4. Raman spectra in the OH/CH stretching region for skin samples from young normal individuals (average of 5), old normal individuals (average of 3), malignant melanoma, MM (average of 5), basal cell carcinoma, BCC (average of 5), pigmented nevi, NV (average of 8), seborrheic keratosis, SK (average of 5).

differences in the total water content were found between malignant, benign skin tumors and normal skin (*26*). The low-wavenumber Raman spectrum in the $R(\bar{v})$-representation showed an increase in water with a bulk-like structure in the malignant tumors, MM and BCC, relative to the benign skin tumors, NV and SK (*26*). Figure 5 shows the $R(\bar{v})$-representation of BCC and SK from our previous work (*26*) and the difference between the two spectra compared to the $R(\bar{v})$-representation of liquid water. Evidently the shape of the exceeding water in the BCC sample is very similar to the overall shape of the water spectrum proving that the excess water in the malignant tumor skin has a bulk-like structure.

Future Directions

Water in skin samples can be characterized by NIR-FT-Raman spectroscopy. The high wavenumber region from 2500 to 3500 cm^{-1} measures the total amount of water. A broad band with a maximum at about 180 cm^{-1} gives an estimate of the presence of water not bound to biomolecules, "free" water with a bulk-like liquid water structure. In the low-wavenumber region

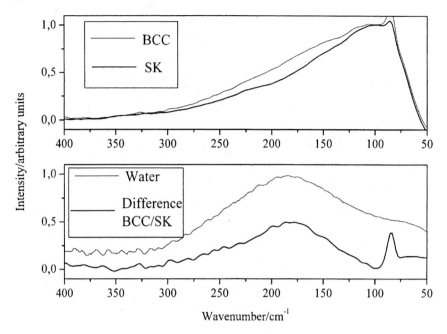

Figure 5. Upper figure: R(ν̄)-representations of skin with basal cell carcinoma, BCC (thinner line) and seborrheic keratosis, SK (thicker line). Lower figure: The difference between the two curves in the upper figure (thicker line) and pure liquid water smoothed with a Savitzky-Golay 9 points smoothing function (thinner line). Adapted from Journal of Molecular Structure, 661-662, M. Gniadecka, O.F. Nielsen, H.C. Wulf, Water content and structure in malignant and benign skin tumours, 405-410 Copyright 2003, reference 26, with permission from Elsevier.

interference from the intense Rayleigh line is overcome by use of the R(ν̄)-representation. Human skin contains approximately the same amount of water as skin from pig and hairless guinea pig, whereas mouse skin contains less water and also less water with a bulk-like structure. In order to improve laboratory modeling of penetration of pharmaceutical products in human skin, water might have a crucial effect. In this context Raman spectroscopy can be a valuable tool in selection of animal skin for the dermatological pharmaceutical industry, although it should be emphasized that Raman spectroscopy in general cannot be justified as a tool for deciding which skin is most suitable.

Before pathological examination of skin samples the samples are often stored at low temperatures. Raman spectroscopy can measure the loss in water content by freezing and thawing and in particular monitor the change in the amount of "free" water.

Our previous Raman investigations of malignant and benign skin tumors revealed a higher amount of "free" water in malignant tumor skin (26). X-ray computer tomography (CT) of liver metastases from malignant melanoma shows changes that might be due to the presence of more "free" water than in a healthy liver (27).

IR spectroscopy is a valuable complementary technique to Raman spectroscopy. Synchrotron based IR micro-spectroscopy is a new technique that shows promising perspectives for studies of skin and human cells (28). At the Max-Lab synchrotron beam line in Lund, Sweden an IR microscope is now being installed (29). This instrument will allow IR-spectra to be recorded with a bolometer detector down to 50 cm^{-1} (around 1,5 THz). In the $R(\bar{v})$-representation of the low-wavenumber Raman spectrum it is very difficult to quantify the amount of "free" water because the water band at 180 cm^{-1} is weak compared to the protein hydrogen bond band at 110-120 cm^{-1}. Hopefully the water band is relatively more intense in the IR spectrum allowing a detection of the "free" water at low concentrations.

Acknowledgements

We wish to thank Dorte Bang Knudsen, Group Manager, Skin Biology, Coloplast Research, Coloplast A/S, Bakkegårdsvej 406A, Humlebæk, Denmark for giving us four *in vitro* samples from breast and belly reduction surgeries. Protocols for collecting the *in vitro* skin samples were accepted by the Ethics Committee of Copenhagen and the Ethics Committee for the Counties of Bornholm, Frederiksborg, Roskilde, Storstøm and Vestsjælland, Denmark. Lykke Ryelund and Mikkel Christensen, University of Copenhagen, are thanked for help in obtaining the Raman spectra. MG and OFN want to thank the Danish Research Academy for financial support (grant no. 51-00-0312). TMG, NRA and KBA want to thank the Ministry of Science, Technology and Innovation for financial support (j.nr. 63781).

References

1. Alberts B.; Johnson A.; Lewis J.; Raff M.; Roberts K.; Walker P. *Molecular Biolology of THE CELL,* Fourth Edition, Garland Science, New York, NY, 2002; p. 55.
2. Levitt M.; Park B.H. *Structure* **1993**, *1*, 223-226.
3. Stanley H.E.; Buldyrev S.V.; Mishima O.; Sadr-Lahijany M.R.; Scala A.; Starr F.W. *J. Physics, Condensed Matter* **2000**, *12*, A403-A412.
4. Olano L.R.; Rick S.W. *J. Am. Chem. Soc.* **2004**, *126*, 7991-8000.
5. Wolthuis R.; van Aken M.; Fountas K.; Robinson Jr J.S.; Bruining H.A.; Puppels G.W. *Anal. Chem.* **2001**, *73*, 3915-3920.
6. Caspers P.J.; Lucasssen G.W.; Carter E.A.; Bruining H.A.; Puppels G.J. *J. Invest. Dermatology* **2001**, *116*, 434-442.
7. Rønne C.; Åstrand P.-O.; Keiding S.R. *Phys. Rev Lett.* **1999**, *82*, 2888-2891.

40

8. Walrafen G.E. in *The Physics and Physical Chemistry of Water;;* Franks F., Ed.; Water. A Comprehensive Treatise, Plenum Press, New York-London, 1972; Vol 1, 151-214.

9. Agmon N. *J. Phys. Chem.* **1996**, *100*, 1072-1080.

10. Knudsen L.; Johansson C.K.; Philipsen P.A.; Gniadecka M.; Wulf H.C. *J. Raman Spectrosc.* **2002**, *33*, 574-579.

11. Leikin S.; Parsegian V.A.; Yang W.-H.; Walrafen G.E. *Proc. Natl. Acad. Sci. USA* **1997**, *94*, 11312-11317.

12. Nielsen O.F. *Chem.. Phys. Lett.* **1979**, *60*, 515-517.

13. Nielsen O.F.; Lund P.-A.; Nicolaisen F.M. *Acta Chem. Scand.* **1981**, *A34*, 749-754.

14. Brooker M.H.; Nielsen O.F.; Praestgaard E. *J. Raman Spectrosc.* **1988**, *19*, 71-78.

15. Murphy W.F.; Brooker M.H.; Nielsen O.F.; Praestgaard E.; Bertie J.E. *J. Raman Spectrosc.* **1989**, *20*, 695-699.

16. Nielsen O.F. *Annual Reports on the Progress of Chemistry;* Phys. Chem. Sect. C; The Royal Society of Chemistry: Cambridge, England, 1993; Vol. 90, pp. 3-44.

17. Colaianni S.E.M.; Nielsen O.F. *J. Mol. Struct.* **1995**, *347*, 267-284.

18. Nielsen O.F. *Annual Reports on the Progress of Chemistry;* Phys. Chem. Sect. C; The Royal Society of Chemistry: Cambridge, England, 1997; Vol. 93, pp. 57-99.

19. Nielsen O.F.; Johansson C.; Jacobsen K.L.; Christensen D.H.; Wiegell M.R.: Pedersen T.; Gniadecka M.; Wulf H.C.; Westh P. In *Optical Devices and Diagnostics in Materials Science;* Andrews D.L.; Asakura T.; Jutamulia S.; Kirk W.P; Lagally M.G.; Lal R.B.; Trolinger J.D., Eds.; Proceedings of SPIE; SPIE-The International Society for Optical Engineering: Bellingham, WA, 2000; Vol. 4098, pp. 160-168.

20. Nielsen O.F. In *Handbook of Raman Spectroscopy: From the Research Laboratory to the Process Line;* Lewis I.R.; Edwards H.G.M., Eds.; Practical Spectroscopy; Marcel Dekker, Inc.: New York, NY, Basel, Switzerland, 2001; Vol 28, pp 593-615.

21. Hetényi B.; Angelis F.D.; Giannozzi P.; Car R. *J. Chem. Phys.* **2004**, *120*, 8632-8637.

22. McIntosh L.M.; Summers R.; Jackson M.; Mantsch H.H.; Mansfield J.R.; Howlett M.; Crowson A.N.; Toole W.P. *J.Invest. Dermatology*, **2001**, *116*, 175-181.

23. Woodward R.M.; Wallace V.P.; Pye R.J.; Cole B.E.; Arnone D.D.; Linfield E.H.; Pepper M. *J. Invest. Dermatology*, **2003**, *120*, 72-78.

24. Gniadecka M.; Nielsen O.F.; Christensen D.H.; Wulf H.C. *J. Invest. Dermatalogy,* **1998**, *110*, 393-398.

25. Gniadecka M.; Nielsen O.F.; Wessel S.; Heidenheim M.; Christensen D.H.; Wulf H.C. *J. Invest. Dermatalogy,* **1998**, *111*, 1129-1132.

26. Gniadecka M.; Nielsen O.F.; Wulf H.C. *J. Mol. Struct.* **2003**, *661-662*, 405-410.

27. L.S. Nielsen, Department of Radiology, Frederikssund Hospital, Frederikssund, Denmark. *Personal Communication.*

28. Diem M.; Romeo M.; Matthäus C.; Miljkovic M.; Miller L.; Lasch P. *Infrared Physics and Technology* **2004**, *45*, 331-338.

29. http://maxsun5.maxlab.lu.se/beamlines/bl73/ (28 March 2006).

Chapter 4

New Approaches Detection Secondary Conformation of Prion Protein in Frozen-Section Tissue by Fourier–Transform Infrared Microscopy

Norio Miyoshi[1], Hiroyuki Okada[2], Masuhiro Takata[2],
Morikazu Shinagawa[2], and Kenichi Akao[3]

[1]Division of Tumor Pathology, Department of Pathological Sciences,
Faculty of Medicine, University of Fukui, Matsuoka, Eiheiji,
Fukui 910–1193, Japan
[2]Prion Disease Research Center, National Institute of Animal Health,
Kannondai, Tsukuba 305–0856, Japan
[3]Spectrometry Instrument Division, JASCO Company, Ishikawa-cho,
Hachioji-City, Tokyo 192-8537, Japan

The spectra and the relative ratio of the secondary conformations (α-helix and β–sheet) of prion infected brains in frozen sectioned tissue were measured by Fourier-transform infrared (FT-IR) microscopy. The tissues were obtained from the Prion Diseases Research Center in the National Animal Health Institute (PDRC/NAHI) of Japan. Both prion infected and normal brain tissues of mice and hamster were embedded adjacently and were frozen-sectioned for the FT-IR microscopy. Spectra from the normal and the prion-infected brain tissue sides were compared. The C-H stretching components in the normal tissue and the amide-I, -II components in the prion tissue were larger than the other sides, respectively. Furthermore, we developed the software to analyze the relative ratios of protein secondary conformation in the tissue by FT-IR microscopy. Results showed that the relative ratio of the β-sheet component was at a higher level (37-40%) in the prion side compared to that in the normal brain tissue. The β-sheet component percentage data were highly contrast against the normal tissue and was small standard deviations comparing the curve-fitting method. We mapped images of the infected brain tissue with respect to the lipid ester, phosphate and protein content.

Introduction

According to the "Prion theory" prion proteins are capable of having more than one secondary conformational structure. This phenomenon can been investigated using several techniques including fluorescence spectroscopy and a fluorescent probe both of which can be problematic in their application. In addition, the aggregation and infection mechanisms remain unclear. The protein biopolymers in brain tissue is not simple at the tissue level. Ideally the brain tissue should be analysed with little or minimal sample preparation such as fixing, embedding in paraffin at high temperaturse, sectioning, de-paraffinising using ethanol or staining with hematoxylin and eosinIt is expected that protein conformation will change as a result of pathological processes. Kneipp, J. et al. (*1*) have reported the molecular changes of preclinical scrapie detected by infrared spectroscopy. Following our work concerning cancer diagnosis (*2*) using FTIR spectroscopic techniques we recently extended our studies to prion-infected brain tissue (without the pathological treatments) using Fourier-transform infrared microscopy (FTIRM) in conjunction with mapping software (IR-SSE developed in collaboration with JASCO Inc., Japan). We report how this approach using a combination of FTIR techniques should prove very useful for explaining protein conformational changes.

Experiments and Methods

(1) <u>Sampling</u>: The infected intra-cerebrally mice for Obihiro strain (*3*) and hamsters for Sc237 (*4*) at each infected stages were sent from PDRC/NAHI (Ibaraki, Japan) after all the safety precautions for the study were fulfilled. Brain tissues (hippocampus domain) were sampled under P2 level conditions and were embedded adjacently with normal brain tissues (hippocampus domain) into an embedding medium for frozen tissue specimens (OCT: Optimal Cutting Temperature; Tissue-Tek, Sakura Fine-technical Co., Ltd., Tokyo, Japan), respectively.

(2) <u>Frozen-sectioning</u>: The frozen compound blocks were sectioning in a 10 μm series at -20 C (Mode: CM3050 S, Leica Co., Germany). One of the

sections was used for the FT-IR spectra measurement and the other one was H&E stained.

(3) FT-IR spectra measurements: Lattice mapping spectra in the 4000-400 cm^{-1} range were collected by a JASCO FT-IRM-410M spectrometer equipped with an Irtron IRT-30M IR microscope (JASCO Co., Ltd., Tokyo, Japan), a motorized XYZ stage, and a liquid nitrogen-cooled mercury-cadmium-telluride-detector. A screen image recorder camera attached to the microscope enabled the acquisition of a photomicrograph of the investigated area. The object area imaged by an individual aperture size was 20x20 μm^2. Sequential spectra were collected from 400 points (20 x 20 points) in the specimen. The area of spectral acquisition amounted to a total of 180 μm^2. For each spectrum, 100 spectra were collected, signal-averaged, and Fourier-transformed to generate spectra with a resolution of 4 cm^{-1} in the transmission mode. All sample spectra of brain tissue were subtracted with the OCT compound frozen section.

(4) Assignments and FT-IRM Image: The peak position of the derived spectra in each necrotic area and undiseased area of the brain tissues were confirmed by Fourier self de-convolution and the second-derivative spectra with a band width of 35 cm^{-1}. The wave numbers of characteristic absorption band in each of the selected spectra were assigned (5-8). To estimate biochemical components, the data obtained were referred to a computerized spectral library data of the FT-IRM-410M software.

(5) Creation of Images by Protein Secondary Structure Analysis: We investigated the conformational changes in protein secondary structure in both the infected and normal hippocampus region brain tissue using a new spectral analytic program (IR-SSE; JASCO Co., Ltd., Tokyo, Japan). This program applies singular value decomposition techniques to estimate the mathematical relationship between the secondary structure types. The program classified the protein conformations into four categories: α-helix, β-sheet, β-turn, and the others included random coil.

Results and Discussion

(1) **Difference spectrum of FT-IR spectra for the infected and the normal brain tissues:** Typical FT-IR spectra for both the infected-prion and the normal hippocampus brain tissue are shown in Figure 1a. The difference spectrum for the normal and infected-prion brain samples are shown in Figure 1b. From the decrease in the lipid (C-H symmetric and anti-symmetric stretching vibration modes; 2,700-3,000 cm^{-1}) and the phosphate (P=O symmetrical stretching vibration mode; 1,080 cm^{-1}) components, it was concluded that the lipid and nuclear DNA components are decreased in the infected tissue.

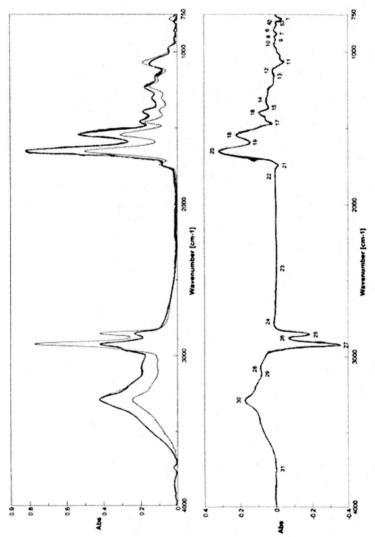

Figure 1.Typical FT-IR spectra of the infected and the normal brain tissue of mice.

(2) **Curve Fitting Results:** Curve fitting (*9*) of the amide-I peak was used for analyzing the secondary protein conformations in the infected and the normal brain tissues. The curve fitting analysis results are shown in Figure 2. 10 components for 4 secondary conformations were separated. Both amide-I absorption spectra for the normal and infected hippocampus brain tissues were each fitted with the original spectra as shown in Figure 2.

(3) **IR-SSE software results and the Comparison:** Akao et al. (*10*) have developed a secondary structural estimation program of protein to be used in conjunction with FT-IR mapping . The protein secondary structure is estimated by a principal component regression (PCR) (*11*) or partial least square (PLS) method. The curve fitting spectra were analyzed using the PCR algorithmic method as shown in (Figure 3) for the same samples in Figure 2. This software included the subtraction analysis of a phosphate buffer and of water (vapor), and we tested for 5% (w/v) solutions of the five types of standard proteins (Lysozyme, Ig-G, Cytochrome-C, Ovalbumin, Pepsin) dissolved in a phosphate buffer. 10 µl of the sample were sandwiched between two CaF_2 plates and then measured using a transmission technique. The instrument of choice for this test was the FT-IR-680 plus, and measurements were repeated three times with 32 integrations, at a of resolution of 4 cm^{-1}, and cosine apodization. The results were as good as can be achieved with respect to reproducibility. In addition, a comparison with the X-ray data reveals a slight shift, and this is thought to be due to the differences between the crystal and liquid states as well as differences due to variations such as purity, species, and tissue types of the measured proteins . Taking these issues into consideration we believe this program is well suited to the secondary structure analysis of protein in brain tissue. The methods most widely used to analyze the secondary structure of proteins involve samples that are crystallized for X-ray crystallography or NMR and CD spectroscopic techniques using aqueous solutions.. However, it is extremely difficult to analyze the secondary structure of multi-component proteins that have anatomy-like structures. On the other hand, since IR spectrascopic techniques can easily measure non-crystalline samples, they are suitable for tissues under sampling conditions as close to *in-vivo* as possible. Furthermore, micro-FTIR enables the surface analysis of heterogeneous samples. Consequently, the IR-SSE mapping program was developed to analyze mapping data The IR-SSE mapping program allows one to use the PCR or PLS technique to analyze each IR spectrum obtained automatically and to display the distribution map of the secondary structure of proteins by color-coding, contour lines, and other means. It also includes a feature that automatically eliminates calculations for areas in which proteins do not exist within a sample. In Figure 3 the four components of protein secondary conformations are shown in the center (calculated spectrum) of right hand side. The experimental results are listed in the table (bottom left hand side),.. The curve fitting results are compared to the PCR data (IR-SSE, JASCO Co., Ltd., Tokyo, Japan) as shown in Table I.

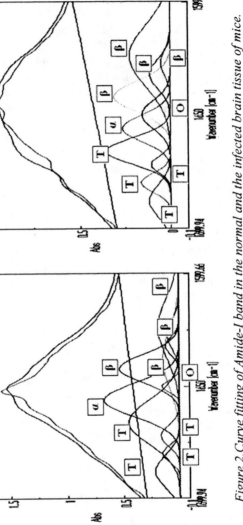

Figure 2.Curve fitting of Amide-1 band in the normal and the infected brain tissue of mice.

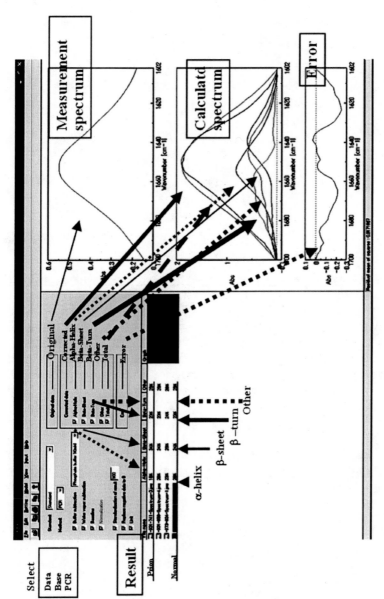

Figure 3 . A typical result of protein secondary conformation analysis in the infected prion side are using the IR SSE software

Table I. The percentages (%) of α–helix and β–sheet components in infected and normal brain tissue

(Normal brain) α-helix: Curve fitting/PCR; β-sheet: Curve fitting/ PCR

===

23.16 / 41	40.87 / 13
17.92 / 41	37.96 / 13
19.98 / 43	35.27 / 12
17.63 / 32	37.51 / 20
(Average values) 20% / 39%	38% / 15%

(Infected brain) α-helix: Curve fitting/PCR; β-sheet: Curve fitting/ PCR

===

16.18 / 16	56.6 / 34
3.22 / 19	66.71 / 33
2.27 / 16	70.19 / 36
19.82 / 24	49.06 / 27
(Average values) 10% / 19%	60% / 33%

(4) Standard deviation values for the β–sheet component using PCR were up to more than half those obtained by the curve fitting method. It appears to us reasonable to use these PCR values to estimate the secondary conformation ratio in those brain tissues; α-Helix and β-sheet components for normal brain tissue were 39% and 15%, for infected brain tissue 19% and 33%, respectively. Typically in infected brain this corresponds to a decrease in the α-helix of 49%. A corresponding more than doubling in the β-sheet component was observed. These data appear to be reasonably in agreement with the values reported for those in the scrapie-associated protein PrP 27-30 in aqueous solution as determined by infrared spectroscopy (*12*). The software used in this analysis has also been applied to a study on nerve toxicity and the physicochemical properties of Aβ mutant peptides from cerebral amyloid angiopathy by Murakami et al (*13*).

(5) Mapping of β-sheet structure ratios in the prion-infected and the normal brain tissues: The percentages of β-sheet component calculated for the FT-IR 400 spectra were plotted against the mapped area of 180 x 180 μm which^2 included both prion-infected and normal hippocampus brain tissues as shown in Figure 4. In the normal (upper) side, the α-helix component percentages were higher than those in the infected brain tissue (lower) side as shown the left side in Figure 1: the higher the percentage the warmer the color (brighter) in the left image. On the other hand, the β-sheet component percentages in the infected brain (lower) side were higher (brighter) than those in the normal one (upper side) as shown in the right side image (the contour representation) of Figure 4, respectively. These images are the first example of an imaging application for prion-infected brain tissues.

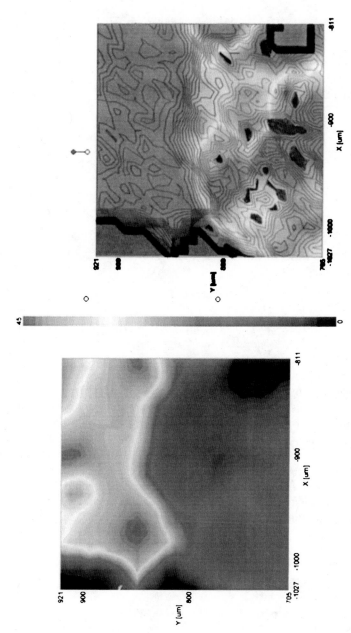

Figure 4. Mapping images of % α–helix and β–sheet components in both sides of the normal and the infected brain tissue of mice

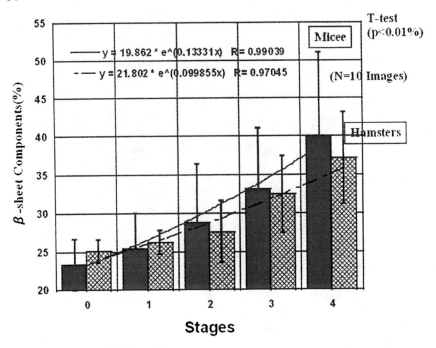

Figure 5. The % β-sheet component in infected brain tissus of hamster and mouse at each stages

(6) β-sheet component percentage (%) in the infected stages of hamster and mice: The average value of the highest percentages for the 10 images (hamster and mouse) of the β-sheet component were plotted against each of the infected stages as shown in Figure 5. The Y-axis scale was expanded from 20 to 55 %. These percentages increased exponentially with the affected stages in both cases. Although the infected speed was different, the relative β-sheet component in hamster brain was lower than that in the mouse one against the stages. Hopefully the combination of infrared spectroscopic techniques and IR SSE software used in this study will prove useful in future, studies for the detection (*14*) of prion diseases.

References

1. Kneipp, J.; Beekes, M.; Lasch, P.; Naumann, D. *J. Neurosci.* **2002**, *22*, 2989-2997.
2. Yamada, T.; Miyoshi, N.; Ogawa, T.; Akao, K.; Fukuda, M; Ogasawara, T.; Kitagawa, Y.; Sano, K. *Clinical Cancer Res.* **2002**, *8*, 2010-2014.

3. Shinagawa, M.; Takahashi, K.; Sasaki, S.; Doi, S.; Goto, H.; Sato, G. *Microbiol. Immunol.* **1985**, *29*, 543-551. *J. Infect. Dis.* **1975**, *131*, 104-110.
5. Manoharan, R.; Baraga, J.J.; Rava, R.P.; Dasari, R.R.; Fitzmaurice, M.; Feld, M.S. *Atherosclerosis*, **1993**, *103*, 181-193.
6. Rigas, B.; Wong, P.T.T. *Cancer Res.* **1992**, *52*, 84-88.
7. Wong, P.T.T.; Goldstein, S.M.; Grekin, R.C.; Godwin, T.A.; Pivik, C.; Rigas, B. *Cancer Res.* **1993**, *53*, 762-765.
8. Rigas, B.; Morgello, S.; Goldman, I.S. Wong, P.T.T. *Proc. Natl. Acad. Sci. USA.* **1990**, *87*, 8140-8144.
9. Dong, A.; Huang, P.; Caugjey, W. S. *Biochem.* **1990**, *29*, 3303-3308.
10. Akao, K. *JASCO Report* **2002**, *44*, 54-57.
11. Sarver, R.W.Jr.; Krueger, W.C. *Anal. Biochem.* **1991**, *194*, 89-100.
12. Caughey, B.W.; Dong, A.; Bhat, K.S.; Ernst, D.; Hayes, S.F.; Caughey, W.S. *Biochem.* **1991**, *30*, 7672-7680.
13. Murakami, K.; Irie, K.; Morimoto, A.; Ohigashi, H.; Shindo, M.; Nagao, M.; Shimizu, T.; Shirasawa, T. *J. Biol. Chem.* **2003**, *278*, 46179-46187.
14. Martin, T.C.; Moecks, J.; Belooussov, A.; Cawthraw, S.; Dolenko, B.; Eiden, M.; von Frese, J.;Kohler, W.; Schmitt, J.; Somorjai, R.; Udelhoven, T.; Verzakov, S.; Petrich, W. *Analyst,* **2004**, *129*, 897-901.

Chapter 5

Proteomic Applications of Drop Coating Deposition Raman Spectroscopy

Corasi Ortiz[1], Yong Xie[1], Dongmao Zhang[2], and Dor Ben-Amotz[1,*]

[1]Department of Chemistry, Purdue University, West Lafayette, IN 47907
[2]GE Advanced Materials, Washington, WV 26181

Recent studies have demonstrated that the drop coating deposition Raman (DCDR) method may be used to obtain high quality vibrational Raman spectra for the identification and quantitiation of proteins, peptides, glycans, and pharmaceuticals. Here we demonstrate applications of DCDR to chromatographic detection, and the identification of peptide and protein structural changes induced by phosphorylation, fibrillation and ligand binding. Such results demonstrate the significant promise of normal (as opposed to surface or resonance enhanced) Raman spectroscopy as a proteomic diagnostics, drug screening and chromatographic detection method with exceptional resolution and reproducibility for high performance chemical identification and quantitiation applications.

Introduction

Although Raman spectroscopy has traditionally been viewed as a method that is primarily applicable to solids and high concentration solutions, new techniques such as the drop coating deposition Raman (DCDR) method have significantly enhanced the utility of Raman in biochemical applications (1-4). Thus, the exceptional structural sensitivity of Raman spectroscopy may now be more broadly applied to the analysis of small quantities of material derived from dilute solutions of relevance to biological, pharmaceutical, environmental and other such applications (5-7). More specifically, the DCDR method facilitates the rapid collection of high quality normal Raman (as opposed to surface or resonance enhanced Raman) spectra of analytes derived from solutions as dilute as 1 μg/mL with volumes as small as 1 nL (8-10). For instance, DCDR Raman spectra have been used to distinguish different conformational and/or post-translational modification states in proteins and peptides, even when the compounds of interest have very similar structure and/or identical mass (8,9).

The key advantages of the DCDR method include its simplicity, rapidity, reproducibility, and high chemical information content (5). In a typical implementation of DCDR, a small volume (in the μL to nL range) of a dilute protein solution is deposited onto a Teflon-coated highly polished stainless steel substrate and a Raman spectrum is collected from the hydrated protein ring which remains after solvent evaporation (either in air or in a vacuum desiccator) (5,6). Important attributes of a good DCDR substrate include high solvophobicity (to a wide variety of solvents), high optical reflectivity and low substrate Raman background. The enhanced detection limits obtained using DCDR arise from the fact that the substrate leads to the formation of a highly concentrated evaporation residue (11,12). In particular, proteins and peptides usually deposit in a ring around the edge of the deposition region, while more highly soluble compounds tend to deposit in the central region (13).

Figure 1 shows typical DCDR spectra of insulin obtained from (a) a 4 mM aqueous solution and (b) a DCDR deposit. The clear similarity of the two spectra suggests that the structure of insulin in the DCDR deposit is very similar to that in an aqueous solution, and thus that the majority of the protein in the deposit remains in a native configuration. More detailed studies have determined that there are about 10-20 water molecules per insulin (6), while other proteins tend to retain a different number of hydration waters in DCDR deposits. Here we present some additional examples of the DCDR method to proteomics, including monitoring protein fibrillation, resolving chromatographic protein bands, and detecting protein-ligand binding events.

Experimental Parameters

Insulin, lysozyme, human serum albumin (Sigma-Aldrich) and tyrosine phosphorylated and unphosphorylated peptides (Bachem) were used with out further purification. Sodium nitrate, sodium dodecyl sulfate (SDS), and suramin were also procured at Sigma-Aldrich. High-purity water (Millipore) or J.T. Baker ultrapure water (Malinchrodt Baker) were used to prepare all the solutions with concentrations of 100 µM, unless stated otherwise. Lysozyme fibrils were formed by preparing a 2mg/mL lysozyme solution in 90% ethanol. Insulin solutions (2mg/mL) used in the chromatographic studies were prepared using an acidified aqueous solution (0.1% trifluoroacetic acid, TFA).

Protein-ligand solutions for the binding studies were prepared at a 1:1 ratio. After the equilibration, 400 µL of protein-ligand solution are placed in a previously rinsed microcon YM-10 (Millipore) ultrafiltration tubes (molecular cutoff of 10,000 g/mol) to remove the excess ligand (6). After 30 min of centrifugation at 1300 rpm, 400 µL of water were added to the ultrafiltration tube and vortexed. The same steps are followed to prepare the control samples.

All protein deposits were produced by pipetting 3 µL of a protein solution onto a spectRIM substrate (Tienta Sciences). The protein deposits were, then, dried in a vacuum desiccator containing anhydrous calcium sulfate (W.A. Hammond Drierite) and connected to a mechanical vacuum pump (model DOA-120-AE, Water Associates). After the samples were visually dry (about 15 min for a 3 µL deposit), DCDR spectra were obtained from the protein ring deposit.

All of the DCDR spectra shown in this work were acquired with a home-built confocal Raman system equipped with a 632.8 nm, 45 mW, HeNe excitation laser [5]. A 100x objective (NA 0.95, Olympus) was used to both focus the laser to a spot size of approximately 1 µm at the protein-coated substrate surface (with about 10 mW of laser power at the sample) and to collect the back-scattered Raman light.

DCDR Spectra of Fibrils

The aggregation of protein fibrils in organs is believed to be the cause of several degenerative disorders including Alzheimer's and Parkinson's disease. Amyloid fibrils are structures which, regardless of the identity of the protein, share a common cross-β-sheet core structure. Several Raman spectroscopic studies have focused on insulin amyloid fibrils (14-19). We have recently used DCDR to confirm the previously reported results and perform a more detailed difference spectroscopic analysis to quantify the principal Raman spectral features associated with insulin fibrillation (20). The following results demonstrate that very similar DCDR difference spectral features are observed upon fibrillation of lysozyme.

Figure 2 shows the DCDR spectra of lysozyme fibrils (a), native lysozyme (b), and their difference spectrum (c). As can be seen in the fibril spectrum, there is a considerable relative intensity increase and shift of the amide I (~ 1660 cm^{-1}) and amide III (~ 1260 cm^{-1}) bands. These blue and red shifts of the amide I and amide III bands, respectively, are consistent with an increase in β-sheet

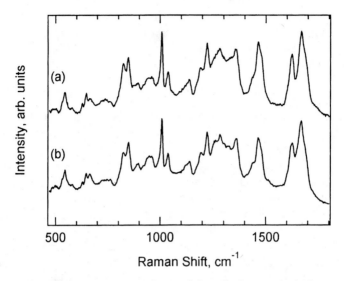

Figure 1. Normal Raman spectra obtained directly from a 4 mM bovine insulin solution. (b) DCDR spectra of a 10 µM bovine insulin solution.

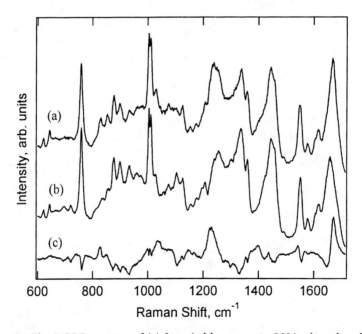

Figure 2. The DCDR spectra of (a) 2 mg/ml lysozyme in 90% ethanol and (b) 2 mg/ml lysozyme in aqueous solution. Spectra (a) and (b) are scaled to the same peak intensities at 1003 cm^{-1}. Spectrum (c) is the difference between spectra (a) and (b). All spectra are plotted using the same scale offset for clarity.

content of the structure (*14-19*). Similar analysis of other proteins (such as insulin and ovalbumin) reveal similar difference spectral features and so suggest that DCDR could be used as a general method for detecting fibril formation.

Detection of Tyrosine Phosphorylation in Peptides

Protein phosphorylation is an important biochemical mechanism used by organisms to control cellular processes. Phosphorylation in eukaryotes occurs predominantly on serine, threonine and tyrosine (which are the only amino acids with hydroxyl groups) (*21-23*). Thus, several publications report the detection of the sites and the degrees of phosphorylation of different peptides (*24-26*). These include our recent demonstrations that DCDR may be used to detect sites of phosphorylation in peptides (*9,10*), as well as a soon to be published demonstration that DCDR may be used to quantify the phosphorylation of α-casein proteins (*27*).

As an example of the sensitivity of Raman spectra to tyrosine phosphorylation, Figure 3 presents the DCDR spectra of a pentamer in its (a) unphosphorylated [IYGEF] and (b) phosphorylated [IY(PO$_3$H$_2$)GEF] states. Since aromatic residues have very high Raman intensities, they produce many of the prominent bands in a protein and peptide spectra. Thus, it is not surprising that tyrosine phosphorylation will induce significant Raman spectral changes. For instance, the intensity of the band at about 852 cm^{-1} from the Fermi-resonance tyrosine doublet is reduced considerably. Further spectral changes induced by tyrosine phosphorylation can be seen in other tyrosine bands at around 1610 cm^{-1} (ring stretch), and 1250 cm^{-1} (ring-O stretch). Given the important role of tyrosine phosphorylation in cell signaling (*28*), it is clear that DCDR could offer a promising method for future tyrosine phosphorylation studies (and such studies are in progress by others).

DCDR Detection of Chromatography Fractions

Previous studies in our lab have demonstrated that the protein concentration detection limit of the DCDR method is comparable to that of standard UV chromatographic detectors (*5,8*). For example, DCDR may be used to detect lysozyme at concentrations as low as 0.1 μM (with a 5 μL deposition volume), which is too low to be detected using UV absorbance at 280 nm (in a 1 cm cell). DCDR has been used to obtain high quality Raman spectra of two insulin variants purified by reverse-phase high-pressure liquid chromatography (RP-HPLC) after direct deposition of the corresponding chromatographic fractions (Figure 4) (*5,8*).

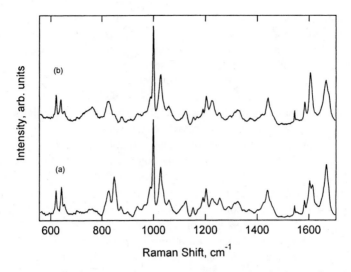

Figure 3. Raman spectra of a pentameric peptide which is either (a) unphosphorylated, or (b) phosphorylated at a single tyrosine residue.

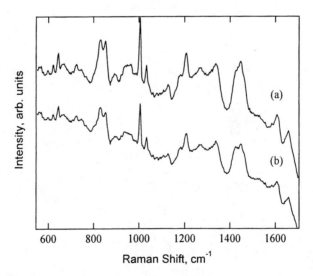

Figure 4. Raman spectra obtained after depositing 10 µL of (a) 21 µM human insulin and (b) 3 µM bovine insulin, each derived from an HPLC fraction (in acetonitrile/water with a 48:52 volume ratio, and using an integration time of 1000s).

58

Current efforts in our lab are focused on using DCDR to enhance chromatographic resolution. In other words, we propose to use the different Raman spectral fingerprints of compounds in a chromatographic peak containing multiple poorly resolved species, to deconvolute and quantify the sub-components. Previous works by Marquardt, et al. and He, et al. have demonstrated the viability of this concept by using conventional Raman and SERS to detect small partially resolved analytes such as alcohols (29,30). However, to our knowledge, no previous such studies have been applied to realistic proteomic mixtures. Figure 5 shows our preliminary DCDR results obtained from sequential HPLC fractions containing two slightly overlapping protein bands. The normalized second derivative DCDR spectra shown in Figure 5 clearly reveal the distinct spectral fingerprints of the two protein sub-bands. DCDR may also be readily combined with reflectance FT-IR and MS performed directly on samples after deposition on a DCDR slide. For example, DCDR may be used in conjunction with standard protein and peptide detection methods such as MALDI-MS or more recent methods such as DESI-MS (5,31). In order to perform MALDI-MS, the DCDR deposited protein or peptide samples are simply mixed with a standard MALDI matrix and directly analyzed using a commercial MALDI-MS instrument (5). Thus, the DCDR chromatographic detection may be combined with UV, FT-IR and MS to derive more information than could be obtained using any of these methods alone.

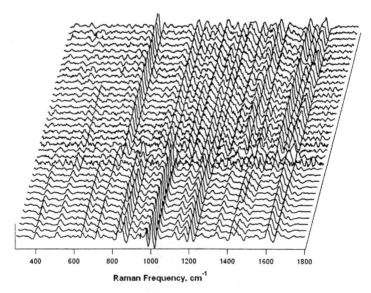

Figure 5. DCDR spectra (in normalized second derivative form) obtained from sequential fractions of a single chromatographic elution containing two partially resolved proteins. Note the clear difference between the spectral signatures of the two protein species: insulin and cytochrome C.

Protein-Ligand Binding Detection using DCDR

Some of the characteristics that make DCDR particularly attractive for the analysis of biologically relevant samples are its relative insensitivity to water interference, as well as its low concentration, small volume and minimal sample preparation requirements. Previous protein-ligand binding Raman studies include several important papers by Callender and coworkers, focused on systems such as the binding of pyruvate to lactate dehydrogenase and other protein-ligand systems (*32-35*). On the other hand, because these studies were performed prior to the development of the DCDR method, they required high protein concentrations (mM) and a carefully balanced Raman difference measurement apparatus to extract ligand spectra from the solution Raman spectra of protein complexes. Carey and co-workers also reported beautiful single crystal studies of protein-substrate complexes (*36,37*). The fact that not all proteins have mM solubilities and/or readily form crystals clearly limits the number of systems which could be studied using previous Raman methods.

Here we report our recent results obtained using a new variant of the DCDR method which facilitates the wider application of Raman scattering to protein-ligand binding studies. The key new feature of this method consists of using ultrafiltration of protein-ligand mixtures, after equilibration at micromolar concentration, followed by DCDR deposition and difference spectral analysis. We thus refer to this new procedure as the ultra-filtration-Raman-difference (UFRD) method (*38*). The following are representative examples of our preliminary UFRD results pertaining to protein binding processes with various different types of ligands.

More specifically, Figures 6 - 8 show UFRD results for three protein-ligand binding systems: i. nitrate ion binding to lysozyme, ii. sodium dodecyl sulfate (SDS), a protein denaturant, binding to lysozyme, and iii. suramin, a drug, binding to human serum albumin. In the case of the lysozyme-nitrate complex (Figure 6), a readily detectable change in the lysozyme-nitrate spectrum (a), as compared to the lysozyme control spectrum (b), is evidenced by the appearance of rather sharp and intense band at about 1040 cm^{-1}. This can be better seen in the difference spectrum (c) which looks remarkably similar to the Raman spectrum a high concentration nitrate solution (d). The major difference between (c) and (d) is the fact that the nitrate band is shifted to lower frequencies which implies that protein-bound nitrate is not fully hydrated.

Since Raman is primarily sensitive to primary and secondary structure, the absence of any protein related bands in the lysozyme-nitrate difference spectrum implies that this binding event does not produce significant secondary structure changes (although it may well affect the tertiary structure of the protein). To better illustrate the sensitivity of the UFRD method to secondary structural changes, we have investigated the effects of SDS, a ligand that is known to denature proteins. Figure 7 shows the UFRD spectrum of the lysozyme-SDS complex. The lysozyme-SDS spectrum (a) shows several changes which can

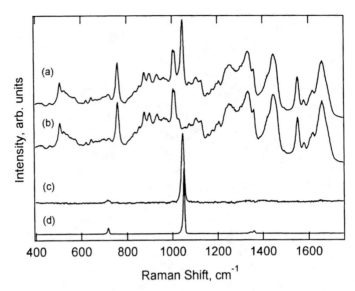

Figure 6. DCDR spectra of the lysozyme-nitrate complex (a) and lysozyme control (b). Spectrum (c) is the difference between (a) and (b) while (d) is the nitrate control spectrum. All spectra are plotted using the same scale and offset for clarity.

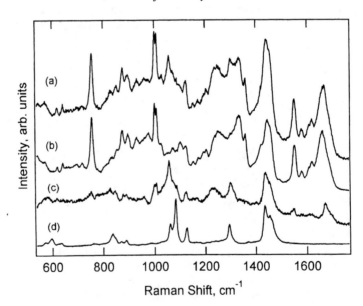

Figure 7. DCDR spectra of the lysozyme-SDS complex (a) and lysozyme control (b). Spectrum (c) is the difference between (a) and (b) while (d) is the SDS control spectrum. All spectra are plotted using the same scale and offset for clarity.

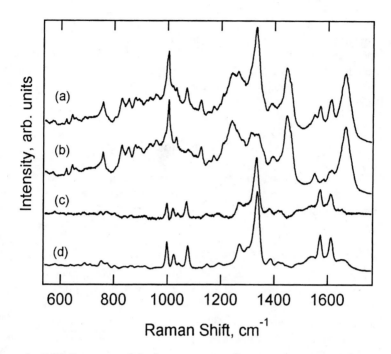

Figure 8. DCDR spectra of the human serum albumin-suramin complex (a) and human serum albumin control (b). Spectrum (c) is the difference between (a) and (b) while (d) is the suramin control spectrum. All spectra are plotted using the same scale and offset for clarity.

best be seen after subtraction of the lysozyme control (b). The resulting difference spectrum (c) clearly shows features which resemble those of the SDS control (d), as well as additional features indicating a change in protein secondary structure. More specifically, the features in the difference spectrum around 1230 cm^{-1} and 1670 cm^{-1} (amide III and amide I, respectively), are consistent with secondary structure changes resulting from protein denaturation (*39*).

Another example of protein-ligand binding is shown in Figure 8, which compares the DCDR spectra of the human serum albumin-suramin complex (a) and the human serum albumin control (b). The resulting difference spectrum (c) clearly reveals an intensity increase broad peak at around 1333 cm^{-1} as well as other peaks that clearly resemble those of the suramin control sample (d). In this case, as with the lysozyme-nitrate complex, there is again no evidence of significant suramin induced protein secondary structure change.

Although, we believe the UFRD technique can be very useful for protein binding studies, it is also important to note that our preliminary results also indicate it is most appropriate for strongly Raman active ligands (such as aromatic compounds) and to systems which have a binding affinity that is stronger than about 100 μM (i.e. a binding constant greater than $1x10^4$ M^{-1}). More specifically, we have tested the method using complexes with affinities between 1 μM and 100 μM (and there is no reason why it should not be applicable to affinities < 1 μM). Our ongoing studies are also focused on using a variant of the UFRD methods to quantify the binding constants of protein-ligand complexes.

Conclusions

DCDR has been used to obtain high quality Raman spectra from small volumes of dilute (low concentration) proteins and peptides, as well as other compounds of biochemical and/or pharmaceutical relevance. This technique has significantly extended the concentration and small volume detection limits of Raman spectroscopy, and thus sets the stage for more widespread use of this technique in biomedical research and discovery. The results we have presented illustrate the use of DCDR in the identification of conformational changes associated with fibrillation, denaturation, post-translational modifications and protein-ligand binding. Moreover, we have shown that DCDR may be used as a chromatographic detection method, either alone or in combination with UV, FTIR and MS, and thus the high chemical information content of Raman spectra could be use to enhance the resolution and chemical identification of compounds, and thus extend the capabilities of current separation, detection and quantitation methods.

References

1. Carter, E. A.; Edwards, H. G. M. Practical Spectroscopy **2001**, *24*, 421-475.
2. Mulvaney, S. P.; Keating, C. D., *Anal. Chem.* **2000**, *72*, 145R-157R.
3. Lyon, L. A.; Keating, C. D.; Fox, A. P.; Baker, B. E.; He, L.; Nicewarner, S. R.; Mulvaney, S. P.; Natan, M. J., *Anal. Chem.* **1998**, *70*, 341R-361R.
4. Carey, P. R. *Biochemical Applications of Raman and Resonance Raman Spectroscopies*; Academic Press: New York, 1982.
5. Zhang, D.; Xie, Y.; Mrozek, M. F.; Ortiz, C.; Davisson, V. J.; Ben-Amotz, D., *Anal. Chem.* **2003**, *75*, 5703-5709.
6. Ortiz, C.; Zhang, D.; Xie, Y.; Ribbe, A. E.; Ben-Amotz, D., *Anal. Biochem.*, in press.
7. Lambert, J. P.; Ethier, M.; Smith, J. C.; Figeys, D., *Anal. Chem.* **2005**, *77*, 3771-3788.
8. Ortiz, C.; Zhang, D.; Xie, Y.; Davisson, V. J.; Ben-Amotz, D., *Anal. Biochem.* **2004**, *332*, 245-252.
9. Zhang, D.; Ortiz, C.; Xie, Y.; Davisson, V. J.; Ben-Amotz, D., *Spectrchim. Acta A.* **2005**, *61*, 471-475.
10. Xie, Y.; Zhang, D.; Jarori, G.; Davisson, V. J.; Ben-Amotz, D., *Anal. Biochem.* **2004**, *332*, 116-121.

11. Deegan, R. D.; Bakajin, O.; Dupont, T. F.; Huber, G.; Nagel, S. R.; Witten, T. A., *Nature* **1997**, *389*, 827-829.

12. Hu, H.; Larson, R. G., *J. Phys. Chem. B* **2006**, *110*, 7090-7094.

13. Zhang, D.; Mrozek, M. F.; Xie, Y.; Ben-Amotz, D., *Appl. Spectrosc.* **2004**, *58*, 929-933.

14. Yu, N.-Y.; Liu, C.; O'Shea, D. C., *J. Mol. Biol.* **1972**, *70*, 117-132.

15. Yu, N.-Y.; Jo, B. H.; Chang, R. C. C.; Huber, J. D., *Arch. Biochem. Biophys.* **1974**, *160*, 614-622.

16. Apetri, M. M.; Maiti, N. C.; Zagorski, M. C.; Carey, P. R.; Anderson, V. E., *J. Mol. Biol.* **2006**, *355*, 63-71.

17. Dong, J.; Wan, Z.; Popov, M.; Carey, P. R.; Weiss, M. A., *J. Mol. Biol.* **2003**, *330*, 431-442.

18. Hiramatsu, H.; Goto, Y.; Naiki, H.; Kitagawa, T., *JACS* **2005**, 127, 7988-7989.

19. Gosal, W. S.; Clark, A. H.; Ross-Murphy, S. B., *Biomacromolecules* **2004**, *5*, 2408-2419.

20. Ortiz, C.; Zhang, D.; Xie, Y.; Ribbe, A. E.; Ben-Amotz, D., manuscript in preparation.

21. Ahn, N.; Resing, A. K., *Nat. Biotechnol.* **2001**, *19*, 317-318.

22. McLachlin, D. T.; Chait, B. T., *Curr. Opin. Chem. Biol.* **2001**, *5*, 591-602.

23. Cohen, P.; *Nat. Cell. Biol.* **2002**, *4*, E127-E130.

24. Marshall, P.; Heudi, O.; Bains, S.; Freeman, H. N.; Abou-Shakra, F.; Reardon, K., *Analyst* **2002**, *127*, 459-461.

25. Adamczyk, M.; Gebler, J. C.; Wu, J., *Rapid Commun. Mass Spectrom.* **2001**, *15*, 1481-1488.

26. Mann, M.; Ong, S.; Gronborg, M.; Steen, H.; Jensen, N. O.; Pandey, A., *Trends Biotechnol.* **2002**, *20*, 261-268.

27. Goodacre, R.; Jarvis, R.; López-Diez, C., private communication.

28. Ullrich, A.; Schlessinger, J., *Cell* **1990**, *61*, 203-212.

29. Marquardt, B. J.; Vahey, P. G.; Synovec, R. E.; Burgess, L. W., *Anal. Chem.* **1999**, *71*, 4808-4814.

30. He, L.; Natan, M. J.; Keating, C. D., *Anal. Chem.* **2000**, *72*, 5348-5355.

31. Takáts, Z.; Wiseman, J. M.; Gologan, B.; Cooks, R. G., *Science* **2004**, *306*, 471-473.

32. Deng, H.; Zheng, J.; Burgner, J.; Callender, R., *PNAS* **1989**, *86*, 4484-4488.

33. Callender, R.; Deng, H.; Gilmanshin, R., *J. Raman Spectrosc.* **1998**, *29*, 15-21.

34. Deng, H.; Callender, R., *J. Raman Spectrosc.* **1999**, *30*, 685-691.

35. Wang, J. H.; Xiao, D. G.; Deng, H.; Webb, M. R.; Callender, R., B*iochemistry* **1998**, *37*, 11106-11116.

36. Carey, P. R.; Dong, J., *Biochemistry* **2004**, *43*, 8885-8893.

37. Altose, M. D.; Zheng, Y. G.; Dong, J.; Palfey, B. A.; Carey, P. R., *PNAS* **2001**, *98*, 3006-3011.

38. Xie, Y.; Zhang, D.; Ortiz, C.; McCready, T.; Ben-Amotz, D., manuscript in preparation.

39. Pelton, J. T.; McLean, L. R., *Anal. Biochem.* **2000**, *277*, 167-176.

Chapter 6

Infrared Spectroscopy of Microorganisms: Characterization, Identification, and Differentiation

E. A. Carter[1], C. P. Marshall[1], M. H. M. Ali[1], R. Ganendren[2],
T. C. Sorrell[2], L. Wright[2], Y.-C. Lee[3], C.-I. Chen[3], and P. A. Lay[1]

[1]Vibrational Spectroscopy Facility, School of Chemistry, The University
of Sydney, Sydney, New South Wales 2006, Australia
[2]Centre for Infectious Diseases and Microbiology and Westmead
Millennium Institute, The University of Sydney at Westmead,
Department of Infectious Diseases, Westmead Hospital, Westmead,
New South Wales 2145, Australia
[3]National Synchrotron Radiation Research Centre, No. 101 Hsin-Ann
Road, Hsinchu, Science Park, Hsinchu 30076, Taiwan

Fourier transform infrared (FTIR) spectroscopy has been used
extensively for characterisation, identification and differentiation of micro–organisms. The technique offers a number
of advantages including: non–destructive analysis, high
reproducibility, rapid sample throughput, and requires minimal
sample amounts and handling. In this paper, examples are
given of the application of the specific FTIR sampling
techniques: attenuated total reflectance and synchrotron–
radiation–based infrared microspectroscopy, to the analysis of
the yeast *Cryptococcus neoformans*, and the acritarchs *Satka
favosa* and *Leiosphaeridia crassa*.

A Brief Overview of Infrared Spectroscopy

Infrared (IR) spectroscopy is a non–destructive analytical technique which provides information about the chemical content (functional groups) of a sample. Other information that can be derived include: chemical structure, molecular environment, bond angle, length, geometry and conformation *(1)*. The sample is irradiated typically with mid–infrared light (4000–400 cm^{-1}) and photons are either absorbed, transmitted, reflected or scattered. A change in the dipole moment during a molecular vibration may result in absorption of IR light (2) .

Molecular vibrations within biological samples typically produce broad, overlapping and complex IR bands that originate from numerous individual components that often have the same chemical functionality. However, the biochemical building blocks such as proteins, lipids and nucleic acids have diagnostic peaks, or marker bands, that often dominate the spectra. Changes in marker band ratios, positions and height are attributed to structural changes that depend on growth status, cell cycle or disease state *(3, 4)* and can provide a wealth of information e.g., protein conformational changes (peak position and lineshape), or increases/decreases in the amount of a specific biomolecule within a sample (peak intensity).

FTIR spectroscopy is an accurate, sensitive and rapid technique for micro–organism identification and differentiation *(4)* that is reproducible and specific for pathogens (microbial cells) at the genus, species and strain level *(5-7)*. This technique overcomes the main weaknesses of conventional diagnostics e.g., traditional biochemical and serological tests, molecular biological, fluorescent antibody and immunological methods. These methods lack accuracy, require time consuming protocols that delay diagnosis, are dependent on resources (kits, reagents, dyes) and/or are expensive *(6, 8)*. The advantages of FTIR spectroscopy over other analytical techniques make the method applicable to: (i) very rapid identification of life–threatening pathogens; (ii) monitoring compositional and structural changes during cell growth; and (iii) characterisation and screening of environmental micro–organisms *(8-10)*.

Sampling Accessories and Techniques

One of the strengths of FTIR spectroscopy is its diverse range of sampling techniques. Examples include attenuated total reflectance (ATR), diffuse reflectance (DRIFT), photoacoustic (PA), grazing angle, microspectroscopy and more specialised techniques such as synchrotron–radiation–based FTIR (SR–FTIR) microspectroscopy *(11-13)*. The following section outlines ATR and SR–FTIR microspectroscopy in more detail and then provides specific applications of their use for the analysis of *C. neoformans*, *S. favosa* and *L. crassa*.

Attenuated Total Reflection (ATR)

Biospectroscopy using an attenuated total reflection (ATR) accessory provides a number of advantages including: good reproducibility, sensitivity, ease of use, rapid sample throughput and the ability to examine biological samples under physiologically relevant conditions, i.e., temperature, pressure, pH, and only microlitres (or micrograms) of sample are required *(14, 15)*.

ATR spectra are generated when infrared light passes through an internal reflection element (IRE) and is internally reflected. The (reflective wave) evanescent field penetrates the sample to a typical depth of 0.5–2 μm, and an absorptions could be detected. Hence, it only interrogates the outer layers of samples i.e., the outer walls and membranes of cells. If the sample absorbs the radiation it will be attenuated before re–entering the IRE *(3, 16)*. Internal reflection will only occur if the angle of incidence is greater than or equal to the critical angle (α_c) as determined by equation 1.

$$\sin \eta_c = \eta_{21} = \eta_2/\eta_1 \tag{1}$$

where η_1 = refractive index of IRE,
η_2 = refractive index of sample, and
$\eta_{21} = \eta_2/\eta_1$

The design of any efficient ATR accessory requires a good sample/IRE interaction to obtain high quality, reproducible spectra. Experimental difficulties are normally related to the IRE in terms of inadequate cleaning, damage and/or irreproducible optical contact.

Microspectroscopy

Infrared microspectroscopy (IR–MSP) is a powerful tool used to probe specific areas of interest in biological samples. However, the spatial resolution of IR–MSP is limited by two factors: aperture size and the diffraction limit. In order to obtain detailed structural information from within biological samples, e.g., individual cells or cell organelles, it would be necessary to use small aperture sizes (below 10 μm) that significantly reduce signal intensity and result in very poor spectral signal–to–noise ratios. Additionally, when the size of the aperture is similar to the wavelength of the incident light, the diffraction effects becomes apparent and further reduce spectral quality *(17)*.

Synchrotron–Radiation–Based Infrared (SR–FTIR) Microspectroscopy

SR–FTIR microspectroscopy takes advantage of the extremely bright, intense, and non–divergent characteristics of the synchrotron source, which is approximately 1000 times brighter than the conventional glowbar source used in a laboratory–based instrument. This allows the use of significantly smaller aperture sizes (3–10 μm), faster data collection, and produces spectra of high quality and good signal–to–noise ratios, and is limited only by diffraction *(11, 12, 17)*. The improved spatial resolution (5–10 μm) that SR–FTIR micro-spectroscopy offers over laboratory FTIR spectrometers (30–50 μm) enables mapping of individual cells in tissues or intracellular features of interest.

Functional Group Mapping and Imaging

Mapping and imaging are specialised vibrational spectroscopic sampling techniques used to visualise the type and distribution of biological constituents (e.g., proteins, lipids and nucleic acids) within a sample. Mapping involves collecting spectra point–by–point, whereas imaging involves simultaneous collection of spectra, i.e., if using a focal plane array of 64 × 64 pixels then 4096 spectra are obtained simultaneously. Both mapping and imaging techniques produce large amounts of data.

When the peak of interest is strong and is not overlapped by other spectral features, data analyses are straightforward and individual wavelength–specific images can be constructed *(18)*. In most cases, the sample complexity and, therefore, that of the spectrum, requires a more rigorous approach to data analyses which is facilitated by multivariate analytical techniques, e.g., cluster analysis, hierarchical analysis and the like that yield high quality detailed images which can reveal anatomical features of interest *(18, 19)*.

Specific Applications

The application of FTIR spectroscopy to the characterisation, identification, and differentiation of biological samples is booming *(4-8, 10, 20)*. This is due to the development of automated high throughput instrumentation and chemometric software packages specifically designed for the analysis of spectroscopic data. The following are specific examples of our research into FTIR spectroscopic analysis of micro–organisms.

Characterisation of a Heat and Chemically Treated Micro–organism

Cryptococcus neoformans is an opportunistic pathogen that primarily affects immunocompromised individuals, particularly patients with T–lymphocyte deficiencies *(21)*. *C. neoformans* is an encapsulated basidiomycetous yeast that is categorised into three biochemically and genetically distinct species (*C. neoformans and C. gattii*) and varieties (*C. neoformans* var. *grubii* and *C. neoformans* var. *neoformans*) and four serotypes (A, B, C and D) *(22)*.

The fungus is surrounded by a thick layer of capsular polysaccharides, glucuronoxlyomannan (GXM) and galactoxylomannan (GalXM) that are external to the cell wall *(21, 23, 24)*. GXM makes up ~90% of the capsular mass and has a backbone of α–1,3–D–mannopyranose units with single β–D–xylopyranosyl and β–D–glucuronopyransosyl residues. GalXM is a minor component, contributing to ~7% of the capsule mass and having a backbone of α–1,6–linked galactose units with side chains composed of galactosyl, mannosyl and xylosyl residues. A third component of the capsule structure, mannoprotein, is a minor fraction of the exopolysaccharide *(23)*.

The rigid cell wall is composed of approximately 90% carbohydrates (β–glucans and α–glucans), 5–10% protein and the remaining 1–2% is attributed to glucosamine and lipids *(23)*. Inside the cell wall is the plasma membrane, composed predominantly of lipids and glycoproteins. The cytoplasm is dominated by proteins and nucleic acids, but also contains lipid droplets.

The desiccated cells and spores that are formed by haploid fruiting, or sexual reproduction, are presumed to serve as infective propagules *(21)*. Infection occurs following inhalation. In most individuals the infection is cleared, or remains dormant, unless the immune system becomes compromised, when the fungus disseminates, with particular trophism for the central nervous system. In severe cases, the infection progresses to potentially fatal meningoencephalitis *(23)*.

The first step in pulmonary infection with *C. neoformans* is adhesion to the lung epithelium. A protein that is necessary for the initial step is secreted phopholipase B (PLB1). PLB1 is a multifunctional protein with three enzyme activities but it is not known which activity is most important in virulence. A recent study has measured the adhesion of the *C. neoformans* var. *grubii* serotype A strain H99, its PLB1 deletion mutant (*Δplb1*) and the reconstituent strain to the human epithelial cell line, A549, and determined that it was PLB1–dependent *(25)*.

The effect of cryptoccal viability on adhesion to a monolayer of A549 cells was examined by comparing live and dead cryptococci. The cryptococci were treated with heat (80 °C for 20 min) or with an antifungal ether–lipid analogue known as miltefosine *(25)*. FTIR/ATR spectroscopy was used to investigate differences in the strains after killing with the heat and chemical treatments.

Figure 1 presents FTIR/ATR spectra of live and treated cryptococci H99 (A, B) and Δplbl cells (C, D). The spectra of the live and heat treated samples are dominated by the amide I (1639 cm^{-1}) and II (1547 cm^{-1}) bands which are generated by the peptide bond formed between amino acid residues within a polypeptide chain or protein (26). These amide vibrations are attributed to the mannoprotein component of the capsule and cell wall of Cryptococcus (23, 27). The other prominent feature is the broad intense peak centred at ~1024 cm^{-1} which is attributed to numerous v(C–O) vibrational modes from polysaccharides also present within the capsule and cell wall (28).

There are several features in the spectra of the heat and miltefosine treated cells that are significantly different from those of the live cells (Figure 1 and Table I). The position of the amide I band in the spectra of the live samples is characteristic of proteins existing in a predominately β–sheet conformation (1). The shift of this band in the spectra of the heat killed cells indicates that the protein secondary structure has undergone a conformational change. However, the most significant changes in this region are observed in the spectra of the miltefosine treated cryptococcal cells. The position of the amide I band remains constant c.f. the live cells; however, the band intensity approximately halves, the lineshape changes dramatically, a large shoulder is prominent at 1596 cm^{-1} and the amide II band completely disappears.

The amide I band was chosen for detailed analysis as its position is sensitive to protein secondary structure. The band arises predominately from v(C=O) stretching of the carbonyl group within the peptide (CONH) bond. Other minor contributing factors arise from v(CN) out–of–plane, δ(CCN) and δ(NH) in–plane vibrations (1). The broad nature of the amide I band is attributed to the presence of a number of secondary structures within the sample. Derivatives were used to deconvolve spectral band widths and positions. This resolution technique, together with deconvolution and curve fitting, is particularly useful for resolving components within a broad band envelope.

Figure 2 indicates that the second derivative spectra of both live strains are primarily composed of two secondary structures (α-helix at ~1653 cm^{-1} and β-sheet at ~1680 and 1631 cm^{-1}) and is further illustrated by the resolved curve fitted amide region presented in figure 3 (Δplbl data not shown). On heating both samples undergo a secondary structure conformational change to form predominately β–sheet protein structures. This is evidenced by a decrease in the α–helical content (1653/1654 cm^{-1}) and an increase in the β–sheet components (1631/1628 and 1680/1691 cm^{-1}) as seen in Figures 2A, 2B, and 3 (1). The position and intensity of the amide II band (1547 cm^{-1}) also changes in the spectra of both heat killed strains. The other peaks resolved from the curve fitting procedure are assigned to amino acid side chain residues, Table II (1, 29).

The spectra of the miltefosine treated samples reveals a loss of virtually all of the amide I and II structure which implies that the mechanism of drug action

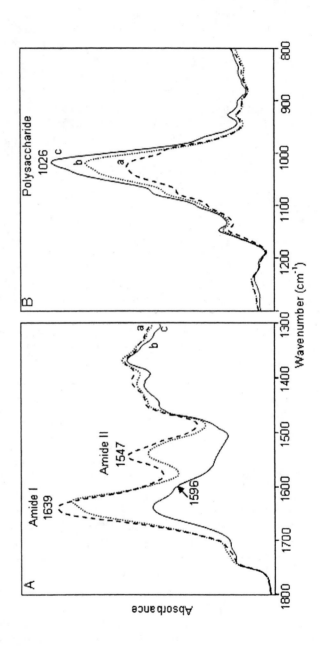

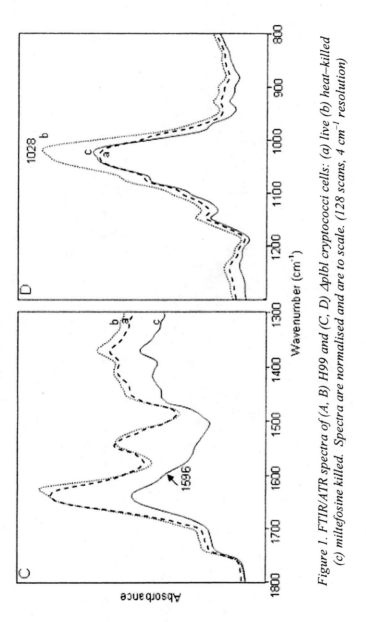

Figure 1. FTIR/ATR spectra of (A, B) H99 and (C, D) Δplbl cryptococci cells: (a) live (b) heat–killed (c) miltefosine killed. Spectra are normalised and are to scale. (128 scans, 4 cm⁻¹ resolution)

Table I. Assignments and Position of Bands from Figure 1

Assignment (1, 26)	Live		Heat Killed		Miltefosine	
	H99	Δplbl	H99	Δplbl	H99	Δplbl
ν(C=O); Amide I	1639	1639	1626	1627	1640	1639
δ(NH); Amide II	1547	1547	1537	1537	–	–
ν(C–O); Polysaccharides	1026	1030	1019	1017	1016	1022

Table II. Assignments and Position of Bands Resolved using Curve Fitting Procedures from Figure 3†

Assignment (1, 30)	Live		Heat Killed		Miltefosine	
	H99	Δplbl	H99	Δplbl	H99	Δplbl
ν(C=O)$_{ester}$	1742	1743	1744	1743	1744	1744
ν(C=O); asp, glu	1723	1725	1724	1725	1723	1727
ν(C=O); Amide I β–sheet/Turns	1676	1678	1680	1677	1679	1680
ν(C=O); Amide I α–helix	1654	1655	1649	1649	1654	1655
ν(C=O); Amide I β–sheet	1631	1631	1623	1623	1632	1633
ν$_{as}$(COO⁻); asp, glu	1597	1597	1596	1597	1598	1597
ν$_{as}$(COO⁻); asp, glu	1571	1567	1572	1571	1571	1571
δ(NH); Amide II β–sheet	1544	1546	1542	1541	1545	1551
ν(C–C)$_{ring}$; tyrosine residue	1517	1518	1516	1516	–	1521

NOMENCLATURE: ν – stretching, δ – deformation, as – anti–symmetric
†NOTE: Curve fitting results not shown for Δplbl spectra.

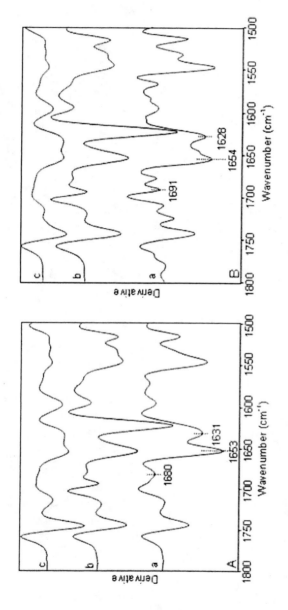

Figure 2. Second derivative spectra of (A) H99 (B) ApIbI cryptococci cells: (a) live (b) heat–killed (c) miltefosine killed.

74

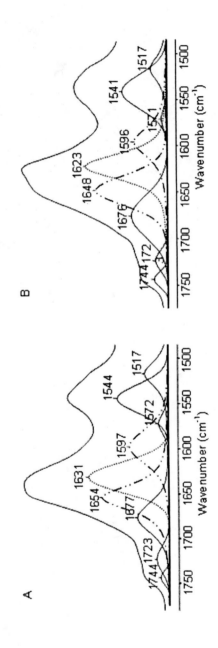

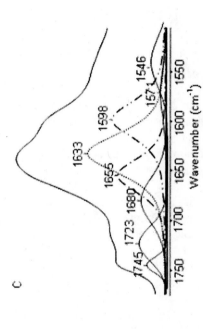

Figure 3. Curve fitting of the amide region (1800–1475 cm^{-1}) of the FTIR/ATR spectra of the H99 strain: (a) live (b) heat–killed (c) miltefosine killed.

involves cleavage of peptide bonds. Miltefosine is structurally similar to a major substrate of PLB1, which is thus a potential drug target. It was hypothesised that miltefosine interferes with the cryptococcal cell wall/membrane biochemistry which is supported by the current results that the protein structure of the cryptococcal cell wall/membrane is affected *(30)*.

Other notable changes are observed in the 1200-900 cm^{-1} region which contains the bands due to numerous $v(C–O)$ vibrations of the polysaccharides of the capsule and cell wall. The position, intensity and lineshape of this band changes as a result of killing the cryptococal cells with both heat and miltefosine. Interestingly, the adhesion ability of these strains to lung epithelial cells differs with the treatment and it has been postulated that glycoprotein adhesins may be exposed during heat-induced capsular changes but are masked by the miltefosine treatment which will be explored in the future *(25)*.

Taxonomic Identification of Fossil Micro–organisms

Micro–algae are diverse and abundant in aquatic environments but most have no, or very limited preservation potential. Nevertheless, a diverse and rich micro–algal fossil record of aquatic palynomorphs exists. This record is an important source of information for solving stratigraphic, palaeoenvironmental (e.g., palaeoclimatological) and evolutionary questions. For this purpose, the morphological characteristics of the fossils have been used almost solely, whereas their potential at the molecular level has not been exploited.

A rich and diverse group of fossil palynomorphs is the Acritarcha. Acritarchs are organic–walled, acid–resistant microfossils of unknown biological affinities, ranging in geological age from the Proterozoic (<2.7 Ga) through to nearly the present. Conventionally, acritarchs have been interpreted as the cysts of eukaryotic micro–algae, but the group probably also includes prokaryotic sheaths, heterotroph protists, vegetative parts of cells or multi–cellular organisms *(31)*, and even animal egg cases *(32)*. Palaeontologists have desired to link diagnostic organic molecules directly to individual microfossil taxa, in part because this might establish the systematic relationships of otherwise problematic fossils. Isolating pure monospecific assemblages is difficult and time-consuming, therefore, the need to develop micro–scale analytical techniques for the successful analyses of acritarch biogeochemistry is desirable an example of which is FTIR micro–spectroscopy *(33)*. A new approach is described here, to not only elucidate organic chemical composition of acritarchs, but in addition to delineate chemical differences within a single acritarch using SR–FTIR microspectroscopy.

Insoluble non–hydrolysable (INH) biopolymers are present in both marine and fresh micro–algae, particularly in Chlorophyceae and Eustigmatophyceae

(34-36). These biomacromolecules, termed algaenans *(37)*, are part of the vegetative cell walls and serve as a structural component consisting of highly aliphatic biopolymers that can be preserved on geological timescales. Algaenans therefore play a major role in kerogen formation via a selective preservation pathway and can be subsequently preserved as ultralaminae *(37)*. In contrast to the algaenans of the motile stage of green micro–algae, dinoflagellates seem to be able to produce a completely different kind of biomacromolecule for their resting cysts, called dinosporin. There is only very limited information on recent dinoflagellate cyst walls. The walls of *Lingulodinium polyedrum* are reported to be relatively condensed and predominantly aromatic, compositionally distinct from 'sporopollenin', unrelated to the walls of green micro–algae (algaenan) *(38)*. Such resistant biopolymeric materials are of special interest to palaeontologists, since they are selectively preserved upon sedimentation and are observed as intact, million–year–old microfossils, which can be used to elucidate biological affinity particularly when morphology alone is not sufficient. Moreover, in contrast to provenance and authenticity issues faced by palaeontologists studying Precambrian (>500 Million year) rocks it is generally straightforward, by virtue of their chemical structures, to recognize when complex hydrocarbon molecules are genuine biogenic remains. Therefore, in the elucidation of biological affinities of fossilized biota, FTIR microspectroscopy is currently the tool of choice in palaeobiology *(33)*.

This study represents the first attempt to develop a model system for the analysis of single acritarchs using SR–FTIR microspectroscopy. The acritarchs used in this study were isolated from two drill cores (Urapunga 6, 230.8 m and Amoco 82/3, 161.7 m) in the Roper Group, Australia (1.5–1.4 billion–year–old). The isolated acritarchs include *Satka favosa* and *Leiosphaeridia crassa*. The geological setting, paleoecological distribution, morphology and wall ultra–structure of these acritarchs have been described previously *(33)*. Figure 4 shows representative SR–FTIR spectra of the two acritarchs investigated. The spectra are of high quality (good signal–to–noise ratio) by comparison with the previously acquired glowbar FTIR spectra *(33)*.

The most prominent features are an intense broad band around 3400 cm^{-1}, assigned to O–H stretching modes (alcoholic OH, phenolic OH, and/or carboxylic OH), a group of overlapped bands around 2900 cm^{-1}, assigned to C–H stretching modes (anti–symmetric and symmetric stretching vibrations from CH_2 methylene groups), moderate to strong bands around 1730–1530 cm^{-1} assigned to the C=O carbonyl stretching modes, a strong band centred at 1600 cm^{-1} due to olefinic/aromatic C=C stretching modes in the spectrum from *L. crassa*, and bands due to C–H deformation modes at 1377 and 1340 cm^{-1} (CH_2 and CH_3). A more detailed peak assignment and comparison between band intensities are given in Table III.

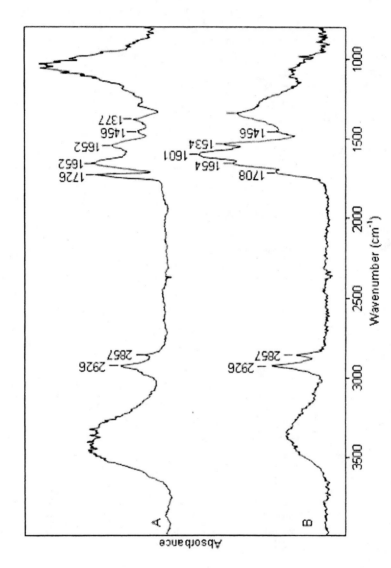

Figure 4. SR– FTIR spectra of (A) S. fabova and (B) L. crassa (32 scans, 4 cm⁻¹ resolution, Aperture size: 20 × 20 μm)

Table III. Assignments, Position (v/cm^{-1}), and Intensities of Absorptions in SR–FTIR Spectra acquired from _S. favosa_ and _L. crassa_

Assignment	v/cm^{-1}	S. favosa	L. crassa
		Intensity	
v(OH)	3600–3000	M	M
v_{as}(CH$_2$)	2926	M	M
v_s(CH$_2$)	2857	M	M
v(C=O)	1726	S	Np
v(C=O)	1708	Np	W
v(C=O)	1654–1652	W	W
v(C=C)	1601	Np	S
v(C=O)	1552	M	Np
v(C=O)	1534	Np	M
δ_s(CH$_2$)	1456	W	W
δ_s(CH$_3$)	1377	W	Np
δ_s(CH$_3$)	1340	Np	W

NOMENCLATURE: S – strong, M – medium, W – weak, Np – not present, v – stretching, δ – deformation, as – anti–symmetric, s – symmetric

Close inspection of the lineshape and relative magnitudes of bands in the aliphatic C–H stretching region (3000–2700 cm^{-1}), reveals a shoulder at 2960 cm^{-1} (v_{as}(CH$_3$)) and in addition, a broad band centered at 2926 cm^{-1} indicating a substantial contribution of polymethylenic chains with moderate branching. There are significant differences, particularly in the 1800–800 cm^{-1} spectral region due to a specific carbonyl group (i.e., ketone or aldehyde), ring stress (the greater the ring stress, the higher the carbonyl frequency), and connection to an aromatic ring or conjugation to a C=C or another C=O group. SR–FTIR spectroscopy shows that the acritarchs, _S. favosa_ and _L. crassa_ are composed of different resistant biopolymers that have been selectively preserved upon sedimentation, which suggests different taxonomic affinities. These biopolymers show differences in their carbon–oxygen functionality (C=O structures in different chemical environments) and degree of conjugation.

The results indicate a relatively short average aliphatic chain–length with a moderate degree of branching and a dominance of aromatic/conjugated moieties in the microfossil walls of _L. crassa_. Comparison with other recent and fossil algal remains _(33)_ shows that almost all dinoflagellates investigated contain a mixed aromatic/aliphatic composition with similar FTIR spectra. This composition is consistent with the resistant biopolymer dinosporin biosynthesised by dinoflagellates. The SR–FTIR spectra of _S. favosa_ however, show a greater component of aliphatic and carbon–oxygen functionality, which

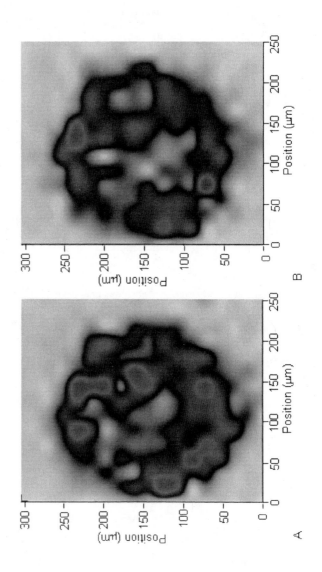

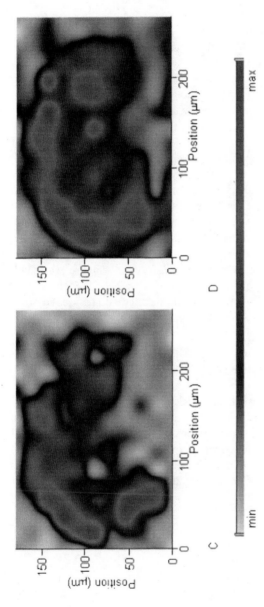

Figure 5. Functional group maps of (A) S. favosa ν(CH) region (3000 to 2800 cm⁻¹) (B) S. favosa ν(C=O) region (1770 to 1705 cm⁻¹) (C) L. crassa ν(CH) region (3000 to 2800 cm⁻¹) (D) L. crassa ν(C=O) region (1760 to 1690 cm⁻¹) (32 scans, 4 cm⁻¹ resolution, Aperture size: 20 × 20 μm, Step Size 18 × 18 μm))

may comprise a new class of biopolymer containing significant aliphatic, branched aliphatic and carbon–oxygen molecular constituents that is different from algaenan, lignin, sporopollenin, or dinosporin macromolecules. To date, little work has been done to elucidate biopolymer compositions of micro–eukaryotes such as euglenids, kinetoplastids, heterokonts, ciliates, red algae, and fungi, indicating a clear need for expanded microchemical and ultrastructural research on living micro–organisms.

S. favosa consists of a complex cyst wall structure composed of interlocking polygonal plates. Figure 5 shows that the C–H (branched hydrocarbon) component is located to the exterior of the cyst wall. Likewise the C=O component is distributed around the exterior of the cyst wall. In addition a few concentrated regions of C=O functionality can be observed within the centre of the cyst. However, the morphology of *L. crassa* is simpler than that of *S. favosa*. In contrast, *L. crassa* is a smooth walled acritarch with no surface ornamentation or surface features *(33)*. Given the uniform morphology it is unexpected to observe chemical differences as revealed by mapping the distribution of C–H and C–O functional groups (Figure 5). Similar to the *S. favosa* the distribution of C–H (branched hydrocarbon) component is located towards the exterior of the cyst wall for the *L. crassa* acritarch (Figure 5). The C=O constituents are also located within the exterior of the cyst wall for *L. crassa*. Future work will expand of this preliminary study of acritarchs in which SR–FTIR data will be compared with SEM morphological and TEM ultra–structural analyses in order to elucidate the observed chemical differentiation.

Acknowledgments

We are grateful to the Australian Research Council (ARC) for Discovery grants an ARC Postdoctoral Fellowship (CPM) an ARC Professorial Fellowship (PAL) and ARC LIEF grants for the FTIR instrumentation; and for National Health and Medical Research Council project grants #211040 and #352354 (TCS). This work was supported by the Australian Synchrotron Research Program, which is funded by the Commonwealth of Australia under the Major National Research Facilities Program.

References

1. Barth, A.; Zscherp, C. *Q. Rev. Biophys.* **2002**, *35*, 369-430.
2. McKelvy, M. L.; Britt, T. R.; Davis, B. L.; Gillie, J. K.; Graves, F. B.; Lentz, L. A. *Anal. Chem.* **1998**, *70*, 119-178.

3. Shaw, R. A.; Mantsch, H. H. In *Encyclopedia of Analytical Chemistry*; Meyers, R. A., Ed. John Wiley & Sons Ltd.: Chichester, UK, 2000; Vol. 1, pp 83-102.

4. Naumann, D. In *Encyclopedia of Analytical Chemistry*; Meyers, R. A., Ed. John Wiley & Sons Ltd.: Chichester, UK, 2000; Vol. 1, pp 102-131.

5. Naumann, D. In *Infrared and NIR Raman Spectroscopy in Medical Microbiology*, Infrared Spectroscopy: New Tool in Medicine; Mantsch, H. H.; Jackson, M., Eds. SPIE-The International Society for Optical Engineering: San Jose, CA, 1998; pp 245-257.

6. Kirschner, C.; Maquelin, K.; Pina, P.; Ngo Thi, N. A.; Choo-Smith, L. P.; Sockalingum, G. D.; Sandt, C.; Ami, D.; Orsini, F.; Doglia, S. M.; Allouch, P.; Mainfait, M.; Puppels, G. J.; Naumann, D. *J. Clin. Microbiol.* **2001**, *39*, 1763-1770.

7. Ngo-Thi, N. A.; Kirschner, C.; Naumann, D. *J. Mol. Struct.* **2003**, *661-662*, 371.

8. Choo-Smith, L. P.; Maquelin, K.; van Vreeswijk, T.; Bruining, H. A.; Puppels, G. J.; Thi, N. A. N.; Kirschner, C.; Naumann, D.; Ami, D.; Villa, A. M.; Orsini, F.; Doglia, S. M.; Lamfarraj, H.; Sockalingum, G. D.; Manfait, M.; Allouch, P.; Endtz, H. P. *Appl. Environ. Microbiol.* **2001**, *67*, 1461-1469.

9. Chenxu Yu, J. I. *Biopolymers* **2005**, *77*, 368-377.

10. Tindall, B. J.; Brambilla, E.; Steffen, M.; Neumann, R.; Pukall, R.; Kroppenstedt, R. M.; Stackebrandt, E. *Environ. Microbiol.* **2000**, *2*, 310-318.

11. Dumas, P.; Miller, L. *Vib. Spectrosc.* **2003**, *32*, 3-21.

12. Dumas, P.; Miller, L. *J. Biol. Phys.* **2003**, *29*, 201.

13. Wetzel, D. L.; LeVine, S. M. In *Infrared and Raman Spectroscopy of Biological Materials*; Gremlich, H.-U.; Yan, B., Eds.; Marcel Dekker, Inc.: NY, 2001; Vol. 24, pp 101-142.

14. Jakobsen, R. J.; Strand, S. W. In *Pract. Spectrosc.*; Mirabella, F. M., Ed. Marcel Dekker, Inc.: NY, 1993; Vol. 15, pp 107-140.

15. Goormaghtigh, E.; Raussens, V.; Ruysschaert, J.-M. *BBA Rev. Biomembranes* **1999**, *1422*, 105.

16. Mirabella, F. M. In *Internal Reflection Spectroscopy: Theory and Applications*; Mirabella, F. M., Ed. Marcel Dekker, Inc.: NY, 1993; Vol. 15, pp 17-52.

17. Diem, M.; Romeo, M.; Matthaus, C.; Miljkovic, M.; Miller, L.; Lasch, P. *Infrared Phys. Techn.* **2004**, *45*, 331-338.

18. Shaw, R. A.; Mansfield, J. R.; Rempel, S. P.; Low-Ying, S.; Kupriyanov, V. V.; Mantsch, H. H. *J. Mol. Struc. Theochem.* **2000**, *500*, 129-138.

19. Levin, I. W.; Bhargava, R. *Annu. Rev. Phys. Chem.* **2005**, *56*, 429-474.

20. Chenxu Yu, J. I. *Biopolymers* **2005**, *77*, 368-377.

84

21. Buchanan, K. L.; Murphy, J. W. *Emerg. Infect. Dis.* **1998**, *4*, 71-83.
22. Kwon-Chung, K. J.; Boekhout, T.; Fell, J. W.; Diaz, M. *Taxon* **2002**, *51*, 804-806.
23. Bose, I.; Reese, A. J.; Ory, J. J.; Janbon, G.; Doering, T. L. *Eukaryot. Cell* **2003**, *2*, 655-663.
24. Doering, T. L. *Trends Microbiol.* **2000**, *8*, 547-553.
25. Ganendren, R.; Carter, E.; Sorrell, T.; Widmer, F.; Wright, L. *Microbes Infect.* **2006**, *8*, 1006-1015.
26. Hopkinson, J. H.; Moustou, C.; Reynolds, N.; Newbery, J. E. *Analyst* **1987**, *112*, 501-505.
27. Cherniak, R.; Sundstrom, J. B. *Infect. Immun.* **1994**, *62*, 1507-1512.
28. Galichet, A.; Sockalingum, G. D.; Belarbi, A.; Manfait, M. *FEMS Microbiol. Lett.* **2001**, *197*, 179-186.
29. Barth, A. *Prog. Biophys. Mol. Bio.* **2001**, *74*, 141-173.
30. Widmer, F.; Wright, L. C.; Obando, D.; Handke, R.; Ganendren, R.; Ellis, D. H.; Sorrell, T. C. *Antimicrob. Agents Chemother.* **2006**, *50*, 414-421.
31. Butterfield, N. J. *Paleobiology* **2004**, *30*, 231-252.
32. Van Waveren, I. M. In *Neogene and Quaternary dinoflagellate cysts and acritarchs*; Head, M. J.; Wrenn, J. H., Eds.; American Association of Stratigraphic Palynologists Foundation: Dallas, TX, 1992; pp 89-120.
33. Marshall, C. P.; Javaux, E. J.; Knoll, A. H.; Walter, M. R. *Precambrian Res.* **2005**, *138*, 208-224.
34. Derenne, S.; Largeau, C.; Berkaloff, C.; Rousseau, B.; Wilhelm, C.; Hatcher, P. G. *Phytochemistry* **1992**, *31*, 1923-1929.
35. Gelin, F.; Volkman, J. K.; Largeau, C.; Derenne, S.; Damste, J. S. S.; De Leeuw, J. W. *Org. Geochem.* **1999**, *30*, 147-159.
36. Versteegh, G. J. M.; Blokker, P. *Phycol. Res.* **2004**, *52*, 325-339.
37. Tegelaar, E. W.; De Leeuw, J. W.; Derenne, S.; Largeau, C. *Geochim. Cosmochim. Acta* **1989**, *53*, 3103-3106.
38. Kokinos, J. P.; Eglinton, T. I.; Go~ni, M. A.; Boon, J. J.; Martoglio, P. A.; Anderson, D. M. *Org. Geochem.* **1998**, *28*, 265-288.

Chapter 7

Probing the Influence of the Environment on Microalgae Using Infrared and Raman Spectroscopy

Philip Heraud[1,2], Bayden R. Wood[1], John Beardall[1,2], and Don McNaughton[1]

[1]Centre for Biospectroscopy and [2]School of Biological Sciences, Monash University, Melbourne, Australia

Since 2000, a diverse range of methods employing vibrational spectroscopy have been developed to study microalgae, mainly focused on probing the effects of environmental change on these ecologically important organisms. These include: FTIR microspectroscopy and Attenuated Total Reflectance (ATR) FTIR spectroscopy of *ex vivo* samples; synchrotron FTIR microspectroscopy of *in vivo* samples; Raman microspectroscopy of living algal cells; and *in vivo* Raman spectroscopy of algal samples using acoustic levitation. It has been demonstrated that *ex vivo* FTIR spectroscopic analysis can track macromolecular changes in microalgae as accurately as conventional methods, whilst requiring much less biological material for the analyses. This work has been extended, through the use of synchrotron sources of IR light, to the measurement of changes in macromolecular composition in living algal cells at the intracellular level. It has also been shown that Raman spectroscopy can detect changes in pigments such as chlorophyll *a* and β-carotene in living microalgal cells, caused by changing light and nutrient conditions, thus providing complementary information on the same processes probed by FTIR spectroscopy. Given the portability of Raman systems using acoustic levitation of samples, it may be possible, in the near future, to extend the laboratory studies to the determination of nutrient status of living algal cells in the field.

Background

Algae in aquatic ecosystems contribute 50% of total planetary primary productivity, fixing about 59 Pg of carbon annually, and hence are not only major components of the global carbon cycle but are also likely to play a major role in global climate change scenarios (1). Understanding the role of microalgae in the planetary ecosystem relies partly on the ability to monitor biochemical changes in cells as they are subjected to perturbations in the environment. Infrared and Raman spectroscopy are proving to be very powerful tools in this respect.

Unlike research into human pathology and microbiology, which have benefited for more than a decade from advances in infrared and Raman spectroscopic analysis, phycological research has only recently been exposed to these powerful techniques. In 2001 the first studies appeared showing that FTIR microspectroscopy could be used to predict the nutrient status of a range of marine microalgal species (2,3). It was demonstrated that FTIR microspectroscopy can determine the relative concentrations of the major macromolecular classes (proteins, lipids and carbohydrates) in algal samples, with measurements matching closely those achieved using conventional, non-spectroscopic methods (3). Moreover, and importantly given the low densities of algal populations, much less biological material is required for the rapid spectroscopic measurements, with measurements even being feasible with single (albeit large; >40 μm diameter) cells. This approach was recently extended to the measurement of intracellular biochemical changes in living algal cells using synchrotron sources of infrared radiation (4). Recently, FTIR microspectroscopy has been used to study field populations of microalgae (5) presenting the possibility for using vibrational spectroscopy as an environmental monitoring tool with microalgae.

In 2005, Raman microspectroscopy was first used to study the effects of environmental changes on suspensions of living algal cells (6). This research showed that Raman spectroscopy can detect changes in pigments such as chlorophyll and β-carotene, the relative levels of which are good predictors of previous nutrient or light stress experienced by the algae. The effects of nutrient changes upon single living algal cells using Raman microspectroscopy were reported the following year (7). This work demonstrated that repeated Raman measurements on living algal cells were possible, allowing environmentally-induced changes in cellular biochemistry to be measured in real time.

This chapter outlines the basic methodologies employed to acquire high quality FTIR and Raman spectra from algal cells using some case studies, the scope of possible research in this area, and provides some possible future research directions.

Methods of vibrational spectroscopy using microalgae

FTIR Microspectroscopy

Preparation of microalgal cell samples for analysis using FTIR microspectroscopy is quite straightforward. There are 2 essential features necessary for obtaining high quality spectra from microalgal samples: cell washing and cell deposition. Cell washing is required, because a number of components commonly found in algal growth media show pronounced spectral bands, which often obscure the bands of biological interest. These include chelating agents such as ethylene diamine tetraacetic acid (EDTA), buffering agents such as Tris[hydroxymethyl] amino methane (Tris), and inorganic nutrients such as potassium dihydrogen phosphate. Two washings of the cells in milliQ water, with a sodium chloride concentration isotonic with the algal cell cytoplasm, has been demonstrated to be sufficient to dilute these components to the extent that bands attributed to them cannot be detected in the infrared spectrum.

The method by which samples are deposited onto the infrared substrate determines, to a large degree, the quality of the spectra acquired. Many algal cell suspensions clump when they are being dehydrated prior to infrared analysis, as a result of cell to cell and cell to substrate adhesions. This leads to inhomogeneity in cell deposition thickness and density. These inhomogeneous deposits cause infrared light to scatter, which manifests as undesirable changes in the baseline of the resulting infrared spectrum. Two methods will produce quite even deposits from most types of algal cell samples. The first involves employing a type of centrifuge designed to deposit lymphocyte cells onto glass slides for haematological analysis (Shandon Cytospin cytocentrifuge, Thermo Electron Corporation, Massachusetts, USA). Infrared reflective Low e slides (Kevley Technologies, Ohio, USA) are used in the place of the normal glass slides to deposit algal samples for FTIR analysis by absorption/reflection. Typically, 0.5 mL of algal cell suspension with a concentration of 10^6 cells mL^{-1} is centrifuged. By varying the speed and acceleration of the Cytospin cytocentrifuge, one can succeed in depositing monolayers of cells from a diverse range of microalgal species onto Low e glass slides. Spectra acquired from these monolayer deposits are much more consistent in terms of spectral intensity and exhibit much less severe base line effects, compared to spectra from deposits laid down via pipette onto Low e slides. This means that less pre-processing (normalisation and baseline correction) of spectra is required prior to analysis.

A second method, developed for straightforward transmission IR spectroscopy, involves using a Bruker silicon multi-well substrate (Bruker Optics Inc., Ettlinger, Germany). These IR transmission substrates, available in 96, 384 & 1536 well formats, were originally developed for analysing bacteria with Bruker's HTS-XT FTIR spectroscopic high-throughput screening system. These substrates enable a much more even drying of cell deposits, compared to conventional IR transmission substrates, because the surface tension properties of the polymer rings surrounding each well, together with the micro-roughened silicon surface, reduce the depth of the meniscus of deposited liquids. Typically, microalgal cell suspensions are concentrated after the second washing stage into a dense cell suspension, the concentration of which varies with cell size up to 10^8 cells per mL, with 5 μL of the cell suspension deposited into each well.

An FTIR microscope is employed to acquire spectra from the IR reflection and transmission substrates. Typically, a 50 μm square aperture is employed which allows regions with the most even cell deposition to be targeted precisely. An FTIR microscope is particularly useful when acquiring spectra from deposits of marine algal species because the microscope can be used to target areas devoid of salt crystals. Although sodium chloride is transparent to mid-infrared wavelength, the small salt crystals left after dehydration of the cell suspension are often of a size that causes scattering of IR photons, leading to baseline effects in the resulting spectra. This problem is minimised in samples prepared using the Cytospin cytocentrifuge system, which uses absorbent pads to remove most of the liquid media when the cell suspension is deposited.

Attenuated total reflectance (ATR) spectroscopy

ATR spectroscopy provides a rapid and easy alternative to transmission and reflection/absorption for the routine analysis of microalgal samples. Excellent results have been achieved using a Golden Gate ATR accessory with diamond window (Specac Inc, GA, USA). Washed algal cells are concentrated into a thick suspension, which is deposited onto the ca 200μm^2 diamond window, and subsequently dried using a stream of warm air. The best results are achieved when the dried deposit is deposited thickly enough so that the underlying diamond window is just occluded from view with the naked eye. This results in spectra with good signal to noise characteristics and minimal baseline sloping.

In vivo synchrotron infrared microspectroscopy

Because the spatial resolution of infrared spectroscopic mapping is wavelength diffraction limited to a range between 3 and 10 μm, a large cell is required if intracellular imaging is to be achieved. A number of genera from the microalgal family Desmidae (desmids) are ideal subjects for *in vivo* infrared microspectroscopy as cell diameters are large, up to 300 µm in some species, but the thickness of the cells is much less, often 10 µm or smaller, allowing infrared photons to penetrate living cells and avoid spectral saturation. Cells from the species *Micrasterias hardyi* can survive for periods up to 24 hours sandwiched lightly between the infrared transparent windows of a modified Bioptecs FCS3 (Bioptechs Inc., PA, USA) flow-through cell (4), allowing long term studies of macromolecular changes to be measured. These flow-through cells were designed for fluorescence microscopy, but can be adapted for infrared microspectroscopy by replacing the glass windows with infrared transparent material such as zinc selenide or calcium fluoride.

A synchrotron source of infrared radiation is required to penetrate the hydrated *Micrasterias* cell, which is about 12 µm thick. Under these conditions, information can be obtained reliably from all spectral regions except the amide I band. Because of the light scattering that occurs through the cell body, spatial resolution is limited to about 20 µm. This is still adequate to differentiate cellular regions containing chloroplasts, the nuclear region, and cytoplasmic areas devoid of chloroplast or nucleus (4).

Raman Spectroscopy

One of the major difficulties with Raman spectroscopy of microalgal samples is that microalgal cells display prominent autofluorescence because they contain high concentrations of chlorophyll and other fluorescent chromophores. This is particularly evident using HeNe lasers, which have an excitation wavelength near the absorbance maximum of the Q_B band of chlorophyll. Best results have been achieved so far using laser excitation wavelengths in the near infrared (7). Moreover, fluorescence is minimized with living cells, due to the existence of intracellular, photochemical and non-photochemical quenching mechanisms present in microalgae. Two methods have been reported to enable high quality Raman spectra to be acquired from algal samples: acoustic levitation which uses massed cells in suspension, and Raman microscopy which can be performed on single cells.

Acoustic levitation

The coupling of an acoustic levitation system with a portable Raman spectrometer provides a novel approach to recording spectra of living algal cells. In addition to portability, the system has the advantage of enabling the measurement of cells in their native environment and in a fixed and contactless position. With this approach, the cells retain their 3-dimensional morphology and total surface area for nutrient uptake. By suspending the cells in air with no attenuation from surfaces, a high signal-to-noise ratio can be achieved compared to conventional Raman measurements performed in macro chambers, cuvettes or microscope systems.

A photograph of the experimental set-up is shown in Figure 1. Aliquots (10 µl) of concentrated algal cells are transferred by pipette into an acoustic node generated by an ultrasonic levitator (Dantec/Invent measurement Measurement Technology, Erlangen, Germany) The instrument uses a piezoelectric vibrator to generate a standing acoustic wave with equally spaced nodes and antinodes. This is achieved through multiple reflections of ultrasonic waves between an ultrasonic transducer and solid reflector. The standard frequency of the ultrasonic levitator is 56 kHz, the standard wavelength is 5.9 mm, the largest theoretical drop diameter is 2.5 mm and the smallest is 15 µm. The samples are levitated below pressure nodes, due to axial radiation pressure and radial Bernoulli stress. Raman spectra are acquired with a Inphotonics portable Inphotote Raman Spectrometer with a 785 nm diode laser and a Inphotonics RamanProbe™ Fibre Optic Sampling Probe. Power at the sample was 50 mW. We recently reported (6) a number of potential applications of this technique including nutrient limitation studies, light exposure experiments and taxonomic identification of algal species.

Raman microspectroscopy

Algal cells are targeted with a 60 × water immersion objective on a Renishaw System 2000 Raman microscope, using the 780 nm excitation line from a diode laser (7). Cells are maintained in nutrient medium at a constant temperature using a temperature-controlled microscope stage, and adhered to aluminium-coated, quartz-glass Petri dishes using a poly-L-lysine coating. Adhesion of cells prevents cells floating and allows even motile cells to be held securely enough to allow spectra to be acquired. About 1 mW power is applied to the cells and each spectrum is acquired during a 10 s exposure. Tests showed that spectral acquisition does not cause any appreciable photodamage to susceptible molecules such as carotenoids.

Experimental cases studies: effects of nutrient perturbations on microalgae

Rationale

Primary production in the world's oceans, lakes and rivers is often limited by the availability of major nutrients, particularly nitrogen, phosphorus, and iron

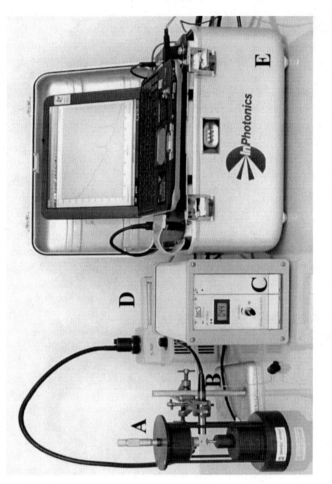

Figure 1. Experimental apparatus A. ultrasonic acoustic levitation device showing cells trapped in a node B. Raman fibre optic probe C. control unit for acoustic levitation device D. Quartz halogen light source E. InPhotonics portable 780 nm Raman spectrometer

(8). Limitation in any of these nutrients can thereby have profound effects on aquatic food chains. Apart from the effect on photosynthesis, nutrient limitation can stimulate microalgae to change their biochemical composition in an attempt to maintain cellular homeostasis. This can result in a change in the food quality of the algae, which may be deleterious to consumers. Although there has been much research involving the measurement of physiological change in microalgal cells subjected to nutrient limitation, much less attention has been paid to changes in cellular composition. Part of the reason for this has been that the conventional methods used to measure biochemical composition in microalgae are quite insensitive, often requiring, at a minimum, millions of cells to achieve a measurement, as well as being very time consuming and subject to large systematic and random errors. Infrared and Raman spectroscopic analyses provide sensitive, non-destructive and rapid new methods for analysing and predicting the effects of environmental change on microalgae, as the following case studies illustrate.

Infrared microspectroscopy of *ex vivo* samples

Microalgae attempt to maintain homeostasis of cellular functioning when subjected to environmental stresses such as nutrient limitation, exposure to high fluxes of visible or ultraviolet radiation, changes in osmolarity of the environment or changes in temperature. Cellular biochemistry is often altered rapidly and substantially in these situations. FTIR spectroscopy has proven to be superior to conventional, non-spectroscopic measurements for monitoring these effects, as it requires much smaller quantities of material to achieve a measurement, conserves the sample, and is rapid and easy to achieve. Moreover, FTIR spectroscopy can measure relative changes in macromolecular concentration in algal samples as accurately as the conventional methods (3).

Figure 2 shows changes in the FTIR spectra of three microalgal species where phosphorus supply had been limited for many days (2). The FTIR spectra are similar between species, except for the prominent band at 1080 cm^{-1} from silicates in the extracellular frustule of the diatom *Nitzschia*. The assignments for the major bands (resulting from proteins, lipids, carbohydrates and phosphorylated molecules) identified in the FTIR spectra of microalgae are given in Table 1. When microalgae are starved of P their ability to carry out energy intensive biochemical processes such as protein synthesis is diminished. Accordingly, carbon that is continued to be fixed by photosynthesis is allocated away from proteins to storage macromolecules such as lipids or carbohydrates (2,3). Figure 2 shows that all three species undergo massive re-organisation of macromolecular composition in response to P limitation, but the nature of the adaptive changes exhibited by the three species are different. For example, the chlorophyte alga *Spherocystis* shows a major allocation of fixed carbon to carbohydrates, evidenced by increases in bands from C-O stretches in the range

Table 1. Band assignments for the IR spectrum of microalgal cell.

Number in Figure 2	Wavenumber values / cm^{-1}	Assignment† (3,4)
1	~ 1740	$\nu_{C=O}$ of ester functional groups primarily from lipids and fatty acids
2	~ 1650	$\nu_{C=O}$ of amides of proteins. Amide I band. May also contain contributions from C=C stretches of olefinic and aromatic compounds
3	~ 1540	δ_{N-H} of amides associated with proteins. Amide II band May also contain contributions from C=N stretches
4	~ 1455	δ_{CH3as} and δ_{CH2as} of proteins
5	~ 1370	δ_{CH3s} and δ_{CH3s} of proteins, and ν_{C-Os} of COO$^-$ groups
6	~ 1320	Amide III band of proteins
7	~1240	$\nu_{P=Oas}$ of the phosphodiester backbone of nucleic acid (DNA and RNA) May also be due to the presence of phosphorylated proteins and polyphosphate storage products
8-10	~1200-900	ν_{C-O-C} of polysaccharides In diatoms a band at 1080 cm^{-1} from $\nu_{Si-O\ s}$ of silica in the frustule is prominent

† ν_{as} = asymmetric stretch, ν_s= symmetric stretch, δ_{as}= asymmetric deformation (bend), δ_s= symmetric deformation (bend)

1200-1000 cm^{-1}. Chlorophyte algae are known to store carbon as starch and the spectral changes observed are evidently due to prominent build up of stored starch in the *Spherocystis* cells.

Conversely, diatoms such as *Nitzchia* allocate fixed carbon during major nutrient starvation to storage lipids rather than carbohydrate. This response can be observed as large increases in the lipid ester carbonyl band at 1740 cm^{-1} in the spectrum of the P starved *Nitzchia* cells (Fig. 2). A strategy that may be described as intermediate to the responses of *Nitzchia* and *Spherocystis* is shown with the cyanophyte, *Phormidium*, which shows increases in stores of both carbohydrate and lipid, when subjected to P starvation.

The ability to measure relative macromolecular composition using very small amounts of sample enables longitudinal experiments that were not

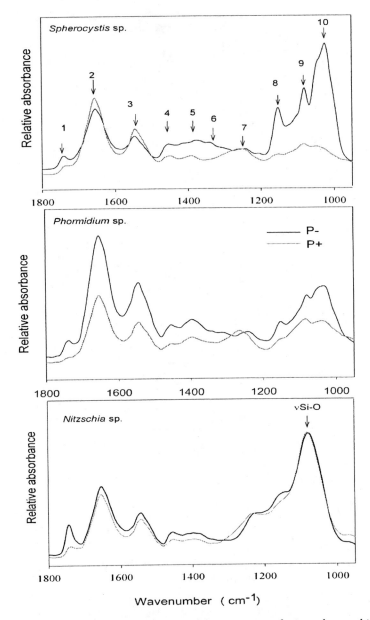

Figure 2 Changes in the FTIR spectra of three species of microalgae subjected to phosphorus limitation. The bands identified by numbers in the spectra from Spherocystis sp. are assigned in Table 1.

possible using conventional methods (3). This is because multiple samples can be drawn from a single algal culture over the course of an experiment without depleting its volume significantly and hence introducing a confounding variable in the experiment. An example of this type of experiment is shown in Figure 3, using the chlorophyte alga *Dunaliella tertiolecta*. At the start of the experiment cells from an exponentially growing culture were re-suspended in a growth medium containing all growth nutrients except for nitrogen. The N starvation is allowed to continue for a number of days after which cells are re-suspended back into a growth medium replete in all nutrients. Recovery is then tracked for a number of days. Under the N starvation, protein synthesis is not supported, evidenced as a decline in amide I and II absorbance, along with a decrease in absorbance from other bands associated with proteins. The onset of the decline in protein is not immediate, with the lag period probably representing a phase during which intracellular N stores are depleted. Coincident with the beginning of the decline in protein absorbance is an increase in C-O stretching bands in the region 1200-1000 cm^{-1}. As mentioned previously, chlorophyte algae are known to store starch in response to stress conditions, and the increase in the carbohydrate bands is evidence for the allocation of carbon fixed by photosynthesis to carbohydrate under the conditions inhibiting protein synthesis. The interpretation of the spectral changes during the starvation phase is corroborated by opposite changes being observed during the recovery phase. Initially, there is a lag phase which appears to represent the engagement of cellular processes involved with the re-initiation of protein synthesis following the re-supply of N to the cells. Following this lag phase, the amide I and II absorbance increases whereas absorbance due to carbohydrate declines, showing that carbon stored in starch is re-allocated back to protein.

There is further scope for these types of experiments. One approach likely to be of value is to examine the interactive effects of a combination of stressors. For example, one might observe the interaction of a nutrient limitation with the exposure of cells to ultraviolet radiation (UVR). Because UVR inhibits photosynthesis we might expect to see different responses to the effects of nutrient limitation, for instance, on the rates of synthesis and relative concentration of macromolecules.

Synchrotron infrared spectroscopy on *in vivo* samples

The previous experiments involving nutrients, described above, all used algal samples consisting of cell populations that were air-dried and fixed. Although capable of tracking changes in populations of cells induced by alterations in nutrient supply, these approaches do not allow changes to be observed in single cells in real time. Recently, experiments employing synchrotron sources of infrared radiation have shown that changes in

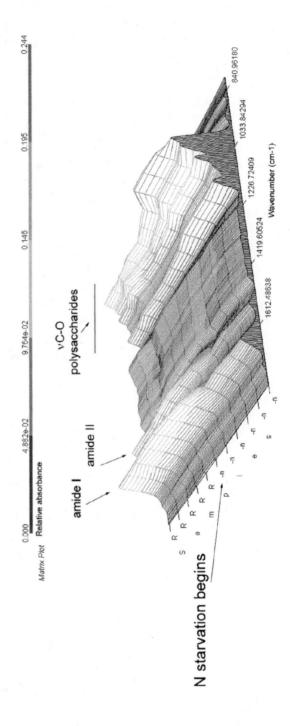

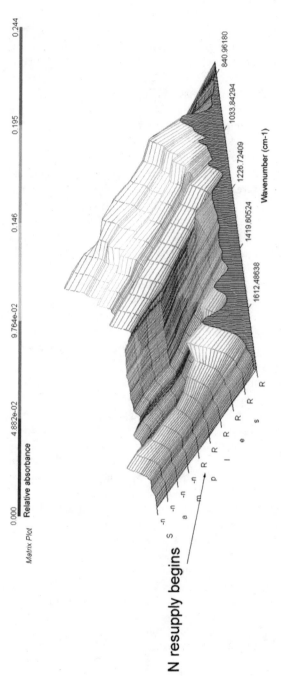

Figure 3. Interpolated plots of FTIR spectra acquired daily from sub-samples drawn from a microalgal culture undergoing starvation of nitrogen (top plot) and then re-supply of nitrogen to the same culture (bottom plot). Spectra taken from the N-starved culture are denoted by n- and spectra acquired during the re-supply of nitrogen by R.

macromolecular composition in single living algal cells can be monitored using FTIR microspectroscopy (4). These experiments maintained "giant" desmid algal cells in IR transparent flow-through cells with spectra recorded during changes in nutrient supply. Apart from being very large, the desmid algal cells had the added advantage of having been extensively studied by cell biologists for several decades, hence the location of major organelles such as the nucleus and chloroplast could be reliably located without staining the cells, and the physiology of these types of cells under a range environmental conditions has also been well documented.

Micrasterias hardyi cells were harvested from exponentially growing cultures and re-suspended in media devoid of either nitrogen or phosphorus. After 3 days of starvation, cells were placed between the windows of an infrared transparent flow through cell and media replete in all nutrients was passed through the flow cell using a peristaltic pump. The size of the spacer used between the IR transparent windows matched the thickness of the *Micrasterias* cells exactly, such that the cells were neither washed away nor was their viability compromised. The high intrinsic brightness of the synchrotron source enabled sufficient IR photons to pass through the cell, liquid medium and windows of the flow cell, to the detector resulting in spectra with acceptable signal to noise and no saturation. The spatial resolution of the technique was limited to ca 20 μm due to the amount of light scattering through the sample; however this did not limit the ability to determine differences in macromolecular composition with the cell because of the large dimensions of the cell (about 300 μm in diameter). The existence of prominent cytoplasmic streaming in the cells after long periods of measurement (up to 24 hours) demonstrated that the *Micrasterias* cells could continue to function normally under the measurement regime. This was due in part to medium supplied to the cells being maintained at growth temperature and light being provided to the algal cells via the stage illumination of the FTIR microscope at approximately the same irradiance as growth conditions. In addition, synchrotron IR light has been shown to have negligible heating effect on samples (9).

Initially, comparisons were made between maps of fixed air-dried cells and those of nutrient-replete, living cells maintained in the IR flow cell. Examples of these maps are shown in Figure 4. These maps showed that the macromolecular distribution was similar in the fixed and living cells. For instance, high proteins and lipid absorbances were observed in the region occupied by the chloroplast. This corresponded with the known composition of the chloroplasts which are rich in phospholipid membranes and proteins associated with photosynthesis. By contrast, the nuclear region was shown to have high protein absorbance but very low lipid absorbance. This corresponded well with the presence of high concentrations of histone proteins in the nucleus, but very little lipid.

For mapping the entire algal cell, the time required for the measurement was ca 6 hours, which is comparable to the time course of nutrient-induced changes in macromolecular concentration deduced from *ex vivo* experiments. During this long period the stability of the microscope stage focus, flow cell and the synchrotron radiation flux is questionable and the washed out protein image of the cell in figure 4 is possibly due to changes in one or all of these. Consequently, spectra were acquired at 10 μm intervals in a line running through the central axis of the cell, through the two chloroplasts, the nucleus and regions of cytoplasm not containing these organelles. Each series of measurements required ca 10 min allowing repeated measurements to be performed every 30 min on two cells. Figure 4 shows a P-starved cell that was re-supplied with P over a period of 10 hours. No change in macromolecular concentration or distribution was seen under these conditions over this time interval. Nevertheless, the results show the consistency of the measurements, and verify the macromolecular distribution observed in the maps of the entire cells. Measurements using N-starved cells re-supplied with N were conducted over much longer time intervals than P-starvation experiments (24 hours). These did show significant changes in macromolecular distribution within the cells, particularly with lipids. A significant (20 - 38%) increase in lipid ester carbonyl absorbance was observed in the chloroplast region in two N-starved cells, over a 24 hour period of re-supply with N.

This experimental approach offers great promise for *in vivo* monitoring of the effects of environmental changes upon algal cell biochemistry. The procedure appears to be limited at the moment to changes that can be induced within a period of about one day. However, this is still long enough to study the effects of many environmental stressors, where knowledge of *in vivo* biochemical changes would shed new light on the processes involved. These, as for *ex vivo* measurements, include photo-inhibition by high fluxes of visible light, exposure to ultraviolet A and ultraviolet B radiation, temperature stress, osmotic stress, poisoning by pollutants, to name a few.

Raman microspectroscopy on *in vivo* samples

Limitation of nitrogen causes a decline in the level of chlorophyll in algal cells, because N is a major constituent of the chlorophyll molecule. Under these conditions there is often a concomitant increase in carotenoids such as β-carotene within cells. Biologists studying the effect of nutrient limitation on aquatic ecosystems would like to measure the levels of these pigments in single cells so that the variability of response within phytoplankton populations to nutrient change can be accessed. Raman spectroscopy offers this potential, and

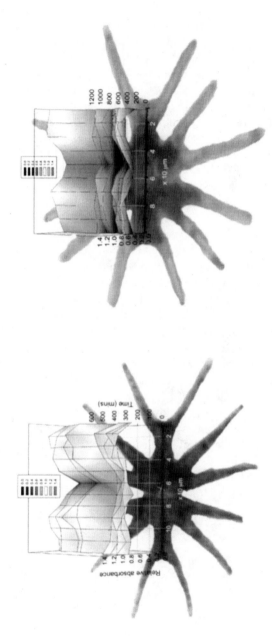

Figure 4. Top is a protein (amide I absorbance) map (left) of an air-dried M. hardyi cell compared with a micrograph of the same cell (right). Middle shows a protein (amide II absorbance) map (left) of a living M. hardyi cell compared with a micrograph of the same cell (right). Bottom shows changes in lipid ester carbonyl absorbance in an N-starved M. hardyi cell re-supplied with N over a 24 h period (right); and changes in lipid ester carbonyl absorbance in an P-starved M. hardyi cell re-supplied with P over a 10 h period

recent studies have shown that *in vivo* Raman spectroscopy can be used to detect β-carotene and chlorophyll in phytoplankton cells (6,7)

Averaged spectra for nitrogen limited and nitrogen replete *Dunaliella tertiolecta* cells are given in Figure 5. All bands in Raman spectra acquired from *D. tertiolecta* cells can be attributed to either chlorophyll *a* or β-carotene (a detailed band assignment is given elsewhere; 6). A number of the prominent bands are highlighted in Figure 5. For example, the band at 1554 cm^{-1} is from CC stretching vibrations, whereas the band at 1328 cm^{-1} results primarily from CH_3 bends, with methine in-plane CH bending contributions and N-C stretching of chlorophyll *a*. The band at 1187 cm^{-1} arises mainly from C-O stretching vibrations of the propionate groups, while the band at 745 cm^{-1} is associated with out-of-plane CH deformations and OCO vibrations from chlorophyll *a*. Bands attributed to β-carotene include that at 1524 cm^{-1} assigned to v_1 from C=C double bonds, the band at 1155 cm^{-1} assigned to v_2 from CC bonds, with v_3 from CC bonds appearing at 1005 cm^{-1}.

Differences between N limited and N replete cells can be observed in Fig. 5. Bands from chlorophyll *a* are less intense in N-limited cells compared with N-replete cells, whereas the opposite is the case for β-carotene bands. PCA scores plots of the spectral data (shown in Fig 5) show a clustering of spectra based on their nutrient status, with nutrient limited cells being positively scored. The spectral basis for the clustering is revealed in both PC1 and PC2 loadings plots (Fig 5) where there is an opposite correlation between bands from β-carotene and those from chlorophyll *a*, with positive loading for the β-carotene bands, illustrated in the figure with the band at 745 cm^{-1} from Chl *a* and the band at 1155 cm^{-1} from □-carotene. This supports the view that N-starved cells have lower levels of chlorophyll a and higher levels of β-carotene, compared with N-replete cells. It has been demonstrated (7) that multivariate models (e.g. Partial Least Squares Regression; PLS-R) can be constructed using this data to compare and classify test samples with very high levels of prediction. Accordingly, these techniques offer the possibility of being able to classify the nutrient status of algal cells based on their Raman spectra.

Conclusions and future direction

Aquatic ecosystems account for approximately the same amount of primary productivity as do terrestrial ecosystems. However, the turnover of carbon in aquatic ecosystems is much faster (0.06 to 0.1 y^{-1}) than that of terrestrial ecosystems (27-59 y^{-1}) (1). Hence, short term disturbance of aquatic primary productivity has the potential to have a major impact on planetary atmospheric carbon dioxide levels. It follows that accurate prediction of aquatic primary productivity requires the development of rapid methods to assess phytoplankton function. Of particular importance are measurements of algal cellular biochemical composition, because these are likely to provide the best indication of the previous recent environmental history undergone by the algal cells, for example whether they have been nutrient limited or exposed to damaging fluxes of light. The newly developed vibrational spectroscopy techniques have the potential to allow rapid monitoring of algal biochemistry in the field. However, interpretation of the results relies on understanding obtained from laboratory studies.

The experimental possibilities using *ex vivo* FTIR microspectroscopy methods are as yet far from the exhausted. Many of the *ex vivo* experiments can be carried out using the FTIR-ATR technique, which is well suited to routine analysis in the algal biology laboratory. ATR has the advantage over microspectroscopy in that it is less expensive in initial cost and consumables compared to microspectroscopy and sample preparation is simpler and more foolproof. There is much more scope for longitudinal studies similar to those begun by Giordano et al (3), by studying the limitation effects of a range of different nutrients, the effects of other environmental stressors, and the interactive effects of different stressors. Eventually, it may be possible to use laboratory studies of this kind to define biochemical states, which reflect certain types of environmental stress. For example, it has already been shown that correlated changes in protein, carbohydrate and/or lipid absorbance can be an indication of limitation of a major nutrient. This knowledge was recently employed in a study where FTIR spectroscopy was used to detect the onset of major nutrient deficiency in microalgal batch cultures, which occurs towards the end of the exponential growth phase (10). This result has a bearing on the industrial growth of microalgae as food for prawn, shellfish and fish culture, because the food quality of microalgae was shown to deteriorate markedly towards the end of the growth cycle as major nutrients become limited. It was suggested that FTIR spectroscopy could be used to optimise harvest time for maximum biomass without a sacrifice in food quality.

In vivo FTIR spectroscopic studies with algal cells are much more specialised, requiring a synchrotron source and the development of sophisticated IR flow-through cells and ancillary equipment. However, because the measurements occur in living cells in real time, they are likely to be more interesting to biologists as well as yielding a deeper insight into cellular composition, structure and function. Refinements in the design of the flow-through cells and the selection of superior cell samples is likely to lead to improved measurements in terms of signal to noise and spatial resolution. The possible processes that could be studied using the technique appear to be limited at the moment only by the imagination of the experimenter. One can envisage being able to measure the effects of other environmental stressors, as well as the actions of pollutants such as heavy metals or hydrocarbons. In addition, basic

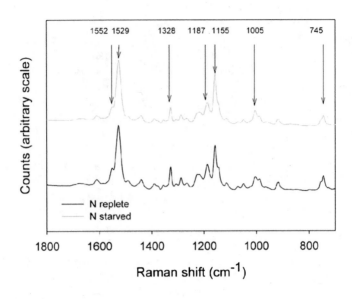

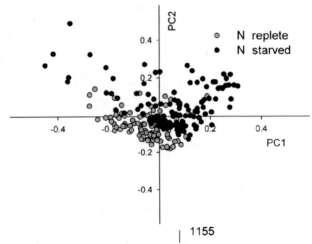

Figure 5. The first panel shows average Raman spectra from 25 N-starved and 25 N-replete Dunaliella tertiolecta cells. The second panel shows the PC1 vs PC2 scores plot for the spectral data. The third panel shows the PC1 and PC2 loadings plots.

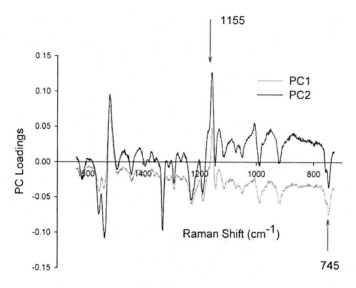

Figure 5. Continued.

cell processes such as protein or lipid synthesis could be probed using this technique by studying the action of inhibitors.

Raman spectroscopy has the advantage over FTIR spectroscopy of being more readily applied to *in vivo* samples. Given the portability of the Raman system using acoustic levitation, described above, it may be possible in the near future to rapidly determine the nutrient status of living cells in the field using Raman spectroscopy. This opportunity would be extremely useful to research focused on monitoring the state of ecosystem productivity using the physiological status of phytoplankton as an indicator. One can also imagine the same knowledge being applied to obtain on-line monitoring of cell status in algal biotechnology applications.

Raman microspectroscopy is best applied to research aimed at understanding basic mechanisms of algal cell function. The latest Raman confocal imaging spectrometers have the potential to rapidly produce three dimensional images of algal cells which may produce new insights into the cellular distribution and functioning of chromophoric molecules. Using different wavelengths to those mentioned in the discussion above, it may be possible to detect pigment types apart from β-carotene and chlorophyll *a*. Interesting candidates in this search are the xanthophyll pigments such as violaxanthin, antheraxanthin and zeaxanthin, which are known to be involved in non-photochemical quenching processes within algal cells essential for adaptation to changing light fluxes in the environment

In summary, the future looks bright for the application of vibrational spectroscopy techniques to microalgal research. The spectroscopic techniques are likely to become powerful tools for both environmental and biotechnology monitoring, as well as yielding new insights into the microalgal composition and function.

References

1. Beardall, J., Raven, J.A. *Phycologia*. **2004**, *43*, 26-40.
2. Beardall, J., Berman, T., Heraud, P.R., Kadiri, M., Light, B., Patterson, G., Roberts, S., Sulzberger, B., Uehlinger, U., Wood B. R. *Aquat Sci*. **2001**, *63*, 44-67
3. Giordano, M.,. Kanzis, M., Heraud, P.,. Wood, B. R., Beardall, J., Burden F., McNaughton, D. *J* Phycol. **2001**, *37*, 271-279,
4. Heraud, P., Wood, B. R., Tobin, M. J., Beardall J., McNaughton, D. *FEMS Microbiol Lett*. **2005**, *249*, 219-225.
5. Sigee, D.C., Dean, A., Levado, E. & Tobin, M.J. *EurJ Phycol*. **2002**, 37, 19-26.
6. Wood, B. R., Heraud, P., Stojkovic, S., Morrison, D., Beardall, J., McNaughton, D. *Anal Chem*. **2005**, *77*, 4955-4961.
7. Heraud, P., Wood, B. R., Beardall J., McNaughton, D. *J. Chemometrics*. **2006**, in press.
8. Harris, G.P. *Phytoplankton ecology, structure , function and fluctuation*. Chapman and Hall Limited, London, 1986.
9. Martin, M., Tsvetkova, N. M., Crowe, J.H., McKinney, W.R. A*ppl. Spectrosc. 2001*, *55*, 111-113
10. Liang, Y., Beardall, J., Heraud, P., *Bot Mar*, **2005**, in press.

Chapter 8

Surface-Enhanced Raman Sensing: Glucose and Anthrax

N. C. Shah[1], O. Lyandres[2], C. R. Yonzon[1], X. Zhang[1], and R. P. Van Duyne[1]

Departments of [1]Chemistry and [2]Biomedical Engineering, Northwestern University, Evanston, IL 60208

This chapter outlines the use of surface-enhanced Raman spectroscopy (SERS) in the development of biological sensors for the detection of anthrax and glucose. In both cases, Ag film over nanosphere (AgFON) surfaces were used as the sensing platform. The localized surface plasmon resonance (LSPR) of AgFON surfaces was tuned to maximize the SERS signal for near infrar-red (NIR) excitations. A harmless analog of *Bacillus anthracis*, namely *Bacillus subtilis*, was quantitatively detected using SERS on AgFON surfaces. Calcium dipicolinate, an important biomarker for bacillus spores, was successfully measured with a limit of detection well below the anthrax infectious dose of 10^4 spores in 11 minutes. For glucose detection, a mixed self-assembled monolayer (SAM) consisting of decanethiol (DT) and mercaptohexanol (MH) was immobilized on an AgFON surface to bring glucose within the zone of the localized electromagnetic field. Complete partitioning and departitioning of glucose was demonstrated. Furthermore, quantitative detection *in vitro* and *in vivo* was achieved.

Introduction

The rapid and accurate identification of biomoleclues is vital for first responders in the event of a biological attack, as well as for patients with a need to monitor chronic or acute conditions. In this chapter, we outline the use of surface-enhanced Raman spectroscopy (SERS) for the detection of calcium dipicolinate (CaDPA), a biomarker for anthrax, and glucose, an indicator for diabetes.

Anthrax is an infection disease caused by *Bacillus anthracis* that requires medical attention within 24-48 hours after initial inhalation of more than 10^4 *B. anthracis* spores (*1*). Therefore, the rapid detection of *B. anthracis* spores in the environment prior to infection is an extremely important goal for human safety. Structurally, a spore consists of a central core cell surrounded by various protective layers. CaDPA is located in these protective layers and accounts for ~10% of the spore's dry weight (*2*) making it is a useful biomarker for bacillus spores (*3*).

Diabetes mellitus is a metabolic disorder where the body either fails to produce insulin or does not respond to insulin, which mediates the uptake of glucose. Failure to regulate glucose levels within tight limits can lead to severe secondary health complications including blindness, kidney disease, nerve damage, and malfunctions related to the circulatory system (*4, 5*). Currently available blood glucose meters are not suitable for continuous monitoring and suffer from low patient compliance, risk of false positives due to interferents, and instability caused by finite protein life cycle. Developing a sensor that measures glucose directly and continuously, in the presence of interfering analytes, will have an enormous impact on the long-term health of diabetics.

We have utilized SERS to design biosensors for both anthrax and glucose. The SERS-based sensor has many advantages: high sensitivity, rapid detection capabilities, and unique spectral fingerprints for detection in the presence of interfering analytes. Normal Raman spectroscopy (NRS), is a very selective technique because every molecular species (e.g. glucose and fructose) has a unique vibrational signature. The limitation of NRS, however, is its weak signal intensity. Higher signal intensity can be achieved by SERS, in which the analyte of interest is brought within a few nanometers of a roughened metal surface, amplifying the signal by 10^6-10^8 and in some cases as high as 10^{15}.

SERS signals can be optimized by tuning the optical properties, known as localized surface plasmon resonance (LSPR), of metal nanostructured SERS-active surfaces to match the experimental parameters. SERS signals are most intense when the LSPR maximum wavelength lies between the excitation wavelength and energy of the Raman scattered photons (*6*). SERS-based sensors utilize Ag film over nanosphere (AgFON) surfaces with highly tunable plasmons, which are optimized to achieve maximum SERS signal. AgFON surfaces are fabricated by depositing silver on hexagonally close-packed polystyrene spheres. The tunability of the LSPR can be easily achieved by changing the sphere diameter (D), as well as the silver metal thickness (d_m).

While CaDPA biomarker in *B. subtilis* spores has an affinity towards the rough silver surface required for SERS detection, many important molecules (e.g. glucose) have low or non-existent binding affinity towards a silver surface (*7*). The work presented herein demonstrates quantitative glucose detection by functionalizing the SERS-active surface with a self-assembled monolayer (SAM). The SAM partitions glucose and concentrates it near the surface of the AgFON (*7-9*). We use a mixed SAM (Figure 1) consisting of decanethiol (DT) and mercaptohexanol (MH) whose dual hydrophobic and hydrophilic properties make it ideal for partitioning glucose (*10*). Although the exact mechanism of DT/MH SAM partitioning has not been well characterized, a space filling computer model shows that combining DT and MH creates a pocket for the analyte. It is hypothesized that glucose partitions into this compartment and approaches closer to the surface than was possible with SAMs used previously (*7-9*).

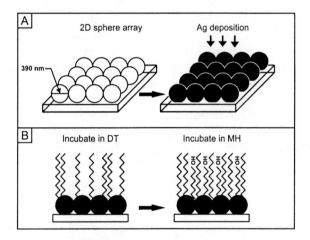

Figure 1: (A) Schematic diagram representing AgFON fabrication. (B) Formation of the DT/MH self-assembled monolayer on the AgFON surface.

Experimental Section

Spore Samples

B. subtilis was purchased from the American Type Culture Collection (Manassas, VA). Spore cultures were cultivated by spreading the vegetative cells on sterile nutrient agar plates (Fisher Scientific), followed by incubating at 30 °C for 6 days. The cultures were washed from the plates using sterile water and centrifuged at 12000 g for ten minutes. The centrifuging procedure was repeated five times. The lyophilized spores were kept at 2-4 °C prior to use. Approximately 1 gram of sample was determined to contain 5.6×10^{10} spores by optical microscopic measurements. The spore suspension was made by dissolving spores in 0.02 M HNO_3 solution and by sonicating for 10 minutes. Calcium dipicolinate was prepared from DPA and calcium hydroxide according to the method of Bailey and co-workers (2).

AgFON Substrate Fabrication

Glass and copper substrates for anthrax detection and glucose detection respectively were pretreated as described previously (10, 11). Approximately 2 µL of the nanosphere suspension (4% solids) was drop-coated onto each glass substrate and 10 µL of the nanosphere suspension was drop-coated onto each copper substrate and allowed to dry in ambient conditions. The metal films were deposited in a modified Consolidated Vacuum Corporation vapor deposition system with a base pressure of 10^{-6} Torr. The deposition rates for each film (10 Å/sec) were measured as described previously (10). AgFON surfaces were stored in the dark at room temperature prior to use.

For glucose detection, AgFON surfaces were first incubated in 1 mM DT in ethanol for 45 min and then transferred to 1 mM MH in ethanol for at least 12 h (Figure 1B). The SAM-functionalized surfaces were then mounted into a small-volume flow cell for SERS measurements.

SERS Apparatus

A titanium-sapphire laser (CW Ti: Sa, model 3900, Spectra Physics, Mountain View, CA) pumped by a solid-state diode laser, λ_{ex} 532 nm (model Millenia Vs, Spectra Physics) was used to generate λ_{ex} of 750 or 785 nm as described previously(10, 11). For the detection of spores, all the measurements were performed in ambient conditions. For the detection of glucose, a small-volume flow cell was used to control the external environment of the AgFON surfaces.

LSPR Reflectance Spectroscopy, quantitative multivariate analysis and time constant analysis was performed as described in our papers (10, 11).

Results and Discussion

Extraction of CaDPA from Spores

CaDPA was extracted from spores by sonicating in 0.02 M HNO_3 solution for 10 minutes. This concentration of the HNO_3 solution was selected because of its capability of extraction and benign effect on the AgFON SERS activity. The sonication procedure was performed because no SERS signal of CaDPA was observed from the spore solution prior to sonication. To test the efficiency of this extraction method, a 3.1×10^{-13} M spore suspension (3.7×10^4 spores in 0.2 μL, 0.02 M HNO_3) was deposited onto a AgFON surface (D = 600 nm, d_m = 200 nm). The sample was allowed to evaporate for less than one minute. A high signal-to-noise ratio (S/N) SER spectrum was obtained in 1 minute (Figure 2A). For comparison, a parallel SERS experiment was conducted using 5.0×10^{-4} M CaDPA (Figure 2B). The SER spectrum of *B. subtilis* spores is dominated by bands associated with CaDPA, in agreement with the previous Raman studies on bacillus spores (*12, 13*). The peak at 1050 cm^{-1} in Figure 2A is from the symmetrical stretching vibration of NO_3^- (*14*). Because of its prominence, this peak is used as an internal standard to reduce sample to sample deviations.

Temporal Stability of AgFON and SAM-functionalized AgFON Substrates

An ideal detection system should run unattended for long periods of time, require infrequent maintenance, and operate at low cost. Previous work has demonstrated that bare AgFON surfaces display extremely stable SERS activity when challenged by negative potentials in electrochemical experiments (*15*) and high temperatures in ultrahigh vacuum experiments (*16*). Moreover, as shown previously by spectroscopic and electrochemical measurements, SAMs adsorbed on metal FON substrates remain stable for 3 to 10 days depending on the metal (*9*). In this work, the temporal stability of AgFON and DT/MH SAM-functionalized AgFON surfaces was studied over a period of 40 and 10 days, respectively.

SER spectra of 4.7×10^{-14} M spores (5.6×10^3 spores in 0.2 μL, 0.02 M HNO_3), well below the anthrax infectious dose of 10^4 spores, were captured on AgFON surfaces of different ages. Both the CaDPA spectral band positions and intensity patterns remained constant over the course of 40 days (*11*). Furthermore, SER spectra of the DT/MH SAM on AgFON surfaces stored in bovine plasma were acquired daily over a period of 10 days. Spectral position of

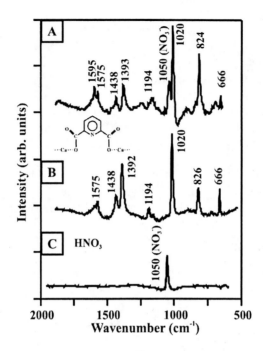

Figure 2. (A) SER spectrum of 3.1 × 10⁻¹³ M spore suspension (3.7 × 10⁴ spores in 0.2 μL, 0.02 M HNO₃) on a AgFON substrate. (B) SER spectrum of 5.0 × 10⁻⁴ M CaDPA. (C) SER spectrum of 0.2 μL 0.02 M HNO₃. λₑₓ = 750 nm, laser power = 50 mW, t = 1 min, D = 600 nm, dₘ = 200 nm. Reproduced with permission from ref. 11. Copyright 2005 J. Am. Chem. Soc.

the C-C stretching vibrational band at 1119 cm⁻¹ remained constant and the intensity decreased by only 2%, which can be attributed to rearrangments in the SAM (*10*). This indicates AgFONs' suitability for potential field sensing applications, as well as *in vivo* detection when functionalized with a SAM.

Adsorption Isotherm and LOD for Bacillus Spores on AgFON Surfaces

The quantitative relationship between SERS signal intensity and spore concentration is demonstrated in Figure 3A. Each data point represents the average intensity at 1020 cm⁻¹ from three samples with the standard deviation shown by the error bars. At low spore concentrations, the peak intensity increases linearly with concentration (Figure 3A inset). At higher spore concentrations, the response saturates as the adsorption sites on the AgFON substrate become fully occupied. Saturation occurs when the spore concentrations exceed ~ 2.0 × 10⁻¹³ M (2.4 × 10⁴ spores in 0.2 μL, 0.02 M HNO₃).

To be practical for long-term health and safety monitoring, a SERS-based detection system has to be capable of detecting less than the life-threatening dose of a pathogen in real or near-real time. In this study, the LOD is defined as the concentration of spores for which the strongest SERS signal of CaDPA at 1020 cm⁻¹ is equal to three times the background SERS signal within a one-minute acquisition period. The background signal refers to the SERS intensity from a sample with a spore concentration equal to zero, which is calculated to be the intercept of the low concentration end of the spore adsorption isotherm (Figure 3A). Although lower detection limits can be achieved using longer acquisition times, these parameters are reasonable for high throughput, real-time, and on-site analysis of potentially harmful species. The LOD for *B. subtilis* spores, evaluated by extrapolation of the linear concentration range of the adsorption isotherms (Figure 3A inset), is found to be 2.1 × 10⁻¹⁴ M (2.6 × 10³ spores in 0.2 μL, 0.02 M HNO₃). It should be noted that previously published NRS studies of anthrax detection were 200,000 times less sensitive and required eight times more laser power (*12*). Similarly, previous SERS studies via the CaDPA biomarker were 200 times less sensitive and required three times more laser power (*13*) than the results demonstrated herein.

To determine the adsorption capacity of extracted CaDPA on an AgFON, the Langmuir adsorption isotherm was used to fit the data (*17, 18*):

$$\theta = \frac{I_{1020}}{I_{1020,\max}} = \frac{K_{spore} \bullet [spore]}{1 + K_{spore} \bullet [spore]} \qquad (1)$$

$$\frac{1}{I_{1020}} = \frac{1}{K_{spore} \bullet I_{1020,\max}} \bullet \frac{1}{[spore]} + \frac{1}{I_{1020,\max}} \qquad (2)$$

where θ is the coverage of CaDPA on the AgFON; $I_{1020\ max}$ is the maximum SERS signal intensity at 1020 cm⁻¹ when all the SERS active sites on the AgFON are occupied by CaDPA; [*spore*] is the concentration of spores (M); and K_{spore} is the adsorption constant of CaDPA extracted from spores on AgFON (M⁻¹). From equation 2, K_{spore} is calculated from the ratio between the intercept and the slope. Slope and intercept analyses of the linear fit (Figure 3B) leads to the value of the adsorption constant $K_{spore} = 1.7 \times 10^{13}$ M⁻¹.

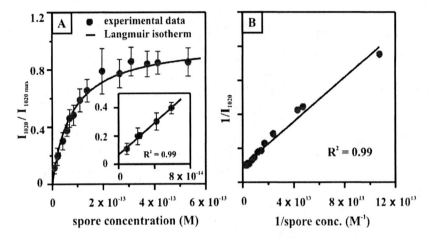

Figure 3. (A) Adsorption isotherm for B. subtilis spore suspension onto a AgFON substrate. I_{1020} was taken from SER spectra that correspond to varying spore concentrations in 0.2 μL, 0.02 M HNO_3 on AgFON surfaces. λ_{ex} = 750 nm, laser power = 50 mW, t = 1 min, D = 600 nm, and d_m = 200 nm. A Langmuir curve was generated using eq 1 with K_{spore} = 1.3 × 10^{13} M^{-1}. The inset shows the linear range that is used to determine the LOD. Each data point represents the average value from three samples. Error bars show the standard deviations. (B) Adsorption data fit with the linear form of the Langmuir model, eq 2. The slope and intercept values are used to calculate the adsorption constant. Reproduced with permission from ref. 11. Copyright 2005 J. Am. Chem. Soc.

Use of Field-Portable Raman Spectrometer for Anthrax Detection

The final goal of this project was to demonstrate the use of SERS as a field-portable screening tool by using a compact Raman spectrometer. Many field-sensing applications require the portability and flexibility not available from conventional laboratory scale spectroscopic equipment. As a first step in this direction, the Raman spectrum from 10^4 B. *subtilis* spores dosed onto a one-month-old AgFON substrate was readily acquired using a commercially available portable Raman instrument. A high S/N spectrum was achieved within five seconds (Figure 4A). The SERS peak positions and intensity pattern for the spore sample was similar to those of CaDPA recorded utilizing the same device (Figure 4B). This is the first example of using a compact, portable Raman spectrometer for the detection of bacillus spores. By coupling the portability and ease of use of this type of device with the molecular specificity and spectral sensitivity inherent to SERS, we open a range of possibilities in the area of detecting bioagents and other chemical threats. In practical field applications of the detection method described, there might be difficulties in collecting B. *anthracis* spores out of the air and dissolving them into a small liquid volume. Most sensor modalities must face this problem.

Reversibility of DT/MH AgFON Glucose Sensor

While CaDPA binds directly to a bare AgFON surface, some molecules such as glucose do not have an affinity to either Ag or Au and need a partition layer to bring them closer to the surface. For glucose detection, we have demonstrated the viability of a DT/MH-functionalized SERS-based sensor. An implantable glucose sensor must successfully monitor the fluctuation of glucose continuously throughout the day. This can only be accomplished if the sensor can completely partition and departition glucose. To simulate real-time sensing, 0 and 100 mM aqueous glucose solutions (pH ~7) were alternately introduced into a flow cell containing the DT/MH-modified AgFON in 20-min intervals without flushing the sensor between measurements. Figures 5A, 5B, and 5C show the first three steps of the pulsing experiment. Figures 5D is the difference spectra that demonstrates partitioning of glucose into the DT/MH functionalized SAM with glucose features at 1461, 1371, 1269, 1131, 916, and 864 cm^{-1}. For comparison to the SERS spectra, Figure 5F shows the normal Raman spectrum of an aqueous saturated glucose solution. Peaks at 1462, 1365, 1268, 1126, 915, and 850 cm^{-1} correspond to crystalline glucose peaks (*19*). The related literature has shown that SERS bands can shift up to 25 cm^{-1} compared to normal Raman bands of the same compound (*20*). Therefore, the features in the difference spectra correspond to the glucose peaks in the normal Raman spectrum of glucose (Figure 5F). Departitioning of glucose is demonstrated by the difference

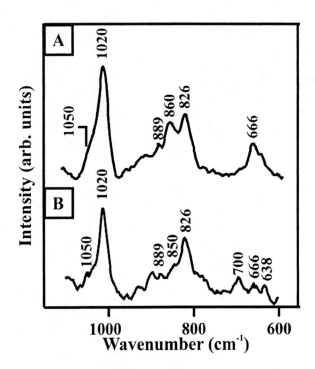

Figure 4. SER spectra obtained by portable Raman spectrometer. (A) SER spectrum of 8.3 × 10⁻¹⁴ M spore suspension (1.0 × 10⁴ spores in 0.2 μL, 0.02 M HNO₃) on 30-day old AgFON. (B) SERS spectrum of 10⁻⁴ M CaDPA in 0.2 μL 0.02 M HNO₃ on 30-day old AgFON substrate. λ$_{ex}$ = 785 nm, laser power = 35 mW, t = 5 sec, resolution = 15 cm⁻¹, D = 600 nm, and d$_m$ = 200 nm. Reproduced with permission from ref. 11 Copyright 2005 J. Am. Chem. Soc.

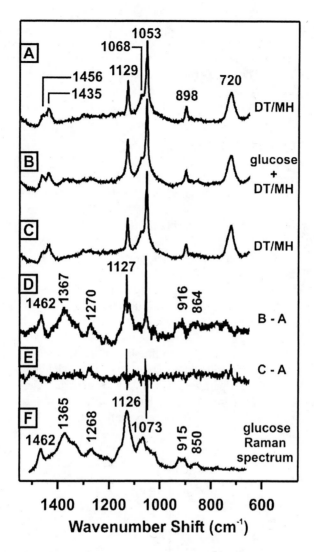

Figure 5. SER spectra demonstrating reversibility of DT/MH functionalized SAM. λ_{ex} = 532 nm, laser power = 10 mW, t = 20 min, pH ~7. (A), (B), and (C) show the first three steps in the pulsing spectrum where 0 and 100 mM aqueous glucose solutions (pH~7) were alternately introduced into a flow cell. (D) shows a difference spectrum demonstrating partitioning. (E) shows a difference spectrum demonstrating departitioning. The normal Raman spectrum of aqueous saturated glucose solution is shown in (F).

spectra shown in Figure 5E. The absence of glucose spectral features clearly demonstrates complete departitioning of glucose. The spectra were normalized by using the nitrate peak at 1053 cm^{-1} as an internal standard to minimize laser power fluctuations. These experiments clearly demonstrate that the DT/MH-functionalized AgFON is a reversible sensing surface that can completely partition and departition glucose.

Time Constant of DT/MH AgFON Glucose Sensor

In addition to reversibility, an implantable glucose sensor must also partition and departition on a reasonable time scale. The real time response was evaluated by calculating the 1/e time constants for partitioning and departitioning. The experiments were conducted in bovine plasma to simulate the *in vivo* environment.

A DT/MH-functionalized AgFON was first incubated in bovine plasma for ~5 hours. The AgFON surface was then placed in a flow cell and a solution containing 50 mM glucose in bovine plasma was introduced at t = 0 to observe partitioning. Departitioning was evaluated by injecting 0 mM glucose in bovine plasma into the flow cell at t = 225 s. Throughout the experiment, SER spectra were collected continuously at 15 s intervals.

The amplitude of the peak at 1462 cm^{-1} was obtained by fitting the spectra to the superposition of three Lorentzian lineshapes and was plotted versus time (Figure 6). The plot was fitted with an exponential curve, yielding the following 1/e time constants: 28 s for partitioning and 25 s for departitioning (*10*). These experiments demonstrate that partitioning and departitioning occur in less than 1 minute, making the SERS based sensor a potential candidate for implantable, continuous sensing.

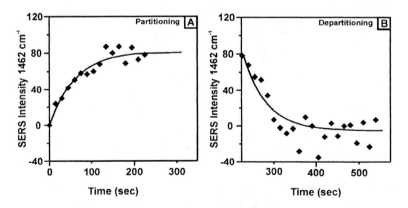

Figure 6. Real-time response to step change in glucose concentration. Partitioning (A) and departitioning (B) of glucose is demonstrated with with λ_{ex} = 785 nm, laser power = 100 mW, t = 15 s. The 1/e time constants were calculated to be 28 s for partitioning and 25 s for departitioning.

Calibration Curve of DT/MH AgFON Glucose Sensor

Futhermore, an ideal glucose sensor must be able to detect glucose in the clinically relevant range 10-450 mg/dL (0.56-25 mM), under physiological pH and in complex media. The DT/MH-functionalized SAM was placed in a flow cell and incubated for 2 minutes with glucose (10 – 450 mg/dl) in filtered bovine plasma. Bovine plasma was used to simulate the *in vivo* environment that the sensor will be exposed to when it is implanted under the skin in interstitial fluid. SER spectra were collected using multiple samples and multiple spots with a near-infrared laser source. Partial least squares leave-one-out (PLS-LOO) analysis was used to construct a calibration model with 92 independent spectral measurements of known glucose concentrations based on 7 latent variables. The resulting calibration model and concentration predictions are presented on a Clarke error grid (Figure 7).

The Clarke error grid is a standard for evaluating the reliability of glucose sensors in the clinically relevant concentration range (0-450 mg/dL) (*21*). Data

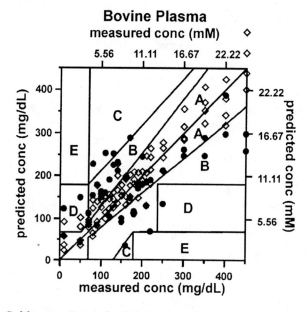

Figure 7. Calibration (◊) and validation (●) plots of glucose in bovine plasma on Clarke error grid. Calibration plot was constructed using 92 data points and validation plot was constructed using 46 data points. Sample was incubated for 2 min in glucose concentrations ranging from 10 – 450 mg/dl with λ$_{ex}$ = 785 nm, laser power = 10-30 mW, t= 2 min. RMSEC = 34.3 mg/dL (1.9 mM) and RMSEP = 83.16 mg/dL (4.62 mM). Reproduced with permission from ref. 10 Copyright 2005 J. Am. Chem. Soc.

points that fall in the A and B ranges are acceptable values. Values outside the A and B range indicate potential failure to detect blood glucose levels properly, usually resulting in erroneous and even fatal diagnosis.

The PLS analysis yielded a root mean square error of calibration (RMSEC) of 34.3 mg/dL (1.90 mM) with 98% of the data falling in the A and B region of the Clarke error grid. The root mean square error of prediction (RMSEP) was calculated to be 83.16 mg/dL (4.62 mM) with 85% of the validation points falling in the A and B range of the Clarke error grid (*10*). Overall, the results show that the DT/MH-modified AgFON glucose sensor is capable of making accurate glucose measurements in the presence of interfering analytes.

Calibration Curve for *in Vivo* DT/MH AgFON Glucose Measurement

Finally, we demonstrated the first successful detection of glucose *in vivo* with an implantable SERS sensor using a rat animal model (*22*). To collect SER spectra, a metal frame containing a glass window was placed along the midline of the back of a Sprague-Dawley rat. A DT/MH-functionalized AgFON surface supported on a Cu mesh was positioned between the skin and the window such that the substrate was in contact with the interstitial fluid. An aliquot of glucose (1 g/mL in saline) was injected in the rat to artificially vary the blood glucose level. An aliquot of blood was drawn from the rat to measure the blood glucose level using the One Touch II blood glucose meter. SER spectra were acquired through the window and showed pronounced Raman band characteristic of the SAM (*22*). Additional bands were observed compared to the spectra collected in the bovine plasma model; however, the Raman bands were not obscured by autofluorescence because near infra-red excitation was used. All SER spectra were analyzed using PLS-LOO and represented on the Clarke error grid. In Figure 8, 21 data points were used to build the calibration model and five data points were used to validate the model. All the data points in both calibration and validation fall in the A and B range of the Clarke error grid with RMSEC = 7.46 mg/dL and RMSEP = 53.42 mg/dL. This is the first report of *in vivo* glucose detection using SERS.

Conclusion

We have successfully demonstrated the applicability of SERS for biological sensing. We have targeted analytes such as CaDPA, a marker for anthrax, as well as, glucose, which plays a crucial role in evaluating metabolic health. In brief, we have shown that SERS is a highly sensitive and selective method allowing rapid quantitative detection of both analytes. In addition, we have

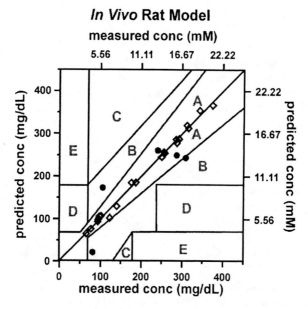

Figure 8. *Glucose calibration (◇) and validation (●) plot using a single substrate and a single spot on a DTMH functionalized AgFON on a mesh. PLS calibration plot was constructed using 21 data points and validation plot was constructed using 5 data points taken over a range of glucose concentrations (10 - 450 mg/dL) in vivo (rat) with λ_{ex}=785nm, laser power = 50 mW, t = 2 min. RMSEC = 7.46 mg/dL (0.41 mM) and RMSEP = 53.42 mg/dL (2.97 mM).*

demonstrated that SERS-active surfaces such as AgFONs are suitable for long term use in the field and as implantable devices in the body. Finally, we demonstrate for the first time the use of a portable Raman spectrometer for anthrax detection and quantitative *in vivo* glucose measurements. Future work will focus on exploring various surface chemistries to improve selective binding of CaDPA to a SERS-active surface, as well as SAMs that minimize spectral overlap with glucose. We will also incorporate a miniaturized spectrometer and a fiber optic probe for transcutaneous measurements of glucose, and perform long-term *in vivo* studies.

References

1. Walt, D. R. *Analytical Chemistry* **2000**, *72*, 738A-746A.
2. Bailey, G. F.; Karp, S.; Sacks, L. E. *Journal of Bacteriology* **1965**, *89*, 984.

3. Goodacre, R.; Shann, B.; Gilbert, R. J.; Timmins, E. M.; McGovern, A. C.; Alsberg, B. K.; Kell, D. B.; Logan, N. A. *Analytical Chemistry* **2000**, *72*, 119-127.

4. Ross, S. A.; Gulve, E. A.; Wang, M. *Chem. Rev.* **2004**, *104*, 1255-1282.

5. Diabetes Overview; National Diabetes Information Clearinghouse (NDIC), http://diabetes.niddk.nih.gov/dm/pubs/overview/index.htm (07-26-05).

6. McFarland, A. D.; Young, M. A.; Dieringer, J. A.; Van Duyne, R. P. *Journal of Physical Chemistry B* **2005**, *109*, 11279-11285.

7. Shafer-Peltier, K. E.; Haynes, C. L.; Glucksberg, M. R.; Van Duyne, R. P. *Journal of the American Chemical Society* **2003**, *125*, 588-593.

8. Yonzon, C. R.; Haynes, C. L.; Zhang, X. Y.; Walsh, J. T.; Van Duyne, R. P. *Analytical Chemistry* **2004**, *76*, 78-85.

9. Stuart, D. A.; Yonzon, C. R.; Zhang, X.; Lyandres, O.; Shah, N. C.; Glucksberg, M. R.; Walsh, J. T.; Van Duyne, R. P. *Analytical Chemistry* **2005**, *77*, 4013-4019.

10. Lyandres, O.; Shah, N. C.; Yonzon, C. R.; Walsh, J. T.; Glucksberg, M. R.; Van Duyne, R. P. *Analytical Chemistry* **2005**, *77*, 6134-6139.

11. Zhang, X.; Young, M. A.; Lyandres, O.; Van Duyne, R. P. *Journal of the American Chemical Society* **2005**, *127*, 4484-4489.

12. Farquharson, S.; Grigely, L.; Khitrov, V.; Smith, W.; Sperry, J. F.; Fenerty, G. *Journal of Raman Spectroscopy* **2004**, *35*, 82-86.

13. Farquharson, S.; Gift, A. D.; Maksymiuk, P.; Inscore, F. E. *Applied Spectroscopy* **2004**, *58*, 351-354.

14. Mosier-Boss, P. A.; Lieberman, S. H. *Applied Spectroscopy* **2000**, *54*, 1126-1135.

15. Zhang, X.; Yonzon, C. R.; Van Duyne, R. P. *Proceedings of SPIE-The International Society for Optical Engineering* **2003**, *5221*, 82-91.

16. Litorja, M.; Haynes, C. L.; Haes, A. J.; Jensen, T. R.; Van Duyne, R. P. *Journal of Physical Chemistry B* **2001**, *105*, 6907-6915.

17. Jung, L. S.; Campbell, C. T. *Physical Review Letters* **2000**, *84*, 5164-5167.

18. Jung, L. S.; Campbell, C. T. *Journal of Physical Chemistry B* **2000**, *104*, 11168-11178.

19. Soderholm, S.; Roos, Y. H.; Meinander, N.; Hotokka, M. *J. Raman Spectrosc.* **1999**, *30*, 1009-1018.

20. Stacy, A. M.; Van Duyne, R. P. *Chem. Phys. Lett.* **1983**, *102*, 365-370.

21. Clarke, W. L.; Cox, D.; Gonder-Frederick, L. A.; Carter, W.; Pohl, S. L. *Diabetes Care* **1987**, *10*, 622-628.

22. Dieringer, J. A.; Lyandres, O.; McFarland, A. D.; Shah, N. C.; Stuart, D. A.; Whitney, A. V.; Yonzon, C. R.; Young, M. A.; Yuen, J.; Zhang, X.; VanDuyne, R. P. *Faraday Discussions* **2005**.

Chapter 9

Surface-Enhanced Raman Spectroscopy Detection of Hyaluronic Acid: A Potential Biomarker for Osteoarthritis

Karen A. Dehring[1], Gurjit S. Mandair[2], Blake J. Roessler[3], and Michael D. Morris[2]

Departments of [1]Biomedical Engineering, [2]Chemistry, and [3]Internal Medicine, University of Michigan Medical Center, Ann Arbor, MI 48109

A novel application of surface-enhanced Raman spectroscopy (SERS) for *in-vitro* osteoarthritis (OA) biomarker detection is described. Hyaluronic acid (HA) is a potential OA biomarker and synovial fluid levels of HA have been correlated with progression of joint space narrowing. However, current immunoassay and chromatographic methods used to identify HA in synovial fluid specimens are cumbersome and often require sophisticated instrumentation. Raman spectroscopy may be an alternative to these analytical methods, providing rapid identification of HA using characteristic Raman biomarker bands. Yet, previous reports of normal (unenhanced) Raman spectroscopy for HA are in aqueous solutions exceed 1000X *in-vivo* concentrations because HA is a weakly scattering polysaccharide. In contrast, SERS could improve the detection limits of HA to below the clinical range and we present, to our best knowledge, the first surface-enhanced Raman spectra of HA. Moreover, the recent commercial availability of gold-coated SERS substrates has enabled rapid the SERS detection of this biomarker at physiological concentrations. Preliminary results show that HA can be readily observed at low concentrations in aqueous solutions and in synthetic models of biofluids, such as artificial synovial fluid, that contain HA at low concentrations. These complex biofluids often contain proteins that will compete with HA for the SERS-active sites, and the resulting spectra are dominated by protein Raman bands. To overcome this problem, we use a simple and validated protein precipitation protocol to artificial synovial fluid prior to deposition onto the SERS substrate. We show that HA can be readily detected in these biofluids at clinically useful levels after protein precipitation.

Introduction

Osteoarthritis (OA), the progressive degeneration of synovial joints affects more than 20 million adults in the United States and is expected to rise to 70 million by 2030.[1] Symptomatic knee OA is one of the most prevalent form of arthritis and affects 6 % of adults 30 years or older.(1) Current methods of detecting OA often rely upon physical examinations and radiographic evaluations with follow-up times of about two years to establish the level of disease progression.(1, 2) However, radiographic assessments are insensitive indicators of short-term OA progression. It is therefore necessary to use additional clinical methods to detect the earliest stages of the disease and to monitor the efficacy of therapeutic interventions.

The evaluation of potential OA biomarkers, such as sulfated and unsulfated glycosaminoglycans (GAGs) has received considerable attention in recent years. It has been demonstrated that the chemical composition of GAGs in synovial fluid is a reflection of arthritic progression in both articular cartilage and the synovial membrane.(3-5) For example, inflammation of the synovial membrane leads to the enhanced secretion of pathological synovial fluids. This fluid contains a lower abundance and molecular weight of the unsulfated GAG hyaluronic acid (HA) compared to normal synovial fluids.(4, 6) These changes are undesirable as the GAG is responsible for the viscoelastic properties of synovial fluid, which aid in the lubrication and protection of articular cartilage from mechanical injury. The decline in HA concentration is caused by infiltration of the plasma fluid and proteins into the synovial fluid, whereas the molecular size reduction is caused by abnormal metabolic processes occurring within the inflamed synovial structures.(4, 7, 8)

We hypothesize that surface-enhanced Raman spectroscopy (SERS) can be used for evaluating the early stages of OA progression by using HA as a biomarker for pathological synovial fluid secretions. However, to fully exploit SERS-based detection in clinical research, it is necessary to use SERS-substrates with reproducible signal enhancements. The recent commercial availability of

SERS-substrates produced by semiconductor techniques has overcome the well-known limitation in reproducibility of enhancement on substrate surfaces.(9, 10) These substrates are based on silicon photonic crystals, which are etched with void architectures and coated with a layer of gold producing a SERS-active surface. In this paper, we explore the feasibility of using these gold-coated SERS substrates for the rapid and reproducible detection of artificial synovial fluid HA at clinically useful levels.

Our approach involves depositing a small volume of liquid containing HA onto a SERS substrate and allowing the droplet to dry naturally under ambient conditions. As a result HA is precipitated in a ring, with residual small molecules located in the center. Ring formation in an evaporating drop is a seemingly simple phenomenon that has recently gained considerable attention in the biomedical and pharmaceutical fields as a way to rapidly measure protein-protein interactions or generate single-crystals for a polymorph crystal screen. The formation of a dried ring onto a solid surface is dictated by capillary flow, and can be affected by variables such as substrate material, analyte concentration, and speed of evaporation.(11-13)

Because most ring formation studies are performed on flat substrates, such as silicon, mica or Teflon©, light microscopy is frequently used to follow solvent evaporation and examine the morphology of the resulting ring formation. The dimensions of the substrates are typically compatible with other microscope-based analytical tools such as cross-polarized light microscopy, fluorescence or microspectroscopy. BenAmotz recently demonstrated the use of normal Raman spectroscopy to detect proteins from dried droplets at concentrations as low as 1 μM in the presence of a buffer.(14) Because the ring formation can function simultaneously as both a low resolution separation and a preconcentration method, it can help overcome the well-known limitations of fluorescence interference and high sample concentration requirements that are inherent in normal Raman spectroscopy. However, normal Raman spectroscopic detection of rings may not be feasible for weakly scattering biomolecules, such as HA. SERS of HA droplets dried on gold-coated SERS substrates can be used in conjunction with the ring formation technique to overcome this problem. Both the preconcentration effect of the dried ring and the surface-enhancement offers an additional improvement in the Raman signal intensity of weakly scattering HA biomolecules.

Materials and Methods

Raman Spectroscopy

Raman spectra were collected with a Raman microprobe, optimized for collection of near-infrared signal, described elsewhere.(15) The system consists of a 400 mW 785 nm laser (Invictus, Kaiser Optical Systems, Inc.) and an epi-illumination microscope (Olympus, BH-2). Laser light was coupled with a 1.0 neutral density filter, Powell lens (StockerYale), and lined-focused through a 20X/0.75 NA Fluar objective (Carl Zeiss, Inc.). A laser power output of ~8 mW was achieved at the objective. Raman scatter was collected using an $f/1.8$ axial transmissive spectrograph (Kaiser, HoloSpec) and detected using an air-cooled, back-thinned deep depletion CCD camera. Raman spectra were acquired with 60 or 120 second integration times. Wavenumber calibration and image curvature correction were performed in Matlab 6.1 (The Math Works, Natick, MA) using built-in and locally-written scripts. Light microscope images of droplets were collected using either a 5X/0.25 NA Fluar (Carl Zeiss, Inc.) or a 10X/0.50 Fluar (Carl Zeiss, Inc.) objective.

Materials

Rooster comb hyaluronic acid (HA, ~2000 kDa), bovine serum albumin (BSA), γ-globulins, and human plasma (*ca.*, 72 % albumin and 15 % γ-globulin) were obtained from Sigma-Aldrich and used as received. All other used reagents and solvents were of analytical grade.

Preparation of Standard Solutions

Aqueous HA standards (4-0.25 mg/mL) were prepared by dilution of a 6-8 mg/mL HA stock solution in water. Artificial synovial fluid standards (ASF) containing human plasma were prepared by dissolution of 25 μL HA standards (4-0.25 mg/mL) in 25 μL solutions containing 11.5 μL deionized water and 13.5 μL human plasma, giving final HA, albumin, and globulin concentrations of approximately 2.0-0.125, 11.8, and 3.8 mg/mL, respectively. A human plasma solution (27% v/v in deionized water) was prepared as an experimental control at the same albumin and γ-globulin concentrations. All solutions were stored at -4 °C.

Protein depletion protocols

50 μL aliquots of each solution (**A-B**) were transferred to 500 μL centrifuge tubes, followed by an equivalent volume of 10 % trichloroacetic acid (TCA) solution. The centrifuge tubes were vortexed for 30 seconds, incubated at -4 °C

overnight, and centrifuged at 9,500 rpm for 10 min. The clear supernatant layers were carefully extracted and stored at -4 °C.

SERS analysis protocols

For all fluids examined, 0.2-0.3 µL of each solution were deposited onto Klarite™ SERS Substrates (Mesophotonics Ltd, Hampshire, UK) and left to air dry at ambient temperature for 30 min. Raman spectra of the ring-like deposits were acquired with 60 or 120 second integration times and ~8 mW laser power. Normal Raman spectra of a 0.5 mg/mL HA deposits dried on bare gold and fused silica slides were also acquired. The resulting Raman spectra were examined between 800 and 1700 cm^{-1}. Pretreatment consisted of subtraction of the detector offset, correction for contributions from the substrate, and baseline subtraction in Grams/AI 7.01 software (ThermoGalactic).

Results and Discussion

Raman bands of HA are shown in Table 1, band assignments are based on literature reports of the spectra of aqueous and solid HA.([16, 17]) The 899, 945, 1050, 1130 and 1410 cm^{-1} bands were used to identify HA in artificial synovial fluid (ASF) preparations. Noise in the measured Raman spectra limited the reproducibility of band positions to ± 2 cm^{-1}.

The effect of substrate surface and HA concentration on droplet shape was studied on fused silica, bare gold, and a SERS substrate. A consistent observation throughout these studies was the asymmetric ring shape when droplets dried on the SERS substrate. As seen in Figure 1a, the droplet shape is similar to an octagon. A similar droplet shape was observed when aqueous HA solutions at various concentrations (0.25-6 mg/ml) are deposited onto the SERS substrate. In addition to the non-spherical shape of the droplet, asymmetric concentric rings at the droplet edge (Figure 1b) were observed when highly concentrated aqueous HA solutions were deposited on the SERS substrate. Previous studies have demonstrated that concentric ring formation is concentration dependent.([11]) Moreover, formation of concentric rings may be related to entanglement of the polymer chains because the concentric rings are not prominent at concentrations below 2 mg/ml. At concentrations greater than 2 mg/ml, it is possible that chain entanglement increases and affects both the hygroscopic nature and mobility of the polysaccharide. Aggregation of HA is

Table 1. Table of Raman shifts for hyaluronic acid (HA)

Raman Shift (cm^{-1})	Band Assignment
899	β-linkages
945	C-C stretch
~1050	C-O,C-C
~1100	C-O-H bend, acetyl group
~1130	C-C
~1150	C-O, C-C, Oxygen bridge
~1210	CH_2 twist
1330	CH bend, Amide III
~1410	CH bend, amide II
~1460	CH_2 bend

more likely at these higher concentrations, this "clumping" of HA may affect the droplet formation.(14) The presence of these concentric rings does not prevent the collection of HA Raman spectra, and may provide additional information about the size distribution of the polysaccharide. Deposition of polygonal rather than circular rings appears to be a result of the interplay between evaporation and the geometry of the SERS substrate. Circular rings were observed when 0.2μl drops of the same HA solutions were deposited on fused silica slides or the bare gold portions of SERS substrates.

Figure 2 shows spectra of a 0.5 mg/ml aqueous HA solution evaporated onto fused silica (a), bare gold (b), and a gold-coated SERS substrate (c). It is clear from Figure 2 that normal (unenhanced) Raman spectroscopy is not possible at low concentrations of HA in solution, even using signal integration times as long as 120 seconds. Raman bands from HA are clearly observed on the SERS substrate, even with an integration time of 60 seconds. The positions of surface-enhanced Raman bands of HA are in good agreement with literature values for bands of the normal Raman spectrum of the molecule.

The use of a SERS substrate enables rapid collection of HA spectra. Even at higher concentrations, attempts to collect unenhanced Raman spectra were unsuccessful. No bands of HA deposited from a 6 mg/ml, 3 mg/ml or 0.5 mg/ml aqueous solution were found on fused silica or bare gold even at integration times of 120 seconds. Previously reported detection limits for HA solutions by normal Raman spectroscopy are in the 40-50 mg/ml range. The limit of detection is reduced by at least two orders of magnitude. The current detection limit of

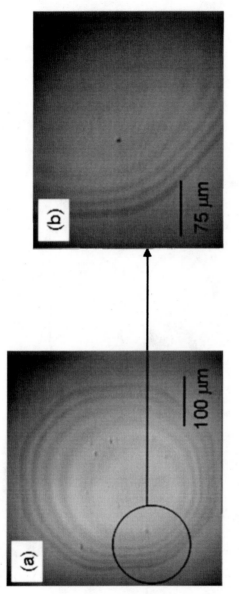

Figure 1. Microscope image of a droplet of 4 mg/ml aqueous solution of HA deposited onto a SERS substrate using an (a) 5X/0.25 NA and (b) 10X/0.50 NA objective. The octagon-like shape of the droplet was observed when deposited onto the SERS substrate, indicating that the surface of the substrate is affecting the formation of the HA droplets.

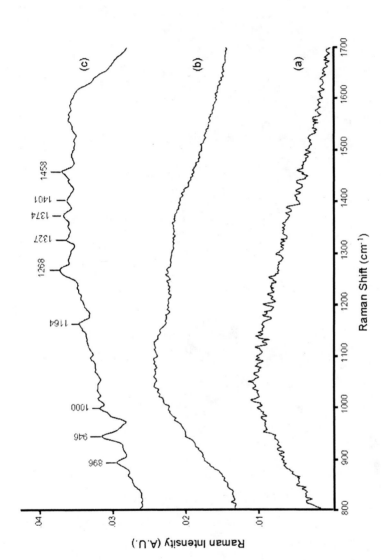

Figure 2. Identification of 0.5 mg/ml aqueous HA solution deposited onto (a) fused silica slide, (b) bare gold and (c) a gold-coated SERS substrate. The measurement time was 120 seconds for spectra (a) and (b) and 60 seconds for spectrum (c). No HA bands were observed when deposited onto the fused silica or bare gold surfaces, but clearly observed when deposited onto the SERS substrate. All spectra have been corrected for the background signal.

about 0.5 mg/mL is within synovial fluid HA levels (0.10-1.14 mg/mL) observed in osteoarthritis patients.(*18*) By contrast, serum HA levels that are correlated with future (approximately 2 years) joint space narrowing are much lower (30.2 ± 19.6 ng/mL).(*19*) To reach such levels a more efficient illumination scheme or an evaporation protocol that produces smaller ring diameters will be need to be examined. The relatively high HA detection limit observed in our studies may be a consequence of using the drop evaporation technique. As HA concentration is reduced the ring of precipitated HA becomes narrower. Because the droplet edge is illuminated by our line-focused laser and most of the deposited HA is not interrogated, the detection limit is relatively high. .

Our current method of producing a droplet allows identification of HA in aqueous solutions at concentrations ranging from 6 mg/ml to 0.25 mg/ml on the SERS substrate. As shown in Figure 3, HA biomarker bands are evident throughout the dilution series. After pre-processing, bands at ~895 cm⁻¹, 945 cm⁻¹, and 1042 cm⁻¹ were fit using a routine in Grams/AI 7.01 software. The effect of HA concentration on band area, band width and band height were examined. The width of the 895 cm⁻¹ band increases with HA concentration, as shown in Figure 4. The heights of the 895, 945 and 1042 cm⁻¹ bands remain almost constant in the 0.25-3 mg/ml range, indicating that monolayer coverage has been reached or perhaps exceeded. The band width is closely related to polymer conformation, which may provide additional information on HA conformation distribution within the ring. The width of the 895 cm⁻¹ band, related to the β-linkages that connect alternating N-acetyl-glucosamine and D-glucuronic acid units, increases approximately linearly with concentration in the 0.25-3 mg/ml range. Raman band width may be a more robust indicator of HA concentration because it is independent of small variations in the substrate surface and can be used at analyte concentrations that yield monolayer or multilayer deposits.

Drop deposition alone is inadequate for separation of HA from the proteins present in synovial fluid. SERS spectra taken from deposits of artificial synovial fluid show protein bands that completely obscure the nearby HA bands. The same problem was encountered canine synovial fluid and canine plasma (data not shown). Further study of the SERS of artificial synovial fluid show that that HA probably binds non-specifically to proteins. Even at higher starting HA concentrations (2 mg/ml) in artificial synovial fluid, Raman spectra taken anywhere in the deposit are dominated by protein bands that obscure any HA signal.

Light microscope images of the deposits from untreated artificial synovial fluid confirm that simple chemical segregation of HA and protein mixtures is inadequate for reduction of interferences from proteins. A few concentric rings are observed, but only protein spectra are seen in all of them. Small molecule impurities are still easily segregated from HA.

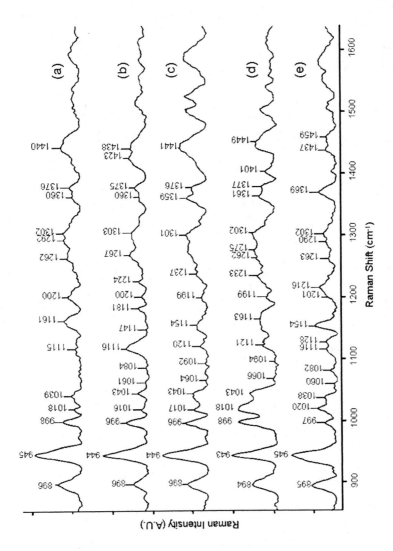

Figure 3 Identification of HA on a SERS substrate was demonstrated at concentrations of (a) 3 mg/ml, (b) 2 mg/ml, (c) 1 mg/ml, (d) 0.5 mg/ml and (e) 0.25 mg/ml. The measurement time was 60 seconds for all spectra, spectra have been corrected for the background substrate signal and baseline corrected.

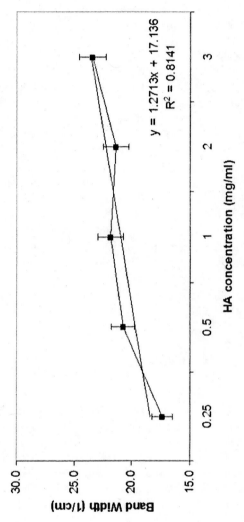

Figure 4. In the range of 0.25-3 mg/ml, the width of the 895 cm^{-1} HA band increases with concentration. The y-axis error bars indicate a 5% error.

Several methods were evaluated for reduction of spectral interference from proteins. We examined filtration, protein precipitation and ultracentrifugation methods to separate protein from HA using protocols from the published literature.(20, 21) Our preliminary conclusion is that filtration of the synovial fluid does not reduce the protein signal, and provides evidence for our hypothesis that hyaluronic acid binds non-specifically with synovial fluid proteins such as albumin or γ-globulin.

Trichloroacetic acid (TCA) protein precipitation followed by ultracentrifugation was successful in removing proteins from HA, rapid and does not interfere with the identification of key Raman biomarker bands associated with HA. This method has been validated against commercially-available protein removal kits and found to provide superior protein-removal efficiency.(22) It should be noted that the TCA protocol dilutes HA in ASF by a factor of two. Although microscope images and Raman spectroscopy show crystalline TCA in the center of the dried droplet deposit, Raman spectra show that some TCA is still contained in the outer HA-rich rings. A broad TCA band is observed between 830-860 cm^{-1} and other bands are found at ~ 945 cm^{-1} and 1365 cm^{-1}. With the exception of the 945 cm^{-1} band, TCA bands do not overlap with HA Raman bands and are not sources of interference.

Most importantly, both albumin and γ-globulin were almost completely removed from the biofluids by precipitation with 10% TCA. As shown in Figure 5b, HA bands can be easily observed at 899, 1040, and 1117 cm^{-1} after treatment with TCA. By contrast, the intense phenylalanine ring breathing band at ~1000 cm^{-1} is reduced to intensities similar to or lower than the intensities of the most intense and characteristic HA bands.

Conclusions

The results demonstrate the potential for hyaluronic acid monitoring of synovial fluid at concentrations indicative of osteoarthritis. Removal of most proteins is necessary for successful measurement. TCA precipitation has been used successfully. Other standard methods will be examined in future studies The incorporation of a simple three-step protein removal protocol reduced the Raman signal from proteins and allowed identification of HA in artificial synovial fluid at clinically relevant concentrations. Other protocols that may provide more complete protein removal are under investigation.

A major weakness of the drop evaporation protocol used in these experiments is that it leaves HA deposited in a ring that is difficult to interrogate efficiently with conventional Raman spectroscopic optics. Optical and optomechanical systems for generation of circular and octagonal laser illumination patterns are under investigation as potential solutions to this problem.

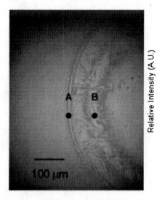

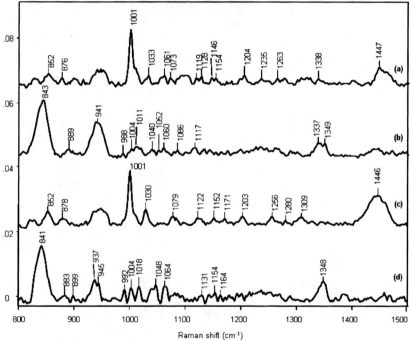

Figure 5. SERS spectra of artificial synovial fluid containing 0.5 mg/mL HA before (a) and after (b) treatment with 10 % TCA solution, taken at point A on the microscope image. Spectra taken at point B on the microscope image are dominated by TCA bands (data not shown). As a control biofluid that does not contain HA, human plasma was diluted 27% v/v in deionized water and also subjected to TCA treatment. SERS spectra of the plasma control before (c) and after (d) TCA treatment indicates removal of protein. There is adequate reduction of the protein signal to enable identification of HA bands, in the 1030-1130 cm^{-1} region, at elevated clinical levels.

Acknowledgements

This work was supported in part by grant #T32 AR07080 from the National Institute of Arthritis and Musculoskeletal and Skin Diseases, NIH, and in part from the University of Michigan Bone Research Center through NIH grant P30 AR46024. KAD acknowledges NIH training grant, T90 DK070071-01, support. We thank Mesophotonics Ltd. for providing the Klarite™ SERS Slides used in this study.

References

1. Leslie, M. *Pain Management in Nursing* **2000**, *1*, 51-57.
2. DeGroot, J.; Bank, R. A.; Tchetverikov, I.; Verzijl, N.; TeKoppele, J. M. *Current Opinion in Rheumatology* **2002**, *14*, 585-589.
3. Fuller, C. J.; Barr, A. R.; Sharif, M.; Dieppe, P. A. *Osteoarthritis and Cartilage* **2001**, *9*.
4. Ghosh, P.; Guidolin, D. *Seminars in Arthritis and Rheumatism* **2002**, *32*, 10-37.
5. Ishimaru, J. I.; Ogi, N.; Mizuno, S.; Goss, A. N. *Osteoarthritis and Cartilage* **2001**, *9*, 365-370.
6. Praest, B. M.; Greiling, H.; Kock, R. *Clinica Chimica Acta* **2003**, *266*, 117-128.
7. Fujimura, K.; Segami, N.; Yoshitake, Y.; Tsuruoka, N.; Kaneyama, K.; Sato, J.; Kobayashi, S. *Oral Surgery, Oral Medicine, Oral Pathology, Oral Radiology, and Endodontology* **2006**, *101*, 463-468.
8. Poortmans, J. R.; S'Jongers, J.-J.; Bidon, G. *Clinica Chimica Acta* **1974**, *55*, 205-209.
9. Netti, C.; Lincoln, J. R. *Microscopy and Analysis* **2005**, *19*, 9-11.
10. Perney, N. M. B.; Baumberg, J. J.; Zoorob, M. E.; Charlton, M. D. B.; Mahnkopf, S.; Netti, C. M. *Optics Express* **2006**, *14*, 847-857.
11. Deegan, R. D. *Physical Review E* **2000**, *61*, 475-485.
12. Deegan, R. D.; Bakajin, O.; Dupont, T. F.; Huber, G.; Nagel, S. R.; Witten, T. A. *Nature* **1997**, *389*, 827-829.
13. Sommer, A. P.; Rozlosnik, N. *Crystal Growth & Design* **2005**, *5*, 551-557.
14. Zhang, D.; Mrozek, M. F.; Xie, Y.; Ben-Amotz, D. *Applied Spectroscopy* **2004**, *58*, 929-933.
15. Timlin, J. A.; Carden, A.; Morris, M. D.; Rajachar, R. M.; Kohn, D. *Analytical Chemistry* **2000**, *72*, 2229-2236.
16. Bansil, R.; Yannas, I. V.; Stanley, H. E. *Biochimica et Biophysica Acta* **1978**, *541*, 535-542.

17. Barrett, T. W.; Peticolas, W. L. *Journal Of Raman Spectroscopy* **1979**, *8*, 35-38.
18. Kvam, C.; Granese, D.; Flaibani, A.; Zanetti, F.; Paoletti, S. *Analytical Biochemistry* **1993**, *211*, 44-49.
19. Pavelka, K.; Forejtova, S.; Olejarova, M.; Gatterova, J.; Senolt, L.; Spacek, P.; Braun, M.; Hulejova, M.; Stovickova, J.; Pavelkova, A. *Osteoarthritis and Cartilage* **2004**, *12*, 277-283.
20. Jiang, L.; He, L.; Fountoulakis, M. *Journal of Chromatography A* **2004**, *1023*, 317-320.
21. Polson, C.; Sarkar, P.; Incledon, B.; Raguvaran, V.; Grant, R. *Journal of Chromatography B* **2003**, *785*, 263-275.
22. Chen, Y.-Y.; Lin, S.-Y.; Yeh, Y.-Y.; Hsiao, H.-H.; Wu, C.-Y.; Chen, S.-T.; Wang, A. H.-J. *Electrophoresis* **2005**, *26*, 2117-2127.

Chapter 10

Surface-Enhanced Raman Spectroscopy- and Surface-Enhanced Infrared Absorption-Based Molecular Sensors

Selective Detection of Polycyclic Aromatic Hydrocarbons Employing Different Silver Nanoparticle Functionalization Methods

C. Domingo[1], L. Guerrini[1], P. Leyton[2], M. Campos-Vallette[2], J. V. Garcia-Ramos[1], and S. Sanchez-Cortes[1]

[1]Instituto de Estructura de la Materia, CSIC, Serrano 121, 28006 Madrid, Spain
[2] Faculty of Sciences, University of Chile, P.O. Box 653, Santiago, Chile

Functionalization of metal nanoparticles usually employed for Surface Enhanced Raman and Infrared Spectroscopies, with adequate host molecules, has allowed the selective detection of some Polyaromatic Hydrocarbons (PAHs), considered as hazardous pollutants. The molecular sensing capabilities of these high sensitivity techniques, SERS and SEIRA, together with their complementarity, are here illustrated for the case of employing calix[4]arenes derivatives as host molecules in order to detect PAHs.

SERS (Surface Enhanced Raman Scattering) and, more recently, SEIRA (Surface Enhanced Infrared Absorption) spectroscopies have rapidly expanded their field of applications. These techniques are based on the use of noble metal nanoparticles (NPs) supporting Localized Surface Plasmons (LSP). They are extremely attractive due to unique high sensitivity and their potential for molecule-specific sensing technologies utilizing vibrational signatures (1,2).

Significant effort, in many research laboratories, is directed towards improving the selectivity and reproducibility of the required nanostructured substrates, in order that these enhanced spectroscopies can be employed as true molecular chemosensing techniques.

The main components involved in the vibrational signal enhancement in the presence of NPs are the metal nanostructure, the adsorbate and the interface between the metal surface and the surrounding medium. The nature of the interface determines the accessibility of the adsorbate to the surface as, firstly, it is necessary a diffusion of the molecule to the surface and, then, the interaction with the metal, directly or through ionic interactions when ions are present.

The metal substrate is essential and the size and morphology of the metal nanostructures will determine the signal intensification. Correspondingly, a substantial part of the research work in our group is dedicated to fabrication and morphological characterization of nanostructured metal surfaces with controlled morphology and optical properties (3-5) using different methods like colloidal aggregation or photoreduction, and further characterization of their SERS and SEIRA properties.

A crucial point in the application of these surface techniques is the nature of the adsorbate. Usually, the adsorption of the active molecules in SERS and SEIRA is driven by some affinity for the metal surface providing a close metal-molecule interaction and allowing for a significant vibrational enhancement. Many molecules are strongly adsorbed undergoing a chemical change at the surface, as we have seen for many polyphenols (6). In contrast, other molecules may not adsorbed at the metal surface due to their poor affinity to the metal. Among the latter are the Polycyclic Aromatic Hydrocarbons (PAHs), considered hazardous pollutants. We are specially interested in the detection of PAHs using enhanced vibrational techniques.

The affinity of adsorbates toward the metal surface may be manipulated by introducing a metal interface. In the case of metal nanostructures the surface is negatively charged, the negative potential or zeta potential value depending on the method employed to prepare the nanostructures (5). This means that positively charged adsorbates could give the most enhanced vibrational spectra due to their high preference to adsorb onto the metal surface. The high polar nature of the interface makes difficult the adsorption of highly apolar molecules such as the PAHs mentioned above and many apolar pesticides, whose detection in trace quantities is of tremendous interest.

But the affinity of the adsorbate toward the metal can be increased by modifying the chemistry of the interface by a proper functionalization of the metal NPs. Lately, we have attempted to modify the surface of metal NPs by the adsorption of molecules displaying a double functionality, i.e. a high affinity to the metal and for the the target molecule (PAHs in our case). The molecules employed in this functionalization, i.e. the host molecules, can be classified into two main groups: natural hosts and specifically designed hosts.

Nucleic acids, proteins and humic acids are among the natural hosts we have employed in the surface functionalization of metal NPs. Since these molecules are the natural targets of many pollutants including PAHs (in fact their biological hazardous effect is due to their ability to interact with them) we have used them as host linkers in order to reach a twofold objective: the pollutant detection and the study of the interaction with these important natural targets (7).

The most successful designed hosts we have found are the calixarenes that can be used in the molecular recognition of PAHs by SERS. Calixarenes are a class of synthetic cyclooligomers formed via a phenol-formaldehyde condensation. They exist in a "cup" like shape with a defined upper and lower rim and a central annulus (see Fig. 1). Calixarenes have interesting applications as host molecules as a result of their preformed cavities (8). By changing the chemical groups of the upper and/or lower rim it is possible to prepare different derivatives with differing selectivities for various guest ions and molecules. Adsorption and self-assembled monolayer formation of calixarenes is a prerequisite for the application of calixarenes in sensor devices (9-12). In previous works we have studied the adsorption and organization of calixarene molecules on metal surfaces, where charge-transfer phenomena can take place with potential for optical sensing.

While the studies carried out by other authors on the adsorption and self-assembly of calixarenes on metal surfaces mainly dealt with thiol functionalized calixarenes (9, 13-15), we have proved that ester-functionalized calixarene derivatives could also be effectively employed as hosts in the selective molecular recognition of PAHs using the SERS technique.

Other kind of host molecules which has rendered good results in the detection pf PAHs are the *quat* molecules (16). These molecules include in their structure a positively charged quaternary nitrogen (quat) and an aromatic moiety. The quat moiety is tightly attached to the surface via an ionic pair formation with addition of chloride ions (17,18) and can interact with PAHs.

As previously indicated, in our functionalization studies, we have centered our attention on the vibrational characterization and detection of PAHs. They constitute a family of different chemical compounds with a condensed multibenzene structure. PAHs are important environmental pollutants formed during the incomplete burning of coal, oil and gas, or other organic substances like tobacco or charbroiled meat (19). PAHs can be found as a mixture of different related molecular compounds in air, soil and water due to both natural processes and the human activity. Because many of them have been reported to be strong carcinogens (20-23) it is important to find an effective and selective method for their trace analysis.

In this chapter, we report the application of several calix[4]arenes with different chemical functional groups in the lower rim, for the selective recognition of large PAHs molecules: anthracene (ANT), dibenzoanthracene

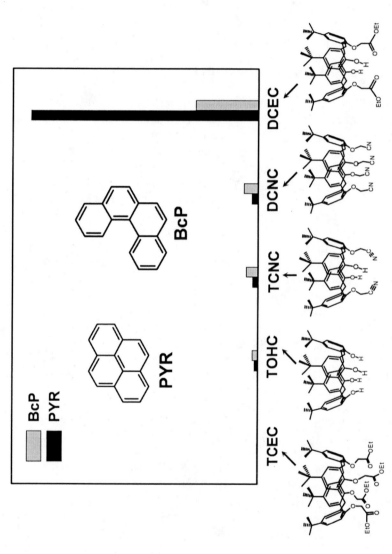

Figure 1. Normalized areas (with respect to the 1405 cm^{-1} band of PYR in the complex with DCEC) of the most intense SERS bands of PYR and BcP (10^{-4} M) on the different assayed calixarenes

(DBA), pyrene (PYR), benzo[c]phenanthrene (BcP), triphenylene (TP), coronene (COR), chrysene (CHR) and rubicene (RUB). The detection is achieved applying both vibrational-enhanced and complementary techniques; SERS and SEIRA. The complementary information provided by both techniques is very helpful in the study of complex systems such as the host/ligand presented in this work.

Experimental Details

Colloidal silver nanoparticles were prepared by using hydroxylamine hydrochloride as reducing agent *(24)*. These nanoparticles have the advantage, with respect to the citrate Ag colloid, of a more uniform distribution of size and shape together with the absence of citrate excess and its oxidation products that could interfere in the SERS measurements *(5)*. In addition, the surface properties of this colloid may be better suited for the detection of PAHs, since the total negative charge, mainly due to the residual chloride ions, should be much lower than in the case of citrate and borohydride colloids. In the latter cases, anions bearing more than one negative charge are adsorbed on the surface increasing the overall nanoparticle negative charge. On the other hand, these chloride-covered nanoparticles present improved adherence properties. This fact facilitates their immobilization on glass, giving rise to films provided with a uniform distribution of metal particles tightly linked to the glass by a simple deposition of an aliquot of the colloidal solution on a glass slide. The immobilized Ag nanoparticles on substrates such as Ge or glass are systems which can be used as substrates for both SERS and SEIRA measurements.

In some cases gold island films deposited on CaF_2 substrates were also employed to obtain SERS and SEIRA spectra. They were prepared by metal evaporation as reported elsewhere *(25)*.

The calixarenes employed for this study were:
25,26,27,28-tetrahydroxy-p-*t*-butylcalix[4]arene (TOHC);
25,27-dicarboethoxy-26,28-dihydroxy-p-*t*-butylcalix[4]arene (DCEC);
25,26,27,28-tetracarboethoxy-p-*t*-butylcalix[4]arene (TCEC)
25,27-dicyano-26,28-dihydroxy-p-*t*-butylcalix[4]arene (DCNC) and
25,26,27,28-tetracyano-p-*t*-butylcalix[4]arene (TCNC).
These calixarenes were synthesized following published procedures *(26)* and are functionalized with four *t*-butyl groups in the upper rim, while the two last calixarenes have a different functionalization in the lower rim with an alternate 1,3 substitution.

SERS spectra at 785 nm were measured with a Renishaw Raman Microscope System RM2000 equipped with a diode laser, a Leica microscope, and an electrically refrigerated CCD camera. The laser power in the sample was

2.0 mW. FT-SERS spectra were obtained with a Bruker RFS 100/S instrument, using the 1064 nm line of a Nd:YAG laser and a Ge detector cooled by liquid nitrogen. The output laser power was 50 mW and a 180° geometry was employed. The spectral resolution was set to 4 cm^{-1} in both cases.

SEIRA spectra were measured on a FTIR Bruker IFS 66 spectrometer provided with a DTGS detector. For the polarized reflection measurements, a variable angle reflectance accessory Specac SC19650 with a KRS5 polarizer was employed. Spectra resolution was 8 cm^{-1}.

Results and Discussion

SERS of calix[4]arene/PAHs complexes: selectivity of the calix[4]arene molecule in function of the lower rim functionalization

The SERS of PAHs cannot be obtained on Ag colloids and in general on other metallic surfaces because these molecules do not manifest any affinity to be adsorbed on the metal surface. However, we succeeded in getting the SERS of calix[4]arene/PAHs complexes at different laser wavelengths excitation (27,28).

Figure 1 shows the normalized areas, corresponding to the most intense SERS bands of PYR and BcP (observed at 1405 and 1338 cm^{-1}, respectively) in the complexes with the different calixarene molecules with a different functionalization in the lower rim. PYR and BcP have a different behaviour with respect to the calixarene selectivity. PYR shows a much higher SERS intensity when it is interacting with DCEC compared with TOHC or DCNC. In contrast, BcP also gives rise to a high SERS intensity when interacting with DCEC (although lower relative intensity than that achieved for PYR), while it shows a significant SERS intensity when interacting with the other calixarenes. This result indicates the existence of a clear selectivity in the molecular recognition of PAHs by calixarenes thanks to the formation of a host/guest complex in which the chemical structure and the size of the PAHs play an important role (26). PYR seems to interact with a high specificity with DCEC. Since the differences between the calixarenes are due to different chemical functionalization on the lower rim, through which the calixarene interacts with the surface, the specificity is attributed to a different adsorption affinity of the host/guest complex.

In Figure 2 the normalized areas of the most intense SERS bands of the different PAHs studied here (the in-plane ring stretching band falling in the 1420-1320 cm^{-1} region) were quantified in their complexes with DCEC. As can be seen, all the 4-ringed PAHs assayed (PYR, TP and BcP) are selectively more enhanced. In particular, PYR, followed by TP, are the PAHs molecules with the highest specificity to DCEC since they lead to the maximum SERS

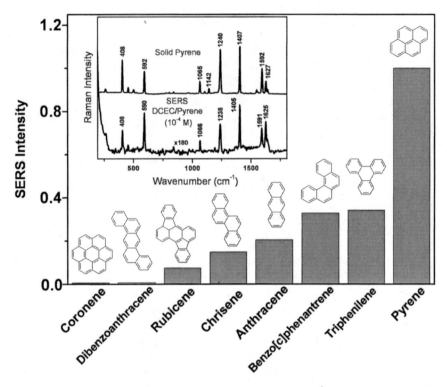

Figure 2. Normalized areas (with respect to the 1405 cm^{-1} band of PYR in the complex with DCEC) of the most intense SERS bands of the studied PAHs (10^{-4} M) on DCEC (10-4 M). All the intensities were registered by exciting at 1064 nm.

intensification. In contrast, the larger PAHs (COR, RUB and DBA) exhibited the lowest SERS intensities. This size specificity is probably determined by the cavity of the calixarene and the *t*-butyl groups placed in the upper rim. It is noted that CHR, another phenanthrene derivative, presented a much lower SERS intensity than BcP. This indicates that the binding properties of each PAH must change depending on their chemical structure, revealing the existence of a size and shape selectivity of DCEC to interact with PAHs molecules.

Calixarenes bearing a dicarboethoxy (-CH$_2$-COO-CH$_2$-CH$_3$) substituents in the lower rim are then good candidates in the design of optical sensors of PAHs based on enhanced vibrational spectroscopies. The presence of these groups in an alternate configuration in the lower rim together with the *t*-butyl groups in the upper rim ensures a strong interaction with both the ligand and the metal surface.

Interestingly, the corresponding tetracarboethoxy calix[4]arene TCEC, which bears four carboethoxy groups in the lower rim, displays a much lower activity in the detection of PAHs.

Complementarity of SERS and SEIRA measurements

Fig. 3 displays the SERS and SEIRA spectra of the DCEC/PYR complex on the same Au/CaF2 substrate, compared to the Raman and infrared spectra of solid samples of PYR and calixarene. The SERS spectrum shows intense PYR bands at 1400 and 1235 cm-1 which correspond to the bands at 1407 and 1241 cm-1 in the solid, and which are assigned to in-plane ring stretching motions and in-plane δ(C-H) vibrations (29). The downwards shift of the latter bands is attributed to the interaction with calixarene (27). Moreover, a marked enhancement of the band at 590 cm-1 and an intensity decrease of that at 408 cm^{-1} are observed. In general, all the intensified bands correspond to symmetric ag modes (30,31) which are enhanced through a Frank-Condon resonance mechanism, most probably associated to a charge-transfer in the complex-surface system, as it also occurs for naphthalene interacting with TOHC (32).

The complementary nature of the information provided by SEIRA and SERS is illustrated in the spectra of the DCEC/PYR complex. SERS spectra are mainly dominated by PYR bands. Thus, SERS can be used to follow the changes occurring in the PAHs as a consequence of the interaction with the calixarene host and for detection purposes.

In contrast, SEIRA spectra display more intense bands corresponding to the more polar groups existing in the calixarene host molecule, while only a weak PYR band is observed at 840 cm^{-1}. Thus, SEIRA seems to be more appropriate to follow structural changes occurring in the host due to the adsorption and the complexation with the analyte and to find a structure/selectivity relationship.

Since SEIRA is a better technique to follow the structural changes undergone by the calixarene host molecule, in the following section we analyze the SEIRA of DCEC and TCEC in order to understand the reasons of the different activity of these molecules in the interaction with PYR and detection of PAHs. *(33)*.

SEIRA of DCEC vs. TCEC: influence of lower rim functionalization on the adsorption and interaction with PAHs

The IR spectrum of solid DCEC dispersed in KBr shows doublets at 1207/1188, 1125/1100 and 1060/1050 cm^{-1} (Fig. 4a). However, when this calixarene is adsorbed on Ag (Fig. 4b) prominent SEIRA bands are seen at 1187, 1127 and 1056 cm^{-1} on Ag and at 1188, 1127 and 1068 cm^{-1} on Au, indicating

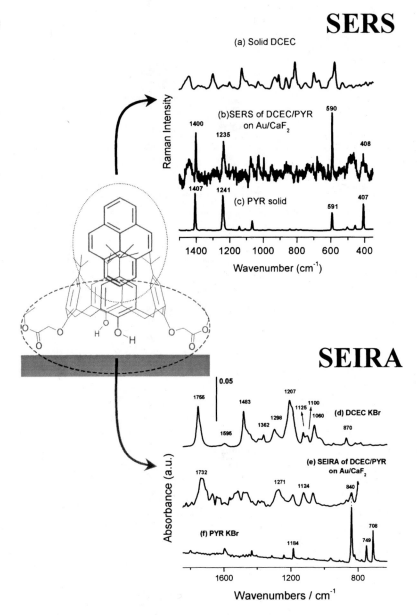

Figure 3. SERS and SEIRA spectra of the DCEC/PYR complex on Au/CaF2 substrate compared with normal Raman and infrared spectra of solid DCED and PYR. Illustration of the configuration adopted by the DCEC-PYR complex on Au.

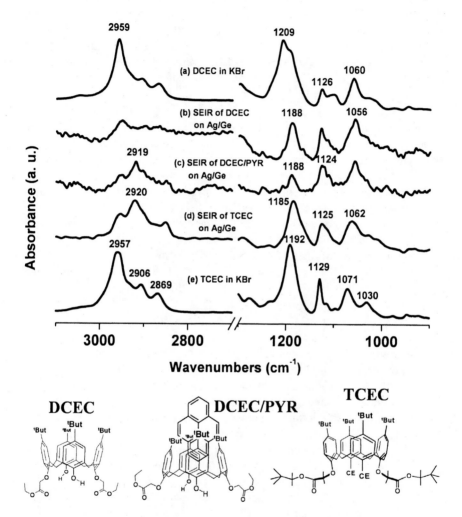

Figure 4. Up: a) IR Absorption spectrum of DCEC in KBr; b) SEIRA of DCEC in Ag/Ge; c) SEIRA of DCE/PYR complex in Ag/Ge; d) SEIRA of TCEC in Ag/Ge; and e) IR Absorption spectrum of TCEC in KBr.

that clear structural changes occur upon adsorption on the metallic surface. In addition, the upwards shift of the ester ν(C-O) band from 1125 to 1127 cm^{-1}, and its intensity increase, is related to the downwards shift of the ν(C=O) ester vibration at 1755 cm^{-1}. The shifts observed for the ester ν(C-O) and ν(C=O) motions indicate that both oxygen groups are interacting with the metal surface through a bidentate complex, as shown in Fig. 4, (bottom, left) due to the electric charge delocalisation occurring in the ester group. Moreover, this interaction induces a rearrangement of the lower molecular rim, responsible for the wavenumber shifts of the 1298, 1207 and 1060 cm^{-1} bands.

The ν(C-O) band at 1207 cm^{-1} seems to be very sensitive to calixarene structural changes involving the aromatic ν(C-O) ether bond through which the carboethoxy group is linked to phenyl rings. This band appears at a lower wavenumber value in TCEC, in comparison to DCEC, due to a more open structure adopted by the four carboethoxy groups in the lower rim of the calixarene induced by the steric hindrance between them. Thus, the shift towards lower wavenumbers in DCEC when adsorbed on the metal can be also interpreted as due to an opening of the carboethoxy groups in the lower rim in order to facilitate the bidentate interaction of DCEC with the surface.

SEIRA spectra of the DCEC/PYR complex on Ag displays a general frequency shifts with respect to the SEIRA of DCEC, mainly the involving carboethoxy group. The ν(C-O) band at 1127 cm^{-1} shifts downwards to 1124 cm^{-1} both on Ag and on Au. Moreover, the ν(C-O) ether aromatic band at 1188 cm^{-1} is weakened, and that at 1298 cm^{-1} shifts downwards to 1271 cm^{-1}. These changes suggest a stronger interaction of carbonyl groups, and a weaker one of the C-O group in the presence of PYR towards a rather monodentate interaction (Fig. 4, bottom, centre). This is attributed to a further opening of the carboethoxy groups in the lower calixarene rim induced by the interaction with PYR. The induced structural change also influences the ether C-O bonds of the aromatic moiety, which could adopt a less perpendicular orientation, thus explaining the 1298 $\rightarrow$ 1271 cm^{-1} shift and the intensity decrease of the 1188 cm^{-1} band.

The structural changes undergone by DCEC upon adsorption and interaction with PYR are also manifested in the 2800-3000 cm^{-1} region corresponding to C-H stretchings (Fig. 4). A relative intensity decrease is seen for the band at 2958 cm^{-1} both on Ag and Au. On the other hand, the interaction with PYR induces an intensity increase of the band at 2917 cm^{-1} on Ag and 2919 cm^{-1} on Au. The changes observed in this region are attributed to the reorientation of the molecule on the surface due to its complexation with PYR. In fact, the most enhanced bands must correspond to C-H stretching vibrations with preferential perpendicular orientation with respect to the surface, according to the SEIRA selection rules *(2)*.

SEIRA spectrum of TCEC on Ag/Ge reveals slight changes in bands falling in the 1250-950 cm^{-1} region, indicating that also TCEC interacts with the metal surface through the ester group as in the case of DCEC. In particular the most intense ν(C-O) band at 1191 cm⁻¹ undergoes a significant intensity decrease and shift to 1184 cm⁻¹, while the band at 1128 cm⁻¹ shift to 1124 cm⁻¹. However, the changes are not as strong as in the case of DCEC, thus indicating a different adsorption of TCEC with respect to DCEC. A detailed analysis of the C-H region and the 1250-950 cm⁻¹ region indicates that TCEC is adsorbed through a monodentate interaction with the metal (Fig. 4, bottom, right), since a bidentate structure with both oxygen atoms interacting with the surface is not possible in TCEC due to the steric hindrance caused by the close carboethoxy groups.

Interestingly, the spectral profile of the C-H stretching region in the SEIRA of TCEC is very similar to that found for the DCEC/PYR complex, since also in DCEC a prominent band at 2919 cm⁻¹ is seen in the complex with PYR. Other similarities between the SEIRA spectra of TCEC and DCEC/PYR complex are found in the lower wavenumber region, mainly corresponding to the position of bands at 1741, 1184 and 1124 cm⁻¹ bands.

The similarity found between SEIRA spectrum of TCEC and that of DCEC/PYR complex could be the key to understand why TCEC is less active binding PAHs molecules, as demonstrated in previous experiments *(27)*. In fact, both the steric hindrance of carboethoxy groups and the interaction with the metal may induce a shutdown of the calixarene upper rim, which may prevent the inclusion of PYR. This structural change does not occur in DCEC, due to the lower steric effect of the two carboethoxy groups. Nevertheless, the interaction of DCEC with PYR could lead to a shutdown of the upper rim and an opening of the calixarene lower rim similar to that occurring for non-complexed TCEC.

Conclusions

The functionalization of metal (silver and gold) nanoparticles with calix[4]arenes derivatives employed as host molecules allows the selective detection of some PAHs using SERS. The corresponding SEIRA spectra provide important information about the structural changes undergone successively by the calix[4]arenes, first when adsorbed onto metal nanoparticles and after when "lodging" the PAH molecule forming a complex. Analysis of SEIRA data allowed to find structure/selectivity relationships, complementing the information given by SERS. Both vibrational enhanced spectroscopies are promising molecular sensing techniques.

Acknowledgements

Ministerio de Educación y Ciencia of Spain (Project FIS2004-00108) and the *Comunidad de Madrid* (MICROSERES Program, S-0505/TIC-0191) are

150

gratefully acknowledged for financial support. Authors also acknowledge support received from the *Convenio Conycit-CSIC (2003/2004 and 2006/2007)*.

References

1. Tian, Z.Q. *J. Raman Spectrosc.* **2005**, *36*, 466-470.
2. Aroca, R.F.; Ross, D.J.; Domingo, C. *Appl. Spectrosc.* **2004**, *58*, 324A-338A.
3. Sanchez-Cortes, S.; Garcia-Ramos J.V.; Morcillo G. *J. Colloid Interf. Sci.* **1994**, *167*, 428-436.
4. Sanchez-Cortes, S.; Garcia-Ramos J.V.; Morcillo G.; Tinti, A. *J. Colloid Interf. Sci.* **1995**, *175*, 358-368.
5. Cañamares, M.V.; Garcia-Ramos, J.V.; Gomez-Varga, J.D.; Domingo, C.; Sanchez-Cortes, S. *Langmuir* **2005**, *21*, 8546-8553.
6. Sanchez-Cortes, S.; Garcia-Ramos, J.V. *J. Colloid Interf. Sci.* **2000**, *231*, 98-106.
7. Sanchez-Cortes, S.; Guerrini; L.; Garcia-Ramos, J.V. *unpublished.*
8. Gutsche, C.D. *Calixarenes, Monographs in Supramolecular Chemistry*; Stoddart J.F., Ed.; Royal Society of Chemistry: Cambridge, 1992.
9. Marenco, C.; Stirling, C.J.M.; Yarwood, J. *J. Raman Spectrosc.* **2001**, *32*, 183-194.
10. Kim, J.H.; Lee, K.H.; Kang, S.W.; Koh, K.N. *Synthetic Met.* **2001**, *117*, 145-148.
11. Zhang, M.H.; Anderson, M.R. *Langmuir* **1994**, *10*, 2807-2813.
12. Troughton, E.B.; Whitesides, G.M.; Nuzzo, R.G.; Allara, D.L.; Porter, M.D. *Langmuir* **1988**, *4*, 365-385.
13. Wehling, B.; Hill, W.; Klockow, D. *Int. Journal of Environ. An. Chem.* **1999**, *73*, 223-236.
14. Faull, J.D.; Gupta, V.K.; *Langmuir* **2002**, *18*, 6584-6592.
15. Faull, J.D.; Gupta, V.K. *Thin Solid Films* **2003**, *40*, 129-137.
16. Guerrini, L.; Garcia-Ramos, J.V.; Sanchez-Cortes, S. *unpublished.*
17. Millan, J.I.; Garcia-Ramos J.V.; Sanchez-Cortes, S. *J. Electroanal. Chem.* **2003**, *556*, 83-92.
18. Millan, J.I.; Garcia-Ramos J.V.; Sanchez-Cortes, S. *J. Raman Spectrosc.* **2003**, *34*, 227-233.
19. http://www.inchem.org/documents/ehc/ehc/ehc202.htm; *Environmental Health Criteria 202*; World Health Organization: Geneva, 1998.
20. van Gijssel, H.E.; Schild, L.J.; Watt, D.L.; Roth, M.J.; Wang, G.Q.; Dawsey, S.M.; Albert, P.S.; Qiao, Y.L.; Taylor, P.R.; Dong, Z.W.; Poirier, M.C. *Mutat. Res-Fund. Mol. M.* **2004**, *547*, 55-62.
21. Falco, G.; Domingo, J.L.; Llobet, J.M.; Teixido, A.; Casas, C.; Muller, L. *J. Food Protect.* **2003**, *66*, 2325-2331.

22. Guo, H.; Lee, S.C.; Chan, L.Y.; Li, W.M. *Environ. Res.* **2004**, *94*, 57-66.
23. Li, D.; Jiao, L. *Int. J. Gastro. Cancer* **2003**, *33*, 3-14.
24. Leopold, N.; Lendl, B. *J. Phys. Chem. B* **2003**, *107*, 5723-5727.
25. Sanchez-Cortes, S.; Domingo, C.; Garcia-Ramos, J.V.; Aznarez, J.A. *Langmuir* **2001**, *17*, 1157-1162.
26. Iwamoto, K.; Shinkai, S. *J. Org. Chem.* **1992**, *57*, 7066-7073.
27. Leyton, P.; Sanchez-Cortes, S.; Garcia-Ramos, J.V.; Domingo, C.; Campos-Vallette, M.; Saitz, C.; Clavijo, R.E. *J. Phys. Chem. B* **2004**, *108*, 17484-17490.
28. Leyton, P.; Sanchez-Cortes, S.; Campos-Vallette, M.; Domingo, C.; Garcia-Ramos, J.V.; Saitz, C. *Appl. Spectrosc.* **2005**, *59*, 1009-1015.
29. Carrasco Flores, E. A.; Campos Vallette, M.M.; Clavijo, R.E.C.; Leyton, P.; Díaz F., G.; Koch. R. *Vib. Spectrosc.* **2004**, *37*, 153-160.
30. Hudgins, D. M.; Sandford, S.A. *J. Phys. Chem. A* **1998**, *102*, 329-343.
31. Shinohara, H.; Yamakita, Y.; Ohno, K. *J. Mol. Struct.* **1998**, *442*, 221-234.
32. Kook, S.K. *B. Kor. Chem. Soc.* **2002**, *23*, 1111-1115.
33. Leyton, P.; Domingo, C.; Sanchez-Cortes, S.; Campos-Vallette, M.; Garcia-Ramos, J.V. *Langmuir,* **2005**, *21*, 11814-11820.

Chapter 11

Protein–Nanoparticle Layer-by-Layer Films as Substrates for Surface-Enhanced Resonance Raman Scattering

Paul J. G. Goulet, Nicholas P. W. Pieczonka, and Ricardo F. Aroca[*]

Materials and Surface Science Group, Department of Chemistry
and Biochemistry, University of Windsor, Windsor,
Ontario N9B 3P4, Canada
[*]Corresponding author: raroca1@cogeco.ca

The development of substrates for surface-enhanced spectroscopy that have additional functionality, beyond optical amplification, is now highly desirable. In this work, we report the fabrication, characterization, and utilization as enhancing substrates of layer-by-layer (LbL) films composed of the glycoprotein avidin and metallic Ag nanoparticles. These nanocomposite films demonstrate strong, bio-specific interactions and are employed as chemically selective substrates for surface-enhanced Raman scattering (SERS) and surface-enhanced resonance Raman scattering (SERRS). Fabricated on glass microscope slides, from solutions of avidin and colloidal Ag, they are characterized by UV-visible surface plasmon absorption and atomic force microscopy (AFM) as a function of the number of bilayers deposited. Their average SERRS enhancement is also monitored as a function of architecture, and is found to plateau after the deposition of ca. 8 bilayers. It is also found that they demonstrate excellent enhancement of SERS/SERRS throughout the visible region of the electromagnetic spectrum (442-785 nm). Finally, biotinylated dye molecules are shown to be selectively adsorbed by these films, containing the natural biotin chelator avidin, relative to untagged analogues. This preferential adsorption, facilitated by biotin capture, yields an additional 'concentration enhancement' of ca. 10^2, and improves detection limits by at least 2 orders of magnitude.

Introduction

Surface-enhanced Raman scattering (SERS) involves immense increases in effective Raman scattering (RS) cross section values of molecules positioned at or near the surface of appropriate nanometric scale metallic particles (*1, 2*). This enhancement, generally of several orders of magnitude, is primarily the result of strengthened local optical fields that arise from the excitation of surface plasmon resonances in the metal. Similarly, resonance Raman scattering (RRS), involving excitation in resonance with an electronic transition of a molecular system, can also be enhanced by metallic nanostructures, yielding the corresponding phenomenon of surface-enhanced resonance Raman scattering (SERRS) (*3*). This effect is defined by its 'double resonance condition', and offers added resonance Raman enhancement, generally on the order of the extinction coefficient of the molecular transition in question. Together, these two phenomena make up a small but important subclass of all surface-enhanced spectroscopies, and are currently receiving a great deal of attention due to their potential for application toward a wide range of problems. In particular, their single molecule sensitivity, high spatial resolution, and high information content make them especially attractive analytical techniques for use in a variety of fields including biology (*4, 5*), medicine (*6*), pharmacy (*7*), environmental analysis (*8*), thin film characterization (*9*), and ultra-trace analysis (*10-12*).

As a result of this, the design of novel nanostructured substrates, that provide functionality beyond simple signal enhancement (*13*), is now of great interest and materials exhibiting strong, bio-specific interactions promise to be among the most widely used for such applications going forward. These include, among others, enzyme/substrate, antibody/antigen, and DNA strand/ complementary strand systems (*14-17*). Of course, for the functionality of these systems to be effectively coupled with the unique optical properties of metallic nanoparticles, to produce the next generation of SERS/SERRS substrates, they will first have to be incorporated together into well designed nanoarchitectures. One of the best strategies for accomplishing this goal is through the use of the layer-by-layer (LbL) technique for thin film fabrication.

The LbL technique was pioneered by Decher (*18*), and is an extremely versatile method for nanodimension film fabrication. It involves the electrostatic assembly of alternating positively and negatively charged layers through a simple and inexpensive procedure, and has been employed using a wide variety of different materials, including, among others, biopolymers (*19*), dendrimers (*20*), and proteins (*21*). Of particular interest, for those employing surface-enhanced spectroscopy, are the recent reports of Ag and Au metallic nanoparticle incorporation into nanocomposite films through the use of the LbL technique (*22*). In fact, several reports have now been published where LbL multilayer films have been directly and successfully employed as substrates for SERS/SERRS (*20, 23, 24*).

In this work, the LbL technique is employed to efficiently couple the strong natural biotin-chelating ability of the glycoprotein avidin with the equally powerful optical enhancing properties of colloidal Ag nanoparticles to produce functional, chemically specific SERS/SERRS substrates. These substrates are characterized by surface plasmon absorption spectroscopy, atomic force microscopy, and SERRS, as a function of the number of bilayers deposited. They are also demonstrated to effectively enhance RS/RRS signals throughout the entire visible region of the EM spectrum. Finally, and perhaps most importantly, they are shown to selectively adsorb biotinylated (tagged) species from solution without loss of enhancing capacity. Biotin-4-fluorescein is found to be preferentially captured by these films, relative to untagged fluorescein, leading to a corresponding 10^2 concentration enhancement and a matching 10^2 increase in detection sensitivity. It is expected that the foundational concepts reported here can be extended toward the implementation of a wide variety of 'smart' SERS/SERRS substrates incorporating other bio-specific systems into LbL architectures.

Experimental

Silver nitrate, sodium citrate, avidin (from egg white), rhodamine 6G (R6G), and fluorescein were purchased from Sigma-Aldrich, while biotin-4-fluorescein (B4F) was purchased from Molecular Probes. All were used without further purification. Colloidal Ag solutions were prepared by citrate reduction of $AgNO_3$ according to the well known method of Lee and Meisel (*25*), and diluted by a factor of 2 before use. All avidin, R6G, fluorescein, and B4F solutions were prepared in 10 mM phosphate buffered saline (PBS) at pH 7.5.

Prior to film fabrication, all glass microscope slides were cleaned and silanized. They were washed in detergent and deionized water, immersed in 20% H_2SO_4, rinsed again in deionized water, and dried. Next, they were immersed into pure acetone for 5 min, immersed into a 2% acetone solution of 3-amino-propyltriethoxysilane for 5 min, and then rinsed thoroughly with water and dried. All LbL films were prepared by immersing these clean, silanized slides into colloidal Ag solution for 30 min, rinsing them with water, drying them with air, then immersing them into a 50 μg/mL avidin solution in PBS for 30 min. Upon removal, these films were again rinsed and dried, then re-immersed into colloidal Ag solution. Through these alternating deposition steps, films were built up until they were composed of up to 14 bilayers.

To monitor their surface plasmon absorption, UV-visible absorption spectra were obtained for all LbL films produced using a Cary 50 UV-visible spectrometer. Atomic force microscopy (AFM) measurements were performed on a Digital Instruments NanoScope IV operating in tapping mode with a n+-

silicon tip (NSC 14 model, Ultrasharp). All images were collected with high resolution (512 lines per scan) at a scan rate of 0.5 Hz. Topographical, error, and phase images were all used for analysis of the surface morphology of the films, and data were collected under ambient conditions. Each scan was duplicated to ensure that any features observed were reproducible.

Raman scattering experiments were conducted employing micro-Raman Renishaw 2000 and InVia systems with laser excitation lines at 442 (HeCd), 514 (argon ion), 633 (HeNe), and 785 nm (diode). All measurements were made in a backscattering geometry using a 50x microscope objective with a NA value of 0.75, providing scattering areas of ~ 1 μm^2. All SERS/SERRS measurements of R6G were recorded from dried 5 μL drops of 10^{-5} M solution (with ~ 5 mm diameters when dry) cast onto avidin/Ag nanoparticle LbL substrates. Fluorescein (results not shown) and biotin-4-fluorescein spectra, on the other hand, were recorded from 14 bilayer films immersed for 30 min into solutions with concentrations ranging from 10^{-4} to 10^{-7} M. After removal from these solutions, films were thoroughly rinsed with deionized water and dried before measurement.

Results and Discussion

Based on the method of Lee and Meisel (*25*), colloidal Ag nanoparticles were prepared in solution through the citrate reduction of $AgNO_3$. The collapse of such colloidal solutions is prevented by the repulsion of negative charges at the surface of different nanoparticles that result from the adsorption of ions and the ionization of surface adsorbates. The extent of the surface charge of colloidal nanoparticles can be monitored through the measurement of the potential across the electrical double layer formed at the particle-liquid interface. This potential is referred to as the zeta potential, ζ, and has recently been measured to be about -50 mV for this colloidal Ag solution, at this pH value (ca. 6) (*26*). Avidin, on the other hand, is cationic at neutral pH values due to its isoelectric point of 10.5 (*27*). Together, these facts make the effective fabrication of LbL films of these materials, through electrostatic interactions, possible.

Surface Plasmon Absorption

After each layer of these avidin/Ag nanoparticle LbL films was deposited, surface plasmon absorption spectra were recorded. In Figure 1, these are shown for films with 2, 4, 6, 8, 10, 12, and 14 bilayers. It can be seen that the extinction of these films grows with increasing number of bilayers, confirming

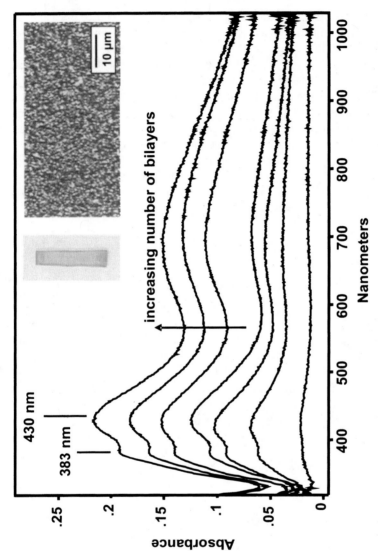

Figure 1. Surface plasmon absorption spectra showing increasing extinction of avidin/Ag nanoparticle LbL films with increasing number of bilayers (2, 4, 6, 8, 10, 12, 14). Insets show digital camera image (left) and 50x microscope image (right) of a 14 bilayer film.

effective film building and the advantage of adsorption over desorption under these conditions. These spectra reveal two well resolved high energy bands at ca. 383 and 430 nm that redshift slightly with increasing number of bilayers. These bands can be assigned to quadrupolar and dipolar particle plasmon resonances, respectively, with the quadrupolar mode being due to the presence of larger spherical particles with diameters up to 70 nm (*28*). The slight redshifting of these bands can be attributed to changes in the dielectric function of the medium surrounding the particles with film growth. Also observed in these spectra is a very broad feature with a maximum at about 700 nm that increases in relative intensity as the number of bilayers deposited is increased. This band can be assigned to dimers and higher order particle aggregates that are known to increase in number as film growth proceeds. Also shown in Figure 1, are digital camera and 50x microscope images of an LbL film made up of 14 bilayers. From these, it can be seen that this film is relatively homogeneous on the macro scale, but becomes less so on the micro scale where Raman measurements are generally made. Large clusters of nanoparticles can be clearly observed in the latter image.

Atomic Force Microscopy

To demonstrate the relationship between the optical and physical properties of the avidin/Ag nanoparticle LbL films produced in this work, their surface morphologies were analyzed by AFM imaging. Selected representative topographical, or height, images are presented in Figure 2 for substrates consisting of 2, 4, 6, 8, 12, and 14 layers. It can be seen that as the number of bilayers deposited is increased, there is a general trend toward greater surface coverage, increased particle-particle interaction, and cluster/aggregate formation, as is expected. This trend is supported by increasing RMS roughness values with film growth, from 7.2 nm for 2 bilayers to 32.6 nm for 14 bilayers. Also, on the basis of the analysis of these images, the diameter of single isolated particles was found to be between ca. 30 and 70 nm, while clusters were generally found to range between ca. 150 and 300 nm. These results are consistent with what is expected on the basis of the surface plasmon results presented for these films.

LbL Film Characterization By SERS/SERRS

In Figure 3, the relationship between the number of avidin and Ag nanoparticle bilayers incorporated into these LbL films, and the mean integrated intensity of R6G SERRS derived from them, is examined. In each case, 5 µL of a 10^{-5} M aqueous solution of the common SERRS dye was cast onto the film,

158

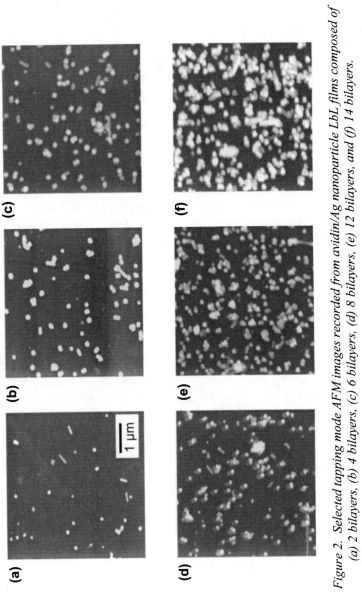

Figure 2. Selected tapping mode AFM images recorded from avidin/Ag nanoparticle LbL films composed of (a) 2 bilayers, (b) 4 bilayers, (c) 6 bilayers, (d) 8 bilayers, (e) 12 bilayers, and (f) 14 bilayers.

allowed to dry, and then probed with laser excitation at 514 nm using identical power, and accumulation time. This laser is in full resonance with the strong electronic transition of the molecular system as well as the surface plasmon resonance of the Ag nanoparticles incorporated into the LbL substrate, fulfilling the double resonance condition of SERRS. To obtain values for mean integrated SERRS intensity, 10 spectra were acquired from different spots on each sample and the 1650 cm^{-1} band of the molecule was fit (Gaussian fitting) using appropriate spectroscopic software. The mean of these 10 values was then determined and plotted in Figure 3. This plot reveals a general increase in mean integrated SERRS intensity up to ca. 8 bilayers, while from 8 to 14 bilayers, intensity appears to level off. This trend generally agrees well with extinction and AFM results, and can be explained on the basis on increasing surface coverage, and particle-particle interaction with increasing number of bilayers. While this result is likely to be concentration and analyte dependent, it does, however, reveal important trends and establish guidelines for further experiment. Therefore, based on these results, it was decided that 14 bilayer films would be used for all further application.

The ability of avidin/Ag nanoparticle LbL films to effectively enhance RS and RRS signals throughout the visible was explored, and the results are shown in Figure 4. Measurements were made from a 5 µL drop of 10^{-5} M R6G cast onto a 14 bilayer film, and recorded with laser excitation at 442 nm, 514, 633, and 785 nm. Notably, strong SERS or SERRS was observed from this substrate at all wavelengths, making it highly flexible for application in a variety of different laboratory settings, with diverse chemical systems. At 442 and 514 nm excitation, this molecular system is resonant, at 633 nm it is pre-resonant, and at 785 nm it is effectively non-resonant. Therefore these results are properly referred to as SERRS, pre-SERRS, and SERS, respectively. Differences between these phenomena are revealed through the variation of the relative intensities in the spectra of Figure 4 as laser excitation is brought in and out of resonance with the molecular electronic transition.

Preferential Adsorption of Biotinylated Species

Finally, the ability of avidin/Ag nanoparticle LbL films to act as 'smart' SERS/SERRS substrates was tested. A film composed of 14 bilayers of avidin and Ag nanoparticles was immersed into a 10^{-4} M solution of biotinylated fluorescein (biotin-4-fluorescein) for 30 min, while another was immersed into a solution of untagged fluorescein with equal concentration, for an equal period of time. Upon removal, both were thoroughly rinsed with water, dried, and SERRS spectra were recorded from them using 514 nm excitation. It was found that the spectra recorded from the two samples were essentially the same in terms of band frequencies, widths, and relative intensities. (The B4F spectrum is shown

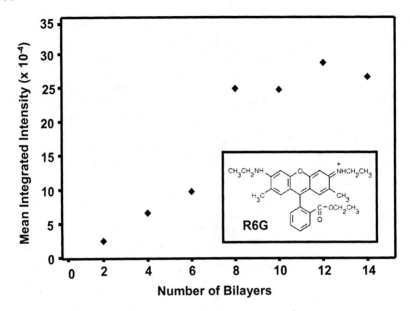

Figure 3. *Mean integrated SERRS intensity of 1650 cm^{-1} band of rhodamine 6G, recorded at 514 nm excitation, as a function of the number of bilayers incorporated into the LbL substrate. Inset shows molecular structure.*

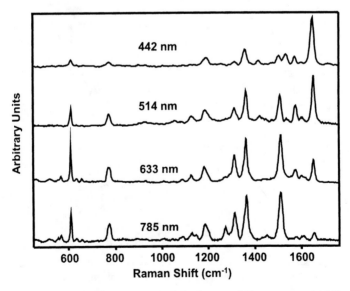

Figure 4. *SERS/SERRS spectra of rhodamine 6G, cast onto 14 bilayer avidin/Ag nanoparticle LbL film, and recorded using laser excitation at (a) 442, (b) 514, (c) 633, and (d) 785 nm.*

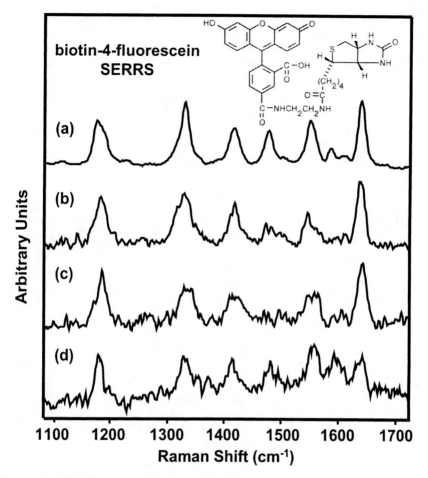

Figure 5. SERRS spectra of biotin-4-fluorescein, recorded at 514 nm excitation, recorded from 14 bilayer avidin/Ag nanoparticle LbL films dipped into PBS solutions with tagged dye concentrations of (a) 10^{-4}, (b) 10^{-5}, (c) 10^{-6}, and (d) 10^{-7} M. All films were thoroughly rinsed and dried following dipping. Inset shows molecular structure of biotin-4-fluorescein.

in Figure 5(a).) However, the biotinylated sample was found to exhibit absolute intensities that were on average 10^2 greater than its non-tagged counterpart. It was also found that fluorescein could not be observed at all spots and was near its detection limit, while biotin-4-fluorescein could be readily observed at all spots with strong intensity. Upon exploration of this, B4F was found to be detectable from substrates dipped, using identical procedures, into solutions with concentrations as low as 10^{-7} M (Figure 5), while fluorescein was undetectable when extracted from solutions with concentrations below 10^{-5} M. Since SERRS bands for each sample arise from the same central chromophore, which is, of course, expected to have an unchanged cross section, the 100 fold increase in intensity, and related 100 fold increase in detection sensitivity, associated with biotinylation can be attributed to a 'concentration enhancement' that arises specifically as a result of the strong, bio-specific interaction between the avidin in the LbL film and the biotin tag of B4F.

Conclusion

In this work, the fabrication, characterization, and application of avidin/Ag nanoparticle layer-by-layer (LbL) films have been reported. The growth of these nanocomposite films has been characterized by surface plasmon absorption, AFM imaging, and SERRS. They were found to exhibit significant Raman/resonance Raman enhancing capability across a broad spectral window, covering essentially the entire visible region. They were also found to demonstrate strong, bio-specific interactions with biotinylated species, thus providing chemical selectivity, without lost sensitivity, in substrates for surface-enhanced spectroscopy. Biotinylated dye molecules were shown to be selectively captured by these substrates, and this preferential adsorption yielded an additional concentration enhancement of ca. 10^2, while detection limits were improved by at least 2 orders of magnitude. It is anticipated that this development will promote further the coupling of the unique optical properties of metallic nanoparticles with systems demonstrating strong bio-specific interactions, particularly through the use of the powerful LbL technique.

Acknowledgements

Financial assistance from NSERC of Canada is gratefully acknowledged.

References

1. Moskovits, M. *Rev. Mod. Phys.* **1985**, *57*, 783.
2. Aroca, R. *Surface-Enhanced Vibrational Spectroscopy*; John Wiley & Sons Ltd: West Sussex, 2006.
3. Johnson, E.; Aroca, R.; Pahapill, J. *J. Mol. Struct., Theochem* **1993**, *293*, 331.
4. Grubisha, D. S.; Lipert, R. J.; Park, H.-Y.; Driskell, J.; Porter, M. D. *Anal. Chem.* **2003**, *75*, 5936.

5. Habuchi, S.; Cotlet, M.; Gronheid, R.; Dirix, G.; Michiels, J.; Vanderleyden, J.; De Schryver, F. C.; Hofkens, J. *J. Am. Chem. Soc.* **2003**, *125*, 8446.

6. Farquharson, S.; Shende, C.; Inscore, F. E.; Maksymiuk, P.; Gift, A. *J. Raman Spectrosc.* **2005**, *36*, 208.

7. Wang, Y.; Li, Y.-S.; Zhang, Z.; An, D. *Spectrochim. Acta, Part A* **2003**, *59A*, 589.

8. Carrasco-Flores, E. A.; Clavijo, R. E.; Campos-Vallette, M. M.; Aroca, R. F. *Appl. Spectrosc.* **2004**, *58*, 555.

9. Jennings, C.; Aroca, R.; Hor, A. M.; Loutfy, R. O. *Anal. Chem.* **1984**, *56*, 2033.

10. Goulet, P. J. G.; Pieczonka, N. P. W.; Aroca, R. F. *J. Raman Spectrosc.* **2005**, *36*, 574.

11. Aroca, R. F.; Alvarez-Puebla, R. A.; Pieczonka, N.; Sanchez-Cortez, S.; Garcia-Ramos, J. V. *Adv. Colloid Interface Sci.* **2005**, *116*, 45.

12. Kneipp, K.; Kneipp, H.; Itzkan, I.; Dasari, R. R.; Feld, M. S. *Chem. Rev.* **1999**, *99*, 2957.

13. Rohr, T. E.; Cotton, T. M.; Ni, F.; Tarcha, P. J. *Anal. Biochem.* **1989**, *182*, 388.

14. Cao, Y. C.; Jin, R.; Nam, J.-M.; Thaxton, C. S.; Mirkin, C. A. *J. Am. Chem. Soc.* **2003**, *125*, 14676.

15. Cao, Y. C.; Jin, R.; Mirkin, C. A. *Science* **2002**, *297*, 1536.

16. Vo-Dinh, T.; Cullum, B. *Fresenius J. Anal. Chem.* **2000**, *3*, 540.

17. Ni, J.; Lipert, R. J.; Dawson, G. B.; Porter, M. D. *Anal. Chem.* **1999**, *71*, 4903.

18. Decher, G. *Science* **1997**, *277*, 1232.

19. dos Santos, D. S., Jr.; Riul, A., Jr.; Malmegrim, R. R.; Fonseca, F. J.; Oliveira, O. N., Jr.; Mattoso, L. H. C. *Macromol. Biosci.* **2003**, *3*, 591.

20. Goulet, P. J. G.; Dos Santos, D. S., Jr.; Alvarez-Puebla, R. A.; Oliveira, O. N., Jr.; Aroca, R. F. *Langmuir* **2005**, *21*, 5576.

21. Anzai, J.; Nishimura, M. *J. Chem. Soc. Perk. T 2* **1997**, 1887.

22. Tian, S.; Liu, J.; Zhu, T.; Knoll, W. *Chem. Mater.* **2004**, *16*, 4103.

23. Li, X.; Xu, W.; Zhang, J.; Jia, H.; Yang, B.; Zhao, B.; Li, B.; Ozaki, Y. *Langmuir* **2004**, *20*, 1298.

24. Aroca, R. F.; Goulet, P. J. G.; Dos Santos, D. S., Jr.; Alvarez-Puebla, R. A.; Oliveira, O. N., Jr. *Anal. Chem.* **2005**, *77*, 378.

25. Lee, P. C.; Meisel, D. *J. Phys. Chem.* **1982**, *86*, 3391.

26. Alvarez-Puebla, R. A.; Arceo, E.; Goulet, P. J. G.; Garrido, J. J.; Aroca, R. F. *J. Phys. Chem. B* **2005**, *109*, 3787.

27. Green, N. M. *Method Enzymol.* **1970**, *18*, 418.

28. Kelly, K. L.; Coronado, E.; Zhao, L. L.; Schatz, G. C. *J. Phys. Chem. B* **2003**, *107*, 668.

Chapter 12

Surface-Enhanced Raman Scattering of Microorganisms

W. R. Premasiri[1], D. T. Moir[2], M. S. Klempner[3], and L. D. Ziegler[1,4]

[1]The Photonics Center, Boston University, 8 Saint Mary's Street, Boston, MA 02215
[2]Microbiotix Inc., 1 Innovation Drive, Worchester, MA 01605
[3]Boston University Medical Center, Boston, MA 02118
[4]Department of Chemistry, Boston University, Boston, MA 02215

A novel metal (Au or Ag) nanoparticle covered substrate has been developed for obtaining surface enhanced Raman scattering (SERS) spectra of vegetative bacterial cells. Species specific vibrational fingerprints are reproducibly observed. These bacterial SERS spectra are metal dependent, exhibit greater species specificity and reduced spectral congestion than nonSERS spectra and show fluorescence quenching. The development of this methodology for the identification and detection of vegetative bacterial cells in human blood is described.

Rapid and accurate techniques for pathogen identification and detection are essential for disease diagnosis and for monitoring disease progression and the intervention efficacies of specific treatments. For example, mortality rates from septicemia, the life-threatening condition resulting from bacteria in the blood, have been cited to be as high as 40 to 50% in the United States and Europe (1-3). Rapid diagnosis and microorganism identification are key for the enhancement of survival rates following infection. In addition, heightened concerns about the use of bacterial pathogens in a terrorist attack, following the intentional release of B. anthracis in the United States in 2001, continue to motivate the development of rapid, reliable and sensitive techniques for bacterial detection in both clinical and environmental samples (4). Traditional methodologies for bacterial identification are generally slow measured against the time scale of disease progression or exposure to environmental release. For example, phenotypic determinations involve culturing, nutrient specific broths and staining or other physical methods for differentiation and require at least 48 hours for identification (5). Currently, the best available methodologies are based on primer mediated polymerase chain reaction (PCR) amplification of extracted DNA (6-9). However, these techniques require highly trained personnel and are expensive. Furthermore, they are subject to risk of contamination since they are sample amplification based techniques, and they are tedious for the identification of components in bacterial mixtures. Thus, methodologies that can provide species and strain specificity at single cell sensitivity and processing times under one hour with relatively simple and inexpensive components offer considerable appeal for pathogen identification and detection.

An alternative approach to these genotypic identification strategies is the use of reagentless spectroscopic methods, such as IR absorption (10, 11), Raman scattering (12-14) and mass spectrometry (15-17) for rapid and sensitive bacterial sensing. Surface enhanced Raman scattering (SERS) studies over the past two decades have amply demonstrated the advantages this spectroscopic technique provides for the identification of biological molecules (18-28). For example, unique vibrational fingerprints may be observed for analytes that are naturally found at low concentrations due to the 3 – 8 order of magnitude Raman cross-section enhancement typically obtained for (ensemble averaged) *molecular* scatterers via SERS (29, 30). Chemical labeling or tagging is not required to obtain SERS signatures because all molecules posses some set of vibrational moieties which can be viewed via Raman excitation. In addition, visible or near IR excited fluorescence due to inherent chromophores or impurities which can spectrally overlap and hence mask nonSERS or bulk Raman signals, is eliminated in SERS (31, 32). Vibrational spectra resulting from IR absorption have also been widely used as a spectroscopic probe of biological molecules and employed as a probe of bacterial identity (10-12, 14, 33). However, in contrast to Raman, the ubiquitous water absorption can limit the accessible spectral range for bacteria in aqueous samples and the spatial resolution and sensitivity of IR methodologies do not allow single cell studies. Furthermore, spectral species differentiation via IR absorption fingerprints can be dependent on quite subtle spectral features in many classes of bacterial species (*12, 33*).

Aside from the attributes of SERS for studies of biological systems in general, significant additional advantages arise for the detection and identification of bacterial pathogens via SERS, particularly when Raman microscopy observations may be exploited. The metal-scattering distance dependence (< ~10 nm) of the SERS enhancement mechanisms (*29, 34, 35*) and the inherent size scale of bacteria (1 μm – 5 μm) account for these additional effects as described in further detail below.

Several normal (non-SERS) vibrational Raman spectra of bacteria have been reported during the past decade (*12-14, 33, 36-39*). When these whole cell Raman signatures are combined with data mining algorithms, such as principal or hierarchical component analysis, neural network analysis or other trained data set analyses, bacterial species, and in some cases strain, identification can be made when the reference library data set has been made sufficiently inclusive (*12-14, 33, 37*). Based on the characteristics of the SERS methodology developed in our laboratory (*40, 41*) and summarized here, the substitution of SERS vibrational fingerprints for normal Raman signatures in such data mining/reference library schemes can only improve the performance of such library based schemes for bacterial determinations.

SERS spectra of bacteria arising from a variety of nanostructured metal surfaces have been previously reported. These include Raman spectra of bacteria in silver and gold colloid solutions (*42, 43*), bacteria placed on electrochemically roughened metal surfaces (*44, 45*), bacteria coated by silver metal deposits (*46, 47*) and bacteria co-deposited with silver colloid aggregates on inert substrates (*48*). However, the bacterial SERS spectra reported in these studies vary considerably in terms of the Raman enhancement efficiency and the relative vibrational band intensities when comparisons are possible for a given species. Furthermore, the more widespread implementation of SERS-based methodologies for microorganism identification has been questioned on the basis of the reproducibility and reliability of SERS vibrational signatures, in general, owing to the very sensitive dependence of this enhancement phenomenon on the microscopic morphology and stability of the SERS active substrates (*43, 48-50*).

Initial efforts in our lab to observe reproducible, strongly enhanced Raman spectra of bacteria near traditional SERS substrates, such as gold colloidal solutions (*51*), or bacteria placed on dried aggregated colloidal precipitates or on metal nanoparticle embedded gels (*50, 52-55*) were largely unsuccessful. Thus, a novel SERS substrate resulting in aggregated gold nanoparticles covering the outer surface of a glassy matrix (SiO_2) grown in an *in situ* sol-gel process, was developed (*40, 41*). These substrates yield reproducible Raman spectra of bacteria that are enhanced sufficiently to observe SERS at the single cell level in ~10 seconds of data accumulation with low excitation laser power.

The scope of this chapter is confined to the characterization of the SERS microscopy of vegetative bacterial cells. The ability to use SERS microscopy for pathogen identification and detection in biological fluids at clinically relevant concentrations requires the development of rapid protocols for enriching and concentrating microorganisms from this fluid, i.e. being able to bring a few bacteria to the focus of the interrogating Raman excitation beam. In addition, it is equally important that the contents of the biological fluid do not poison or "foul" the SERS active substrate by preventing the bacterium from physi-adsorbing to the nanostrucutured surface, destroying the integrity of the SERS substrate surface or preventing the observation of bacterial vibrational bands due to masking Raman contributions from residual fluid components. A preliminary procedure for enriching and concentrating bacteria from spiked human blood is described here and SERS substrate fouling does not prevent the observation of Raman spectra of recovered vegetative bacterial cells. When SERS spectra, however, do show changes in their vibrational signatures as a function of biological fluid exposure or any other environmental change, these vibrational spectra are a sensitive, molecular level probe of cell surface layer structural changes in such media. In blood, such changes could result from the activity of proteases, surface adsorption or release of serum components including proteins, nutrients or metal ions.

Experimental Methods

Substrate Preparation

The central experimental component responsible for the successful development of a reliable and sensitive SERS based platform for bacterial identification and detection is the SERS active substrate. In our initial attempts to obtain bacterial SERS fingerprints, we were unable to observe SERS signals from bacteria in metal colloidal solutions or placed on Au nanoparticle embedded gel matrices (*52, 53*). When bacteria were placed on aggregated Au nanoparticles that had precipitated from Au colloidal solution onto a glass surface, reasonably intense SERS spectra could be observed. The 785 nm excited SERS spectrum of *B. thuringiensis* on such a substrate is shown in Fig. 1c. The corresponding SEM of this nanostructured substrate in the vicinity of the bacterium is displayed in Fig. 2c. However, SERS bacterial signatures could not be reliably obtained from these substrates because the strongly enhancing

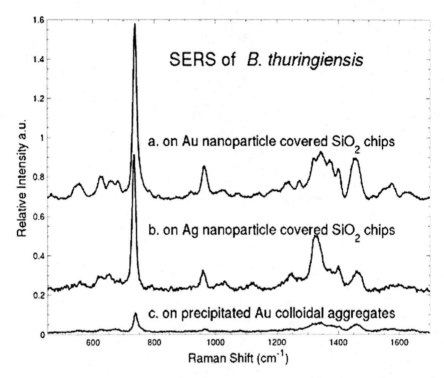

Figure 1. (a) SERS spectrum of B. thuringiensis on a gold nanoparticle covered SiO₂ substrate, (b) SERS spectrum of B. thuringiensis on a silver nanoparticle covered SiO₂ substrate, (c) SERS spectrum of B. thuringiensis on a gold aggregates that have precipitated from a colloidal solution and dried onto a clean glass slide.

metal clusters are randomly distributed and insufficiently uniform to provide concsistently strong SERS enhancement over the entire substrate surface area. Thus, a scheme for the production of a novel metal nanostructured substrate that provided large SERS enhancements for micron sized analytes was developed. The bacterial spectra reported here were obtained on an aggregated metal (Au or Ag) nanoparticle coated SiO_2 SERS substrate produced by a multi-step *in situ* growth procedure. SERS spectra of *B. thuringiensis* on these gold and silver nanostructured substrates are shown in Figs. 1a and 1b. The relative intensity of the SERS spectrum on the metal nanoparticle covered substrates is about an order of magnitude stronger than that due to the precipitated gold aggregates.

Recently, various sol-gel based recipes incorporating metal colloids have been shown to result in successful SERS substrates (*49, 50, 52-57*). Large Raman enhancements have been found for nano-sized metal particles dispersed in the resulting gel structure, in part, due to the large stabilized metal surface areas available to molecular-sized scatterers. In contrast, the sol-gel derived SiO_2 SERS substrate produced by the procedure described here is covered by immobilized clustered aggregates of monodispersed sized gold nano-particles that have been grown *in-situ*.

The recipe for the production of gold nanaoparticle covered chips has been described previously (*40, 41*), and is only briefly summarized here. A gold ion doped sol-gel is formed by the hydrolysis of tetramethoxysilane in an acidic methanol solution (10 ml HPLC grade methanol/5 ml H_2O/3 ml $Si(OCH_3)_4$ 99.99% Sigma) of $HAuCl_4$ (50 µl of ~1M $HAuCl_4$). After ~ 3 hours of stirring to complete the hydrolysis, 25 µl aliquots are drawn into polypropylene microcentrifuge tubes and dried in a glovebox for ~ 24 hours at ambient temperature and highpurity Nitrogen gas flow (relative humidity ~40%). The resulting gel pellets are exposed to water saturated air for ~ 1 hour and then mixed vigorously (~30 sec) with 0.66 mM aqueous sodium borohydride. This first reduction step provides Au seeds for subsequent nano-structured metal particle growth. The solution is drained, and 50 ml of H_2O are added to the resulting gel pellets providing a greatly diluted $NaBH_4$ solution for an additional reduction stage. Following 30 minutes of gentle shaking, the substrates are covered with a paraffin film and incubated for ~24 hours at room temperature. This second reduction step promotes the slow growth of gold nano-particles on the exposed SiO_2 outer surface during this period. This recipe is adapted for the production of silver nanoparticle covered SiO_2 SERS substrates by substituting $AgNO_3$ for $HAuCl_4$, carefully limiting the first reduction step to 5 seconds and using a five-fold lower reducing agent ($NaBH_4$) concentration. The SEM image of the SERS substrate viewed with higher magnification (Fig. 2b) reveals that clusters of 2 to 20, ~90 nm gold particles cover the solid SiO_2 surface and are not found in the interior of this substrate (*40, 41*).

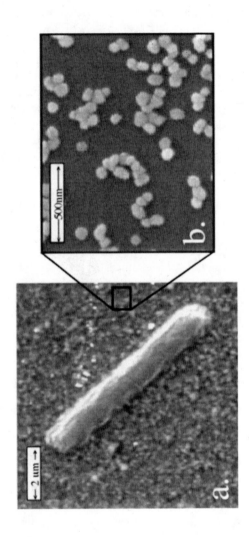

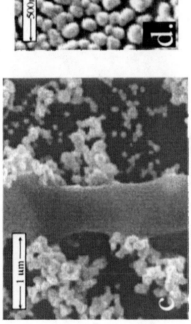

Figure 2. (a) A scanning electron micrograph (SEM) of a two-cell chain of B. anthracis Sterne bacteria on a gold nano-particle covered SiO_2 SERS chip. The relative size scales of this 2-cell bacterium (1 μm x 6 μm) and the nano-structured surface roughness are displayed here. (b) An SEM image of this SERS substrate at a ~10 fold enhanced magnification. Clusters of 1 – 15 ~80 nm Au particles are evident on the surface of the SiO_2 SERS substrate. (c) An SEM of B. thuringiensis on gold aggregates precipitated onto a glass slide. The corresponding SERS spectrum is shown in Fig. 1c. (d) An SEM of the silver nano-particle covered SiO_2 chip that gives rise to the maximum bacterial SERS enhancement. The SERS spectrum shown in Fig. 1b is due to this SERS substrate.

Bacterial Sample Preparation and Handling

B. anthracis Sterne strain was obtained from the Colorado Serum Company and verified to be attenuated. *E. coli* K12 (12-4500), *S. typhimurium* (15-5351A), *B. cereus* (15-4869), and *B. thuringiensis* (15-4926) were obtained from Carolina Biological Supply (catalog numbers noted). Human blood and blood plasma were obtained from SeraCare Life Sciences. The *in vitro* bacterial samples were grown in 5 ml of LB (Sigma) broth (~ 4-5 hours) to an $OD_{600} = $ ~1, washed five times with Millipore water and re-suspended in 0.25 ml of water. An Eppendorf pipette is used to place ~ 1 μl of the bacteria suspension on the SERS active substrate. Raman measurements are usually acquired within minutes after placing the bacterial samples on the SERS surface. Bacterial cells were enriched and concentrated about 10-fold from human blood samples as follows. The red cells and bacteria were pelleted by centrifugation in a microfuge, and the plasma supernatant together with most of the white cell buffy coat were removed by pipetting. The red cells were lysed selectively by addition of a 20-fold volume of 0.08% Na_2CO_3 and 0.005%Triton X-100 (*58*). Bacteria were pelleted by centrifugation, resuspended in phosphate buffer solution (PBS) and analyzed by SERS.

Raman Instrumentation

A Renishaw Raman microscope (model RM-2000) capable of ~2λ spatial resolution was used to obtain the 785 nm diode laser excited data presented here. Typically, SERS spectra of bacteria on the metal nanoparticle covered substrates are obtained with an incident laser power of 1 – 3 mW and a spectral data acquisition time of ~20 seconds. Spectral resolution is usually set to 3 cm^{-1} for this cooled CCD (400x578 array size) detection system (0.25 m spectrometer fitted with a 1200 groove/mm grating). The 520 cm^{-1} vibrational band of a silicon wafer provides frequency calibration.

Results and Discussion

SERS Enhancement Factors, Fluorescence Quenching and Spectral Reproducibility

Typical Raman cross-section enhancements due to the proximity of nano-structured metal surfaces are in the range of 10^3 to 10^8 per molecule from ensemble averaged SERS measurements although single molecule enhancements as large as 10^{14} (59-61) have been reported at selected nanoparticle cluster "hot spots". Raman enhancement factors sensitively depend on the nature of the SERS active substrate, particularly in terms of surface morphology and surface layer charge distribution, the nature of the analyte electronic structure and the excitation wavelength (29, 61). As a first step in characterizing the SERS activity of these novel metal aggregate covered SiO_2 substrates, the Raman cross-section enhancement of a molecular scatterer on a gold nanoparticle covered substrates was determined. The scattering cross-section of the 912 cm^{-1} C-C stretching mode of the simple amino acid glycine is enhanced by a factor of $5x10^7$ due to 785 nm excitation on the gold substrate. After normalization for incident intensities and data accumulation times, an estimate of the relative number densities of a 1M aqueous glycine solution and a monolayer of 1mM glycine solution on the SERS substrate in the scattering volume, are used to derive this per molecule enhancement factor (62). The magnitude of this ensemble averaged, scattering amplification indicates that these nanoparticle covered surfaces are effective SERS promoting substrates for molecular samples.

However, the inherent orientational and distance constraints arising from the structure of bacterial outer cell layers as well as the typical cell dimensions offer additional potential limitations for the metal surface enhancement of bacterial Raman signals. Thus the ability of a SERS substrate to provide significant enhancement for molecules in liquid solution does not guarantee that the same substrate will be effective in promoting species specific SERS fingerprints for bacteria with large and reproducible enhancement factors. As we reported previously (40), the Raman scattering cross-sections (i.e. scattered power per bacterium) of Gram-negative, E. coli, and Gram-positive, B. anthracis Sterne are amplified by factors of 2×10^4 and 5×10^4 respectively on the gold particle covered substrates when the intensities of the strongest vibrational band in SERS and bulk Raman spectra are compared and normalized for excitation intensity, number density and data acquisition time. This enhancement factor is so large that SERS spectra of single bacterial cells can be obtained in ~20 seconds of data acquisition time (40, 41).

The per bacterium enhancement factor reported here ($2 - 5 \times 10^4$) is not a measure of the per molecule SERS enhancement more typically quoted in the SERS literature because only those correctly oriented molecules close to the SERS active substrate are enhanced. Thus, presumably only some fraction of cell surface components account for these bacterial SERS spectra. Vibrational bands due to these outer surface components may contribute only weakly to the observed bulk Raman spectrum excited at 785 nm which is determined by both cytoplasmic and surface layer components. The SERS bacterial enhancement factors reported here are thus, necessarily, lower estimates of the per molecule

Raman enhancement effect arising from these gold covered SiO_2 substrates consistent with the glycine solution based results.

A comparison of the SERS spectra displayed in Fig. 1 (a and b) reveals that the SERS spectra on these nanoparticle covered substrates are metal dependent. We attribute this differences to the distinct nanoscale morphologies of the corresponding gold and silver surfaces (see Figs. 2b and 2d). The Raman amplification is nearly the same for the two metal nanoparticle covered substrates as seen in Fig. 1. Significant differences in the SERS activity of bacterial spores on these Au and Ag substrates are observed and will be described elsewhere.

The Raman scattering of *Bacillus* vegetative bacteria excited at 785 nm illustrates an additional important attribute of SERS. The non-SERS or bulk Raman spectrum of *B. anthracis* Sterne excited at 785 nm sits on top of a broad structureless fluorescence feature (*40*). A relatively weak and noisy Raman spectrum can be identified at this experimentally convenient wavelength overlapping the much more intense fluorescence of the bulk Bacillus emission specta. In contrast, only a strongly enhanced Raman emission, lacking the broad fluorescence feature, is evident in the SERS spectrum of this species excited at the same wavelength. The proximity of the metal surface results in the decreased fluorescence quantum yield from the fluorophore (*31, 32, 34, 35*) due to the rapid energy transfer of the excitation to the nearby metal substrate. Consequently, no analogous broad fluorescence is observed in the 785 nm excited SERS spectra of *Bacillus* species. However, whatever the origin of this overlapping fluorescent feature, the close proximity of the metal structure surface, required for the observation of SERS, concurrently eliminates any fluorescence relative to the SERS signal. Thus, SERS is essential for the observation of high quality Raman vibrational signatures of some pathogenic microorganisms, such as those of the *Bacillus* family, excited by red to visible radiation.

As discussed above, producing sensitive and reproducible substrates has been a major obstacle to the more wide spread use of SERS for analytical applications. We have previously demonstrated (*40, 41*) the excellent reproducibility obtained for SERS spectra of a Gram-positive (*B. anthracis* Sterne) and Gram-negative (*S. typhimurium*) bacteria acquired from different locations on the same substrate and from different substrates. The *absolute* scattering intensities of spectra for a given species normalized for identical excitation and data collection parameters vary by less 15 % at ~735 cm^{-1} (signal maximum). One source of observed spectral differences is ascribed to the inhomogeneity of the metal particle coated SiO_2 substrate surfaces produced by our current growth procedure. Work is in progress to further reduce this source of spectral inhomogeneity. An additional evaluation of SERS reproducibility on these substrates is illustrated below.

Bacterial SERS and nonSERS Raman Comparison

As discussed above, the normal (non-SERS) Raman and IR absorption spectra of different bacterial species and strains are often very similar to one another (*12, 13, 33*). By contrast, more distinct species-specific vibrational spectral signatures are observed in the SERS spectra than in the corresponding normal Raman spectra (*40*). Only bacterial molecular components close to the SERS active surface (within ~10 nm) with the correct orientation are preferentially enhanced due to the distance and orientation dependence of the electromagnetic and charge-transfer Raman enhancement mechanisms (*29, 63*). The greater species differentiation evident in the bacterial SERS spectral "fingerprints" than in the corresponding normal Raman or IR spectra, implies that bacterial cells of different species are chemically more distinct at their cell surface layers than in their cytoplasm. Thus, these results argue that bacterial identity is more dependent on outer layer components than on cytoplasmic cellular machinery. Greater bacterial species differentiation at the cell surface is consistent with the view that bacteria have evolved to interact with specific environmental niches to create new species.

The relative Raman intensities and frequencies of the SERS and non-SERS spectra *for a given species* are generally quite different (*40*). In addition to enhanced species vibrational specificity, SERS spectra of bacteria generally exhibit fewer vibrational bands, i.e. reduced spectral congestion, than their corresponding bulk Raman spectra. The simplified and more species distinct vibrational signatures arising from SERS observations, makes the SERS technique potentially a more powerful spectroscopic tool than normal Raman (or IR absorption) for bacterial identification, in addition to the advantages derived from the enormously enhanced Raman scattering efficiency. These observed differences in the bacterial Raman spectra are also attributed to the proximity and orientation effects of the SERS enhancement mechanisms.

Species Specificity of SERS

The SERS spectra of seven bacterial species acquired on gold nanoparticle covered substrates are shown in Fig. 3. These spectra result from the scattering of approximately 25 – 100 bacterial cells, (depending on the sample species), illuminated by ~ 2 mW of incident 785 nm laser power and 10 seconds of data accumulation time. They are single scan spectra that have been corrected only for the spectral response of the system and some baseline effects. Bands in the hydrogen stretching (2900 – 3200 cm^{-1}) region are not observed in these SERS spectra and thus only the 400 to 1700 cm^{-1} fingerprint region are displayed here. These spectra demonstrate two of the most important results of our SERS studies

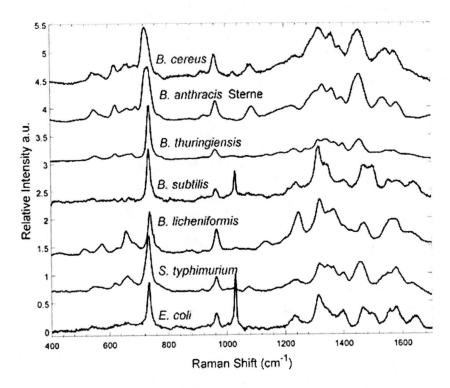

Figure 3. SERS spectra of seven bacterial species obtained on gold aggregate coated SiO₂ chips. An incident laser power of 2 mW and a data accumulation time of 10 sec were used to obtain these spectra. Spectra are offset vertically for display purposes and top-to-bottom ordered according to their phylogenetic relationship.

up to this point: (1) SERS spectra of bacteria excited by low laser power radiation at 785 nm exhibit excellent signal/noise when cells are placed on the aggregated gold nanoparticle coated substrates, and (2) each bacterial species is characterized by a unique SERS vibrational fingerprint. These features are central to the development of a SERS-based platform for bacterial pathogen detection and identification.

The SERS spectra in Fig. 3 are ordered from top to bottom according to their phylogenetic relationship, i.e. the more closely related species are adjacent to each other in this figure (*64*). However, the observed vibrational signatures do not consistently exhibit systematic trends that correlate with this lineage. For example, SERS spectra of the Gram-positive species, *B. anthracis* Sterne, *B. cereus*, *B. thuringiensis*, *B. subtilis* and *B. licheniformis* and the Gram-negative species, *E. coli* and *S. typhimurium*, do not show obvious spectral differences attributable to these two classes of bacterial outer surface structures (*65*). However, spectral regions with considerable homology can be discerned between some closely related species. For example SERS spectra of the closely related members of the *B. cereus* group; *B. anthracis* Sterne, *B. cereus* and *B. thuringiensis* (*66*), all exhibit a very similar pattern of relative intensities in the 1250 – 1650 cm^{-1} region and, as discussed further below, the *B. anthracis* Sterne and *B. cereus* SERS spectra are extremely similar. Qualitatively, the SERS bacterial fingerprints of all seven species exhibit a finite number of vibrational bands in common with each other in the 500 – 1250 cm^{-1} region. For example, bands at ~ 735, 965, 1030, 1080 and 1240 cm^{-1} are evident in all these SERS spectra although the relative intensities of these features vary as a function of bacterial species. Furthermore, when these spectra are viewed at higher resolution, reproducible small frequency shifts and bandwidth changes are evident for some of these recurring vibrational bands. The relatively small number of vibrational features that characterize these SERS spectra argues that robust algorithms for bacterial identification should be readily achievable using these markers.

As an additional example of the ability of these SERS spectra to distinguish closely related species/strains, we explicitly consider the SERS signatures of *B. anthracis* and *B. cereus* which have the most similar SERS spectra of the species shown in Figure 3. These members of the *Bacillus cereus* group of bacteria exhibit very different pathological effects. *B. anthracis* and *B. cereus* are respectively a highly lethal human pathogen and an ubiquitous soil bacterium which infrequently causes human disease such as diarrhea or conjunctivitis. In a recent study (*66*), these bacteria were judged to be members of the same species and found to be virtually genetically indistinguishable by their 16s rDNA sequences, generally considered to be the gold standard for bacterial identification. The nearly identical SERS signatures of *B. anthracis* Sterne and B. cereus are evident in Figs. 3 and 4. To highlight the specificity of the SERS signatures obtained on the gold nanoparticle substrates however, SERS derivative difference spectra of two *B. cereus* spectra, two *B. anthracis* and a *B. anthracis*, *B. cereus* SERS derivative spectrum are shown in Fig. 4, as well. The inter-species *B. anthracis*, *B. cereus* derivative difference spectrum (Fig. 4f) shows much larger intensity features as compared to the intra-species difference spectra (Figs. 4d and 4e). These single species difference spectra are a measure of the reproducibility of the vibrational fingerprints. The unique SERS signatures of these bacteria unequivocally distinguish the identity of these closely related organisms in contrast to the results from several established DNA sequence dependent methods (*66*).

SERS Spectra of Bacteria in Biological Fluids

All of the bacterial spectra described so far in this chapter are from samples grown in laboratory culture medium without any contact with human biological fluids. The practical application of SERS for *in vivo* bacterial identification, however, will require the generation of spectra from cells isolated from human body fluids. A central question to be addressed for this application of SERS is whether the biological fluid will cause fouling of the surface, i.e. occupy sites on the surface that might preclude close (< 10 nm) approach of the micron sized vegetative bacterial cells, and/or result in masking SERS signals due to prominent features from other components in biological fluid. Our results thus far indicate that these factors are not a problem on the Au and Ag nanoparticle covered SERS substrates employed here, at least for the biological fluids we have investigated thus far. A SERS spectrum of human blood spiked with 3×10^7 *B. anthracis* Sterne per ml and not subjected to our bacterial enrichment procedure is shown in Fig. 5(d). The concentration of red and white cells, which are 50 to 100 times larger (by area) than the bacterial cells is $>10^9$ per ml. Red blood cells completely dominate the reflective image of this sample. However, spectral features of the *B. anthracis* Sterne SERS spectrum are clearly evident in this vibrational spectrum. A corresponding SERS spectrum of *B. anthracis* Sterne exposed only to standard LB broth is shown in Fig. 5 (a) for comparison. The characteristic SERS bacterial bands at 735 cm^{-1}, 970 cm^{-1} and 1350 cm^{-1} are clearly evident in the blood spiked sample. These data illustrate that while our planned protocol for bacterial identification and detection (*vide infra*) will employ a series of rapid bacterial enriching procedures, the metal nanoparticle surface acts as a "getter" to effectively provide some inherent level of bacterial enrichment. We ascribe this to favorable electrostatic interactions between the outer bacterial cell wall components and the native double layer charge distribution on the prepared SERS substrate.

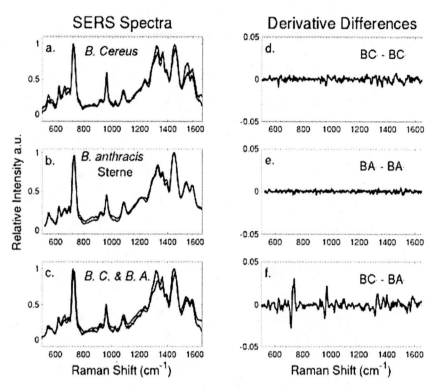

Figure 4. a) Two normalized SERS spectra of B. cereus, b) Two normalized SERS spectra of B. anthracis Sterne, c) Normalized B. cereus and B. anthracis Sterne spectra are overlayed, d) The difference between the derivatives of the two B. cereus spectra shown in (a) are plotted as a function of frequency (derivative difference spectrum), e) Derivative difference spectrum corresponding to the two B. anthracis spectra shown in (b), f) The B. cereus, B. anthracis Sterne derivative difference spectrum.

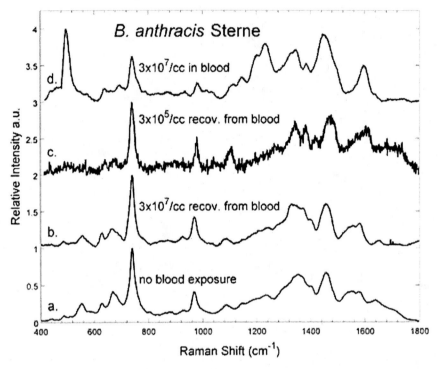

Figure 5. SERS spectra of B. anthracis Sterne on Au nanoparticle covered substrates. a) grown in LB broth, no blood exposure, b)recovered from human blood spiked with 3x10⁷ bacteria per cc, c) recovered from human blood spiked with 3x10⁵ bacteria per cc, d)10⁸ bacteria in human blood

However, in order to enhance these bacterial signatures and suppress the contribution of the biological fluid to achieve detection and identification at biologically meaningful concentration levels ($10^4 - 10^2$ per cc) in whole blood a bacterial enrichment and concentration protocol is required. We have developed a procedure for this purpose which results in ~30 % bacterial recovery from human blood samples spiked with laboratory-grown bacteria. The key steps in our current method are high speed centrifugation to remove white blood cells and most of the plasma, followed by red cell lysis by addition of a solution of 0.08% Na_2CO_3 and 0.005% Triton X-100 (*58*). After the resulting solution is centrifuged, supernatant removed and resuspended in phosphate buffered saline, a µl of this solution is placed on our SERS substrate and SERS spectra nearly completely dominated by the bacterial signals are obtained. A SERS spectrum of *B. anthracis* Sterne cells spiked at 3 x 10^7 per ml in blood and enriched by this procedure is shown in Fig. 5(b). Comparison with the SERS spectrum of vegetative *B. anthracis* Sterne exposed only to LB growth medium shows that spectral changes due to blood exposure are minimal. Maybe more important for the future development of this methodology, fouling or biological fluid "solvent" bands do not prevent the observation of this clearly recognizable SERS fingerprint. We have been able to detect and identify *B. anthracis* Sterne cells in human blood samples spiked with concentrations as low as 3 x 10^5 per cc (Fig. 5(c)) by the current enrichment and concentration procedure and SERS substrates. In addition we find that the SERS bacterial signature is independent of blood exposure times over the time scale tested here thus far (15 - 90 minutes).

Conclusions

The sensitivity, selectivity, specificity and spatial discrimination that SERS microscopy can offer, as our results on *in vitro* bacterial samples demonstrate (*40, 41*), should thus greatly enhance the implementation of a practical, portable, rapid SERS-based procedure for microorganism detection and identification. We estimate that, ultimately, within a time frame of minutes, spectra of bacteria enriched and concentrated from clinical isolates can be acquired and identified with the use of current data mining strategies. In contrast to other bacterial identification techniques, this simple, sensitive, bacterial diagnostic platform requires no labeling, hybridization, growth, or PCR amplification, but relies on the large Raman signal enhancements obtained on the novel metal nanostructured SERS substrates we have developed. The prospects are encouraging that bacterial identification in human blood is possible on a time scale of ~ 30 minutes at clinically relevant concentrations via this SERS approach when combined with bacterial concentrating protocols.

In addition to fully achieveing this goal, future challenges associated with this research effort include providing molecular level assignments for the observed SERS bacterial vibrational features, optimizing software data mining procedures for bacterial identification and the extension of this methodology to pathogen detection in other biological fluids. Efforts in these directions are ongoing in our laboratory.

Acknowledgment

The support of the Army Research Laboratory (Cooperative Agreement DAAD19-00-2-0005, Task 17) and the National Institute of Health (STTR Grant # 1 R41 AI066641-01) are gratefully acknowledged.

References

1. Bernhardt, M.; Pennell, D. R.; Almer, L. S.; Schell, R. F. *J Clin Microbiol* **1991**, *29*, 422.
2. Washington, J. A., 2nd; Ilstrup, D. M. *Rev Infect Dis* **1986**, *8*, 792.
3. Renaud, B.; Brun-Buisson, C. *Am J Respir Crit Care Med* **2001**, *163*, 1584.
4. Hughes, J. M.; Gerberding, J. L. *Emerg Infect Dis* **2002**, *8*, 1013.
5. Jernigan, J. A.; Stephens, D. S.; Ashford, D. A.; Omenaca, C.; Topiel, M. S.; Galbraith, M.; Tapper, M.; Fisk, T. L.; Zaki, S.; Popovic, T.; Meyer, R. F.; Quinn, C. P.; Harper, S. A.; Fridkin, S. K.; Sejvar, J. J.; Shepard, C. W.; McConnell, M.; Guarner, J.; Shieh, W. J.; Malecki, J. M.; Gerberding, J. L.; Hughes, J. M.; Perkins, B. A. *Emerg Infect Dis* **2001**, *7*, 933.
6. Bell, C. A.; Uhl, J. R.; Hadfield, T. L.; David, J. C.; Meyer, R. F.; Smith, T. F.; Cockerill, F. R., 3rd *J Clin Microbiol* **2002**, *40*, 2897.
7. Hoffmaster, A. R.; Meyer, R. F.; Bowen, M. P.; Marston, C. K.; Weyant, R. S.; Barnett, G. A.; Sejvar, J. J.; Jernigan, J. A.; Perkins, B. A.; Popovic, T. *Emerging Inf. Dis.* **2002**, *8*, 1178.
8. Oggioni, M. R.; Meacci, F.; Carattoli, A.; Ciervo, A.; Orru, G.; Cassone, A.; Pozzi, G. *J Clin Microbiol* **2002**, *40*, 3956.
9. Head, I. M.; Saunders, J. R.; Pickup, R. W. *Microb Ecol* **1998**, *35*, 1.
10. Naumann, D.; Helm, D.; Labischinski, H. *Nature* **1991**, *351*, 81.
11. Helm, D.; Labischinski, H.; Schallehn, G.; Naumann, D. *J Gen Microbiol* **1991**, *137 (Pt 1)*, 69.
12. Maquelin, K.; Kirschner, C.; Choo-Smith, L. P.; Ngo-Thi, N. A.; van Vreeswijk, T.; Stammler, M.; Endtz, H. P.; Bruining, H. A.; Naumann, D.; Puppels, G. J. *J Clin Microbiol* **2003**, *41*, 324.

13. Maquelin, K.; Choo-Smith, L. P.; Endtz, H. P.; Bruining, H. A.; Puppels, G. J. *J Clin Microbiol* **2002**, *40*, 594.

14. Maquelin, K.; Kirschner, C.; Choo-Smith, L. P.; van den Braak, N.; Endtz, H. P.; Naumann, D.; Puppels, G. J. *J Microbiol Methods* **2002**, *51*, 255.

15. Vaidyanathan, S.; Winder, C. L.; Wade, S. C.; Kell, D. B.; Goodacre, R. *Rapid Commun Mass Spectrom* **2002**, *16*, 1276.

16. Goodacre, R.; Kell, D. B. *Curr Opin Biotechnol* **1996**, *7*, 20.

17. Lay, J. O., Jr. *Mass Spectrom Rev* **2001**, *20*, 172.

18. Grabb, E. S.; Buck, R. P. *J Am Chem Soc* **1989**, *111*, 8362.

19. Cotton, T. M.; Kim, J. M.; Chumanov, G. D. *J. Raman Spectrsoc.* **1991**, *22*, 729.

20. Chumanov, G. D.; Efremov, R. G.; Nabiev, I. R. *J Raman Spectrosc* **1990**, *21*, 43.

21. Herne, T. M.; Ahern, A. M.; Garrell, R. L. *J Am Chem Soc* **1989**, *113*, 846.

22. Wantanabe, T.; Maeda, H. *J Phys Chem* **1989**, *93*, 3258.

23. Vo-Dinh, T.; Houck, K.; Stokes, D. L. *Anal Chem* **1994**, *66*, 3379.

24. Weldon, M. K.; Morris, M. D.; Harris, A. B.; Stoll, J. K. *J Lipid Res* **1998**, *39*, 1896.

25. Seibert, M.; Picorel, R.; Kim, J. H.; Cotton, T. M. *Methods Enzymol* **1992**, *213*, 31.

26. Kneipp, K.; Kneipp, H.; Manoharan, H.; Hanlon, E. B.; Itzkan, I.; Desari, R. R.; Feld, M. S. *Phys Rev E* **1998**, *57*, R6281.

27. Cavalu, S.; Cinta-Pinzaru, S.; Leopold, N.; Kiefer, W. *Biopolymers* **2001**, *62*, 341.

28. Shafer-Peltier, K. E.; Haynes, C. L.; Glucksberg, M. R.; Van Duyne, R. P. *J Am Chem Soc* **2003**, *125*, 588.

29. Moskovits, M. *Rev Mod Physics* **1985**, *57*, 783.

30. Haynes, C. L.; Van Duyne, R. P. *J Phys Chem B* **2003**, *107*, 7426.

31. Chance, R. R.; Prock, A.; Silbey, R., *Molecular Fluorescence and Energy Transfer near Interafaces.* ed.; Wiley: NYC, 1978; 'Vol.' 37, p 1.

32. Pineda, A. C.; Ronis, D. *J Chem Phys* **1985**, *83*, 5330.

33. Kirschner, C.; Maquelin, K.; Pina, P.; Ngo Thi, N. A.; Choo-Smith, L. P.; Sockalingum, G. D.; Sandt, C.; Ami, D.; Orsini, F.; Doglia, S. M.; Allouch, P.; Mainfait, M.; Puppels, G. J.; Naumann, D. *J Clin Microbiol* **2001**, *39*, 1763.

34. Gersten, J.; Nitzan, A. *J Chem Phys* **1980**, *73*, 3023.

35. Weitz, D. A.; Garoff, S.; Gersten, J. I.; Nitzan, A. *J Chem Phys* **1983**, *78*, 5324.

36. Maquelin, K.; Choo-Smith, L. P.; van Vreeswijk, T.; Endtz, H. P.; Smith, B.; Bennett, R.; Bruining, H. A.; Puppels, G. J. *Anal Chem* **2000**, *72*, 12.

37. Choo-Smith, L. P.; Edwards, H. G.; Endtz, H. P.; Kros, J. M.; Heule, F.; Barr, H.; Robinson, J. S., Jr.; Bruining, H. A.; Puppels, G. J. *Biopolymers* **2002**, *67*, 1.

38. Puppels, G. J.; de Mul, F. F.; Otto, C.; Greve, J.; Robert-Nicoud, M.; Arndt-Jovin, D. J.; Jovin, T. M. *Nature* **1990**, *347*, 301.
39. Puppels, G. J.; Colier, W.; Olminkhof, J. H. F.; Otto, C.; de Mul, F. F.; Greve, J. *J. Raman Spectrsoc.* **1991**, *72*, 5529.
40. Premasiri, W. R.; Moir, D. T.; Klempner, M. S.; Krieger, N.; Jones II, G.; Ziegler, L. D. *J. Phys. Chem. B* **2005**, *109*, 312.
41. Premasiri, W. R.; Moir, D. T.; Ziegler, L. D. *Proceed SPIE* **2005**, *5795*, 19.
42. Sockalingum, G. D.; Lamfarraj, H.; Beljebbar, A.; Pina, P.; Allouch, P.; Manfait, M. In *Direct on-plate analysis of microbial cells: a pilot study by sursface-enhanced Raman spectroscopy*, Europ Conf Spectrosc Biol Mol, Boston, 1999; Kluwar Academic Publ: Boston, 1999; 599.
43. Fell Jr., N. F.; Smith, A. G. B.; Vellone, M.; Fountain III, A. W. *Proceed SPIE* **2002**, *4577*, 174.
44. Guzelian, A. A.; Sylvia, J. M.; Janni, J.; Clauson, S. L.; Spenser, K. M. *Proceed SPIE* **2002**, *4577*, 182.
45. Grow, A. E.; Wood, L. L.; Claycomb, J. L.; Thompson, P. A. *J Microbiol Methods* **2003**, *53*, 221.
46. Efrima, S.; Bronk, B. V.; Czege, J. *Proceed SPIE* **1999**, *3602*, 164.
47. Zeiri, L.; Bronk, B. V.; Shabtai, Y.; Czege, J.; Efrima, S. *Colloid Surf A* **2002**, *208*, 357.
48. Jarvis, R. M.; Goodacre, R. *Anal Chem* **2004**, *76*, 40.
49. Olson, L. G.; Lo, Y.-S.; Beebe Jr., T. P.; Harris, J. M. *Anal Chem* **2001**, *73*, 4268.
50. Lee, Y.-H.; Farquharson, S.; Rainey, P. *Proc SPIE* **1999**, *3857*, 76.
51. Creighton, J. A., Metal colloids. In *Surface enhanced Raman Scattering*; Chang, R. K.; Furtak, T. E., Eds. Plenum: NYC, 1982; 315.
52. Volkan, M.; Stokes, D. L.; Vo-Dinh, T. *J Raman Spectrosc* **1999**, *30*, 1057.
53. Premasiri, W. R.; Clarke, R. H.; Londhe, S.; Womble, M. E. *J Raman Spectrosc* **2001**, *32*, 919.
54. Garcia-Rodriguez, F. J.; Gonzales-Hernandez, J.; Perez-Robles, F.; Vorobiev, Y. V.; Manzano-Ramirez, A.; Jimenez-Sandoval, S.; Chao, B. S. *J Raman Spectrosc* **1998**, *29*, 763.
55. Akbarain, F.; Dunn, B. S.; Zink, J. I. *J Raman Spectrosc* **1996**, *27*, 775.
56. Lee, Y. H.; Dai, S.; Young, J. P. *J Raman Spectrosc* **1997**, *30*, 635.
57. Farquharson, S.; Smith, W.; Lee, Y.; Elliot, S.; Sperry, J. F. *Proc SPIE* **2002**, *4575*, 62.
58. Sullivan, N. M.; Sutter, V. L.; Finegold, S. M. *J Clin Microbiol* **1975**, *1*, 30.
59. Nie, S.; Emory, S. *Science* **1997**, *275*, 1102.
60. Kneipp, K.; Wang, Y.; Kneipp, H.; Perelman, L. T.; Itzkan, I.; Desari, R. R.; Feld, M. S. *Phys Rev Lett* **1997**, *78*, 1667.
61. Markel, V. A.; Shalev, V. M.; Zhang, P.; Huynh, W.; Tay, L.; Haslett, T. L.; Moskovits, M. *Phys Rev B* **1999**, *59*, 10903.

62. Suh, J. S.; Moskovits, M. *J Am Chem Soc* **1986**, *108*, 4711.
63. Picorel, R.; Bakhtiari, M.; Lu, T.; Cotton, T. M.; Seibert, M. *Photochem. Photobiol.* **1992**, *56*, 263.
64. Cole, J. R.; Chai, B.; Marsh, T. L.; Farris, R. J.; Wang, Q.; Kulam, S. A.; Chandra, S.; McGarrell, D. M.; Schmidt, T. M.; Garrity, G. M.; Tiedje, J. M. *Nucleic Acids Res* **2003**, *31*, 442.
65. Lengler, J. W.; Drews, G.; Schegal, H. G., *Biology of the Prokaryotes.* ed.; thieme: Stuttgart, 1999; 'Vol.' p.
66. Helgason, E.; Okstad, O. A.; Caugant, D. A.; Johansen, H. A.; Fouet, A.; Mock, M.; Hegna, I.; Kolsto *Appl Environ Microbiol* **2000**, *66*, 2627.

Chapter 13

Surface-Enhanced Raman Spectroscopy-Based Optical Labels Deliver Chemical Information from Live Cells

Janina Kneipp[1,2], Harald Kneipp[1], and Katrin Kneipp[1,3]

[1]Wellman Center for Photomedicine, Harvard Medical School,
Boston, MA 02114
[2]Federal Institute for Materials Research and Testing, FG I.3,
D–12489 Berlin, Germany
[3]Harvard–MIT Division of Health Sciences and Technology,
Cambridge, MA 02139

We demonstrate the application of surface-enhanced Raman scattering (SERS) in robust and sensitive optical probes for measurements in living cells. The hybrid probes consist of gold or silver nanoaggregates with an attached reporter species, e.g. a dye. They are detected based on the SERS signature of the reporter molecule. This approach results in several advantages compared to other optical labels, such as improved contrast, high spectral specificity, multiplex capabilities, and photostability. SERS probes do not only highlight targeted cellular or other biological structures through the specific reporter spectrum, SERS in the local optical fields of the gold or silver nanostructures also provides sensitive and spatially localized molecular structural information on the cellular environment.

Introduction

Since more than a decade ago, it has repeatedly been proven that the fingerprint-like information in Raman microspectra can add to our understanding of the biochemical background of various regular, induced or pathological changes in eukaryotic cells(*1 - 3*). This is a basic research goal and a major prerequisite for progress in the areas of molecular medicine and nanobiotechnology. However, Raman investigations of living cells are not complication-free if damage to the cells is to be avoided, because the applicable maximum intensity of the excitation laser is limited, but probe volumes are small and analyte concentrations low.

In addition to the new challenges for Raman probing in live cells, current cell biology research generates a strong need for better optical labels(*4*). Although state-of-the-art fluorescence labels using dyes and quantum dots provide high sensitivity, the information they can deliver on chemical composition or molecular structure is very limited(*5, 6*). Therefore, improving optical labels regarding sensitivity, specificity, molecular structural information content, and spatial localization is another demanding subject in current biophotonics research.

Here, we will show that both issues can be addressed by applying surface enhanced Raman scattering (SERS) in the local optical fields of noble metal nanoparticles. (for a more recent overview on SERS see refs (*7 - 12*)). Nanostructures from gold or silver lead to significant improvements of the Raman signals from the molecules in their nanometer-scaled environment. This suggests optical labels based on SERS signals of reporter molecules attached to gold and silver nanoparticles instead of using fluorescence tags (*13 - 15*). Moreover, aside from providing the specific SERS spectrum of the reporter dye, gold nanoparticles also enhance the Raman signatures of their environment and in this way enable sensitive chemical probing inside biological structures, such as inside living cells (*16 - 19*) .

In the following we investigate silver and gold nanoclusters and dyes commonly used in biological studies for their potential use in SERS hybrid probes. From two biocompatible components, the dye indocyanine green (ICG), complexed with serum albumin protein, and gold nanoparticles, we constructed a SERS probe and introduced it into cultured cells.

Experimental

Silver and gold nanoparticles were produced in aqueous solutions by chemical preparation as described in ref (*20*). This preparation process results in isolated metal nanoparticles and small clusters comprised from 3-10 nanoparticles. Gold colloidal solution containing 60 nm gold spheres was also purchased from Polysciences Inc. (Eppelheim, Germany/ Warrington, PA). Both kinds of gold nanoparticles gave the same SERS spectra and provide almost the same enhancement factors. Methylene Blue, Hoechst33342, and indocyanine green (ICG) (all purchased from Sigma) were prepared in 10^{-5} - 10^{-7} M stock solutions in water or, for applications in cells, in 5 mg/ml aqueous solution of human serum albumin (HSA). These stock solutions were added to the aqueous solution of silver and gold nanoclusters for final concentrations between 10^{-7} and 10^{-10} M.

For application in cells, a SERS nanoprobe consisting of 10^{-7} M indocyanine green (ICG) complexed with human serum albumin (HSA) on 60 nm gold nanoparticles was delivered into cultured cells of a metastatic Dunning R3327 rat prostate carcinoma line (MLL) (donated by Dr. W. Heston, Memorial Sloan-Kettering Cancer Center, New York, NY). After overnight incubation, nanoparticles that were endocytosed must be included in lysosomes. In accordance with this assumption we found gold accumulations in the range of 100–1000 nm in lysosomes by light and electron microscopy after incubation for 20 hours and longer. The presence of gold particles in the cells could also be verified by the appearance of SERS signals collected from the cells. Microscope inspection provided evidence that the cells were dividing after incubation with the ICG-SERS probes. While incubated with the nanosensors, the cells were visibly growing, and no evidence of cell death was found. A slightly lower density after 20 hours incubation with the diluted medium was observed when compared with control cells growing in undiluted medium, probably resulting from the dilution of nutrients. Controls in diluted medium with ICG/NaCl and NaCl-diluted culture medium showed growth rates similar to those incubated with the gold particles and with the ICG-SERS nanoprobes.

For testing the SERS probes, SERS spectra were measured from aqueous solutions (25 µl droplets) of silver and gold nanoclusters loaded with different concentrations of the reporter dye. Excitation at different wavelengths (680 nm, 786 nm and 830 nm) was applied. (For detailed experimental parameters see Figure captions.)

A 60× microscope water immersion objective was used for focusing the excitation laser to a probed volume of ~50 fL. The same microscope objective was also used to collect the Raman scattered light. A single stage spectrograph with holographic edge or notch filters in front of the entrance slit and a liquid nitrogen-cooled CCD detector were used for spectral dispersion and collection of the scattered light.

Raman spectra were acquired from the living cells in the physiological environment (PBS buffer) using laser intensities of ~2 mW in accumulation times of 1 sec and less.

Raster scans over single living cells were carried out with a computer-controlled x,y-stage in 2 μm steps at a laser spot size of ~4x10^{-8} cm^2. All experiments on the cell culture were carried out at 830 nm excitation.

Results and Discussion

A key question in creating hybrid SERS probes for biological applications is a nanostructure that provides a high level of electromagnetic field enhancement. This is the prerequisite for an intense and stable spectroscopic signature of a reporter but also for an efficient and sensitive probing of the cellular environment. Moreover, the nanostructure should be biocompatible in order to enable probing of cells and other biological structures without influencing their physiological state or inducing any damage. So far, aggregates formed by silver or gold nanoparticles provide the highest enhancement level observed in SERS experiments, yielding enhancement factors up to 14 orders of magnitude (*12, 21, 22*). Figure 1 shows typical nanoclusters formed by silver and gold nanoparticles and their plasmon absorption. Interestingly, and very useful for biological applications, extremely high enhancement factors for aggregates can be obtained at near infrared excitation despite the fact that the plasmon resonance for isolated nanoparticles can be found around 400 nm for silver and 520 nm for gold.

As reporter molecules to be attached to these nanoaggregates we chose dyes known and approved in biological and medical applications as fluorescence markers. However, instead of using the broad and relatively non-specific fluorescence signals of these dyes, here we rely on their specific (surface enhanced) Raman signature.

Figure 2a displays SERS spectra measured from two biocompatible dyes, Methylene Blue and Hoechst33342, attached to silver nanoaggregates. Both dyes show very specific SERS spectra comprised from several narrow lines. The idea to create an optical label based on these SERS signals provides several advantages over other labels. The characteristic, fingerprint-like pattern of many narrow Raman bands of the label molecule rather than, e.g., one broad fluorescence band, provides a high specificity, which has been used in multiplexing experiments unprecedented so far by fluorescence studies (*15*). Moreover, the near infrared excitation intensity is not in resonance with the electronic transitions of the two dyes. This prevents photobleaching and results in a high label stability.

SERS experiments performed on silver and gold nanoaggregates at NIR excitation show that nanoclusters of both metals exhibit comparably good SERS enhancement factors (Figure 2b). Figure 2b compares SERS spectra of the dye indocyanine green on silver and gold nanoaggregates. The favorable enhancing

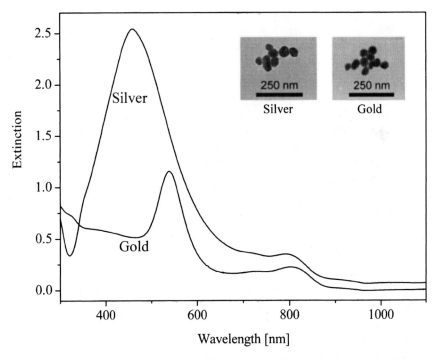

Figure 1. Extinction spectra and EM images of typical silver and gold nanoaggregates used for SERS studies in cells.

properties along with their biocompatibility suggest gold nanoaggregates as very promising SERS nanosensors for intracellular studies (*12*).

Figure 2c illustrates the detection limits for the reporter dye ICG bound to HSA in gold colloidal solution. The dye is widely used in medical imaging and surgical applications based on its fluorescence signal. Usually in these applications, local ICG concentrations are in the micromolar (10^{-6}M) range. Spectra A to D in Figure 2c were collected from gold colloidal solutions containing ICG in 10^{-7} M to 10^{-10} M concentration, respectively. Spectrum D demonstrates that the strong Raman line at 945 cm^{-1} is still detectable at an ICG concentration of 10^{-10} M. This corresponds to ~3 dye molecules in the probed 50 fL volume. Surface enhanced Raman lines of ICG in such low concentration experiments can be understood in terms of very high SERS enhancement factors in the hot spots of the nanoaggregates (*12, 22*). Comparing the signal count rate per molecule for the strongest surface enhanced Raman lines of ICG in such low concentration experiments with the count rate of the fluorescence per "free" ICG molecule, the number is one order of magnitude higher for the SERS lines.

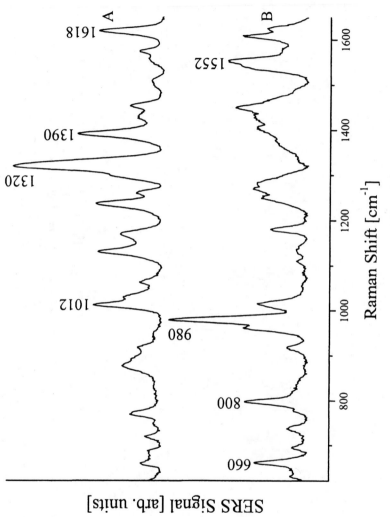

Figure 2. a) SERS spectra of the biocompatible dyes Methylene Blue (A) and Hoechst 33342 (B) on silver nanoaggregates. The signals were collected from ~ 5000 molecules in the probed volume, 3 mW 830 nm excitation, collection time 1 second.

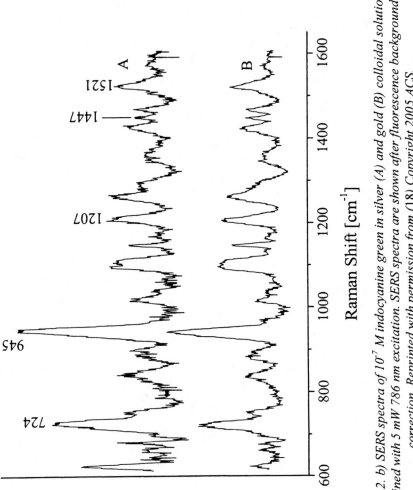

Figure 2. b) SERS spectra of 10^{-7} M indocyanine green in silver (A) and gold (B) colloidal solutions obtained with 5 mW 786 nm excitation. SERS spectra are shown after fluorescence background correction. Reprinted with permission from (18) Copyright 2005 ACS.

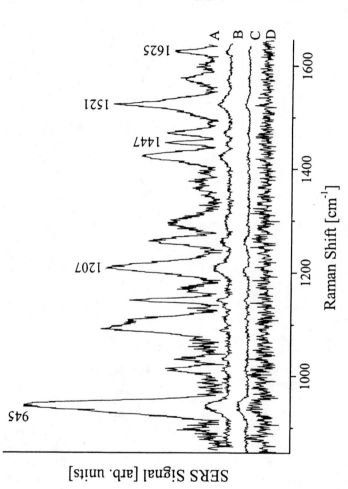

Figure 2. c) SERS spectra of ICG bound to HSA in gold colloidal solution. Spectra A to D were collected from gold colloidal solutions containing ICG in 10^{-7} M to 10^{-10} M concentration, respectively, at 15 mW, 830 nm excitation, 1 second collection time Reprinted with permission from (18) Copyright 2005 ACS

As has long been practiced in various cell biology applications (*23 - 25*) gold nanoparticles can be delivered into live cells (see also Experimental section). Figure 3 illustrates the basic concept of intracellular SERS probes. Large nanoparticle aggregates can easily be visualized by standard bright field microscopy inside the cells, here, in the lysosomal structures of a fibroblast cell.

Raman measurements with laser intensities of $\sim 10^5$ W cm^{-2} and accumulation times of 1 sec and less, and excitation with 830 nm (out of resonance) preclude from the observation of non-SERS spectra from the cells. However, at positions in the cells where gold nanoparticles are present, surface-enhanced Raman spectra can be measured in the living cells in their physiological environment.

Figure 4 shows the detection and imaging of a SERS label made from gold nanoparticles and ICG in a single live cell using the 945 cm^{-1} SERS signal (see also Figure 2). In addition to the SERS signals of the reporter ICG, at these positions, also the Raman signatures from the cellular components in the immediate surroundings of the gold nanostructures experience surface enhancement. This enables sensitive chemical probing of the particles' vicinity in very short times.

Figure 5 illustrates this ultrasensitive molecular probing using SERS nanosensors made from gold nanoparticles and ICG. Spectrum A represents the spectrum of pure ICG in the physiological environment. Spectrum B shows the SERS signature of the reporter ICG along with SERS lines that originate from the cellular surroundings of the gold nanoparticles. After subtracting the SERS signal from the reporter ICG (trace A), trace C displays the SERS spectrum of the cellular components. The Raman lines in spectrum C can be assigned to vibrations characteristic for protein and nucleotide molecular groups, such as C-H deformation/bending modes at 1450 cm^{-1}, C-N deformation at 1166 and 1229 cm^{-1}, possible contributions from Phe and Tyr $\sim$1207 cm^{-1}, as well as cytosine and adenosine ring vibrations and/or protein amide II contributions around 1540 cm^{-1} (tentative assignments after refs. (*26 - 29*).

As described above, SERS takes place in the local optical fields of metal nanostructures and is therefore restricted to the immediate vicinity of the gold nanoparticles (see also Figure 3 for a schematic). Thereby, SERS probes enable the acquisition of Raman signals not only at high sensitivity but also from nanometer scaled volumes. This local confinement of the SERS effect has several advantages over regular Raman experiments: For SERS studies in cells, SERS nanoprobes can be positioned at discrete locations, e.g., in a specific cellular compartment, and the spectral information is obtained only from the nano-environment of these probes and hence that particular compartment. This is different from the spectral information collected in a "normal" Raman microspectroscopic experiment, where all positions in a whole cell are probed. This also means that the maximum lateral resolution in such a Raman experiment

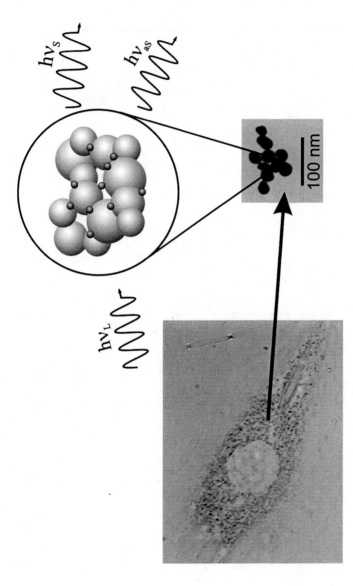

Figure 3. Basic concept of intracellular SERS probes: Gold nanoparticles are transferred into cells. Aggregates, which provide optimum SERS enhancement and are typically utilized in the live cell experiments are shown in the transmission electron micrograph and the schematic drawing. During excitation with laser light in the near-infrared ($h\nu_L$), such gold nanoaggregates provide enhanced local optical fields in their nm-scaled vicinity, leading to surface-enhanced Stokes ($h\nu_S$) and anti Stokes Raman signals ($h\nu_{aS}$) of the attached reporter molecules as well as of different cellular molecules (see also Figure 5).

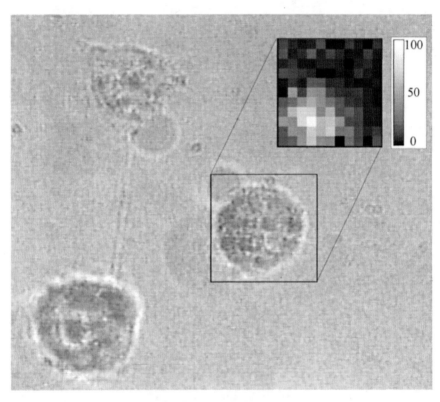

Figure 4. Spectral map of ICG in a cell based on the 945 cm⁻¹ SERS line of the molecule (see also Figure 2b). Intensities are shown in gray scale (highest value in white). A photomicrograph of the cell, indicating the mapped area, is shown for comparison. Scale bar: 28 microns.

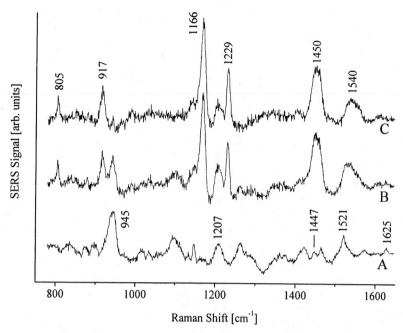

Figure 5. SERS spectra measured inside living cells incubated with ICG on colloidal gold: SERS spectrum of ICG (A), SERS spectrum of ICG and cell environment (B), difference spectrum C displaying SERS features of cell components. For the assignment of the bands see text.

is no longer limited by the excitation wavelength; instead, it is influenced by the metal nanostructure used to provide the enhancement.

Conclusions

To summarize, we have demonstrated a new type of optical probe based on SERS that is stable, specific, and biocompatible, and can be used for applications in live cells.

SERS nanosensors can be detected and imaged based on the unique spectroscopic signature of the SERS signal of a reporter molecule attached to the gold nanostructure. In addition to its own detection by the characteristic SERS spectrum of the reporter, the probe we are proposing delivers surface enhanced Raman signatures of the cell components in its vicinity. This provides the capability of ultrasensitive chemical characterization of nanometer scaled units in single live cells. Due to the large effective Raman scattering cross section, SERS probes fulfill the requirements of dynamic *in vivo* systems –the use of very low laser powers and very short data acquisition times.

Acknowledgement

This work was supported in part by DOD grant # AFOSR FA9550-04-1-0079, NIH grant # PO1CA84203, and by the generous gift of Dr. and Mrs. J.S. Chen to the optical diagnostics program of the Massachusetts General Hospital Wellman Center for Photomedicine.

References

1. Bakker Schut, T. C.; Wolthuis, R.; Caspers, P. J.; Puppels, G. J. *Journal of Raman Spectroscopy* **2002**, *33*, 580-585.
2. Feofanov, A. V.; Grichine, A. I.; Shitova, L. A.; Karmakova, T. A.; Yakubovskaya, R. I.; Egret-Charlier, M.; Vigny, P. *Biophys. J.* **2000**, *78*, 499-512.
3. van Manen, H.-J.; Kraan, Y. M.; Roos, D.; Otto, C. *PNAS* **2005**, *102*, 10159-10164.
4. Weissleder, R. *Nature Biotechnology* **2001**, *19*, 316-317.
5. Mader, O.; Reiner, K.; Egelhaaf, H. J.; Fischer, R.; Brock, R. *Bioconjugate Chemistry* **2004**, *15*, 70-78.
6. Kim, S.; Lim, Y. T.; Soltesz, E. G.; De Grand, A. M.; Lee, J.; Nakayama, A.; Parker, J. A.; Mihaljevic, T.; Laurence, R. G.; Dor, D. M.; Cohn, L. H.; Bawendi, M. G.; Frangioni, J. V. *Nature Biotechnology* **2004**, *22*, 93-97.
7. Kambhampati, P.; Child, C. M.; Foster, M. C.; Campion, A. *Journal of Chemical Physics* **1998**, *108*, 5013-5026.
8. Kneipp, K.; Kneipp, H.; Itzkan, I.; Dasari, R. R.; Feld, M. S. *Chemical Reviews* **1999**, *99*, 2957-2975.
9. Kneipp, K.; Kneipp, H.; Itzkan, I.; Dasari, R. R.; Feld, M. S. *Journal of Physics-Condensed Matter* **2002**, *14*, R597-R624.
10. Moskovits, M. *Journal of Raman Spectroscopy* **2005**, *36*, 485-496.
11. Haynes, C. L.; Yonzon, C. R.; Zhang, X. Y.; Van Duyne, R. P. *Journal of Raman Spectroscopy* **2005**, *36*, 471-484.
12. Kneipp, K.; Kneipp, H.; Kneipp, J. *Accounts of Chemical Research* **2006**, *39*, 443-450.
13. Isola, N. R.; Stokes, D. L.; Vo-Dinh, T. *Analytical Chemistry* **1998**, *70*, 1352-1356.
14. Ni, J.; Lipert, R. J.; Dawson, G. B.; Porter, M. D. *Analytical Chemistry* **1999**, *71*, 4903-4908.
15. Nam, J. M.; Thaxton, C. S.; Mirkin, C. A. *Science* **2003**, *301*, 1884-1886.
16. Kneipp, K.; Haka, A. S.; Kneipp, H.; Badizadegan, K.; Yoshizawa, N.; Boone, C.; Shafer-Peltier, K. E.; Motz, J. T.; Dasari, R. R.; Feld, M. S. *Applied Spectroscopy* **2002**, *56*, 150-154.

17. Zeiri, L.; Bronk, B. V.; Shabtai, Y.; Czege, J.; Efrima, S. *Colloids and Surfaces a-Physicochemical and Engineering Aspects* **2002**, *208*, 357-362.
18. Kneipp, J.; Kneipp, H.; Rice, W. L.; Kneipp, K. *Analytical Chemistry* **2005**, *77*, 2381-2385.
19. Premasiri, W. R.; Moir, D. T.; Klempner, M. S.; Krieger, N.; Jones, G.; Ziegler, L. D. *Journal of Physical Chemistry B* **2005**, *109*, 312-320.
20. Lee, P. C.; Meisel, D. *Journal of Physical Chemistry* **1982**, *86*, 3391-3395.
21. Kneipp, K. *Experimental Technique of Physics* **1988**, *36*, 161.
22. Kneipp, K.; Wang, Y.; Kneipp, H.; Perelman, L. T.; Itzkan, I.; Dasari, R. R.; Feld, M. S. *Phys. Rev. Lett.* **1997**, *78*, 1667.
23. Feldherr, C.; Kallenbach, E.; Schultz, N. *J. Cell Biol.* **1984**, *99*, 2216-2222.
24. Tkachenko, A. G.; Xie, H.; Liu, Y. L.; Coleman, D.; Ryan, J.; Glomm, W. R.; Shipton, M. K.; Franzen, S.; Feldheim, D. L. *Bioconjugate Chemistry* **2004**, *15*, 482-490.
25. Chithrani, B. D.; Ghazani, A. A.; Chan, W. C. W. *Nano Letters* **2006**, *6*, 662-668.
26. Thomas Jr., G.; Prescott, B.; Olins, D. *Science* **1977**, *197*, 385-388.
27. Parker, F. S. *Applications of Infrared, Raman, and Resonance Raman Spectroscopy in Biochemistry*; Plenum Press: New York and London, 1983; Vol.
28. Peticolas, W. L.; Patapoff, T. W.; Thomas, G. A.; Postlewait, J.; Powell, J. W. *Journal of Raman Spectroscopy* **1996**, *27*, 571-578.
29. Puppels, G. J.; De Mul, F.; Otto, C.; Greve, J.; RobertNicoud, M.; Arndt-Jovin, D. J.; Jovin, T. *Nature* **1990**, *347*, 301-303.

Chapter 14

Photonic Explorers Based on Multifunctional Nanoplatforms: In Vitro and In Vivo Biomedical Applications

Yong-Eun Lee Koo[1], Rodney Agayan[1], Martin A. Philbert[2], Alnawaz Rehemtulla[3], Brian D. Ross[4], and Raoul Kopelman[1,*]

Departments of [1]Chemistry, [2]Environmental Sciences, [3]Radiation Oncology, and [4]Radiology, University of Michigan, Ann Arbor, MI 48109

Photonic explorers for biomedical use with biologically localized embedding (PEBBLEs) have been developed to explore, image and monitor, microspectroscopically, live cell processes, with minimal physical and chemical interferences. The design, employing a nanoparticle made of a biologically inert matrix as a platform for loading the active molecular components by encapsulation or covalent attachment, is universal, flexible and allows for facile interchange of its components. The same design concept has been applied to build a multifunctional nanoplatform for cancer diagnosis and therapy. This nanoplatform has a polyacrylamide (PAA) core containing photosensitizers and MRI contrast agents, with a surface-coating consisting of both molecular targeting groups and polyethylene glycol (PEG), for the recognition of the tumor vasculature and for controllable particle residence time, respectively. The PEBBLEs have been successfully applied *in vitro* for intracellular measurements and imaging of important ionic and molecular species. *In vivo* studies in a rat 9L gliosarcoma model showed significant MRI contrast enhancement and photodynamic therapeutic efficacy with targeted nanoplatforms, demonstrating significant promise for their use as a combined therapeutic and diagnostic tool for cancer.

Introduction

Intra-cellular imaging of the biochemistry and biophysics of live cells has been of prime interest for decades. To visualize, track and quantify molecules and events in living cells, the analytical sensors to be introduced to the cells should be small and inert enough not to cause any physical and chemical interference with normal cell behavior. The use of traditional chemical (electrochemical or optical) sensors for intracellular measurement has been hampered by physical invasiveness due to their large size, despite efforts towards miniaturization (1-8). Instead, elegant fluorescent molecular probes that are sensitive to intracellular analytes of interests have been developed and have become a common and powerful approach for these tasks (9). These molecular probes, however, tend to suffer from the following problems, thus limiting the indicator dyes available for reliable intracellular measurements: 1. The indicator molecules have to penetrate the cell's membrane, which often requires derivatized indicator molecules. 2. Once inside the cell, they tend to be sequestered unevenly into various cell compartments. 3. The measurement is often affected by non-specific proteins and other cell components. 4. The best available dyes are sometimes cytotoxic to cells and the mere presence of the dye molecule may chemically perturb the cell. 5. Often the dye is not "ratiometric," i.e., has only a single spectral peak, which then requires technologically more demanding techniques, such as picosecond lifetime resolution or phase sensitive detection. We note that just loading into the cell a separate reference dye, for ratiometric measurements, is not a solution, because of the factors 2. and 3. listed above. We note that most of the above problems have never been an issue for traditional chemical (electrochemical or optical) sensors, where the sensing components are separated from the interfering environment but are exposed to the analytes through their semipermeable membrane.

A new type of sensor called PEBBLE (Photonic Explorer for Biomedical use with Biologically Localized Embedding) has been developed as a non-invasive real-time sensing and imaging nano-tool for cells. It combines the advantages of both traditional sensors and molecular probes by adopting the nanoparticle as a platform to be loaded with proper fluorescent molecular probes. The usual invasiveness of the traditional chemical sensor, fiber sensor or microelectrode, is minimized as the size of the PEBBLE is reduced to that of the tip of the traditional sensor, but still, the fluorescent dyes are protected from the chemical interference due to the cellular environment by the nanoparticle matrix. The size of PEBBLE sensor has been mostly in the range of 20-600 nm and the single PEBBLE nanosensor takes up only 1 PPM to 1 PPB of a mammalian cell volume (10). The small size of the PEBBLEs results in minimal physical perturbation to the cell and allows simultaneous non-invasive delivery of a number of distinct PEBBLEs into a cell. Like fluorescent molecular probes, these sensors can be used with conventional microscopy techniques, enabling high spatial and temporal resolution. The small size of the PEBBLE also shortens the response time of the sensor.

The PEBBLEs are introduced into cells by standard delivery methods that were adapted for this purpose, such as gene gun, pico injection, liposome incorporation or endocytosis. These PEBBLE delivery methods circumvent the issues related to "naked" molecular probes, such as problem 1. listed above. Also, the inert protective matrix of the nanoparticles eliminates interferences such as protein binding and/or membrane/organelle sequestration (*11*); thus getting around problems 2. and 3. listed above. The pores of the PEBBLE matrix are sufficiently large to allow diffusion of the smaller analyte ions but small enough to exclude the diffusion of larger proteins into the core matrix. This exclusion of macromolecules is quite significant, since many "naked molecule" dyes change their fluorescence properties in the presence of proteins; on the other hand, the PEBBLEs containing the very same dyes were not affected by protein presence (*12-14*). The nanoparticle matrix also provides protection for the cellular contents, enabling dyes that are sometimes toxic to cells to be used for intracellular sensing, thus obviating problem 4. listed above. There is thus negligible physical and chemical perturbation to the cell, and, indeed, cell viability after PEBBLE delivery is about 99%, relative to control cells (*15*). Loading of multiple components per each nanoparticle is possible due to the size of the nanoparticle being much larger than that of the molecular dyes. This also allows the PEBBLEs to perform as a ratiomentric sensor by co-encapsulation of the sensing and reference component, obviating problem 5. above. The ratiometric mode of operation assures that the measurements will be unaffected by excitation intensity, absolute concentration and sources of optical loss, which seems to be essential for intracellular measurement where there are many interfering factors (*14,16*).

The same nanoplatform concept has been extended to prepare a multifunctional nanomedicine for cancer diagnosis and therapy, by loading the nanoparticle with contrast imaging agents and therapeutic agents, instead of sensing elements.   The application of nanoparticle technology to cancer detection and therapy has many advantages: 1) It can improve the efficacy of existing imaging and treatment through efficient targeting and high payload per nanoparticle. Compared with the blood vessels in normal tissues, those of growing tumors are frequently more 'leaky' to circulating macromolecules and larger particles, allowing them to access easily the tumor's interior and then to be trapped inside (*17-19*). This phenomenon, known as the enhanced permeability and retention effect (*20*), is not applicable for small molecular drugs being used today for chemotherapy as they do not discriminate tumor tissue from normal tissue. Introduction of molecular targeting groups and/or hydrophilic coating on the nanoparticle surface can further enhance the selective targeting toward specific cell types, with many targeting groups per particle. Each nanoparticle can also carry a large amount of molecular therapeutic or contrast agents. When the nanoparticles packed with these agents are delivered selectively to tumor sites, with enhanced surface properties (targeting and/or hydrophilic coating), a high localized density of the agents can be achieved to provide inherent signal amplification in the signature detection as well as a coherent, critical mass of destructive power for intervention. 2) The nanoparticles can reduce immunogenicity and side effects. The maximum tolerated dose of the drug or contrast agents can be increased as the non-toxic (biocompatible) polymer reduces the exposure to toxicity. Furthermore, there is much less possibility of multiple-drug-resistance, i.e., the cancer cell's ability to pump back out the drug molecule. 3) The multi-functional nanoplatform that carries both imaging agents and drugs can integrate the efforts for detection, treatment and follow-up monitoring of tumor response, leading to decisions about the need for further treatment. It is an innovative and efficient strategy for cancer diagnosis and treatment compared to the current clinical practice where tumor detection and therapy are mostly performed separately.

Below, the design and applications of the platforms are given in detail.

Nanoplatform Design

The nanoparticle can effectively serve as a chemically inert and non-toxic platform for sensing, imaging and therapy with its engineerability to accommodate proper components. The components of desired functionalities are loaded into nanoparticles by standard procedures, i.e., either encapsulation or covalent linkage inside and/or on the surface of the nanoparticles. The type of loaded components may depend on the nanoparticle matrix but the number (single or multiple) and amount of the components are controllable.

PEBBLE Design

Figure 1 shows a schematic diagram of PEBBLE for intracellular sensing. The basic design of nanoplatform for biosensing, PEBBLE, has fluorescent indicator and reference dye entrapped in or covalently linked to a nanoparticle. PEBBLEs to sense H^+, Ca^{2+}, Mg^{2+}, Zn^{2+}, O_2, $Cu^{+/2+}$, Fe^{3+} and OH radicals have been developed based on this design. The analyte permeates the matrix and interacts with the indicator molecule selectively, causing the fluorescence to be enhanced or quenched, while the fluorescence of the reference molecule remains constant, resulting in a ratiometric measurement. The type of nanoparticle matrix is selected on the basis of the permeability and solubility of the analytes as well as

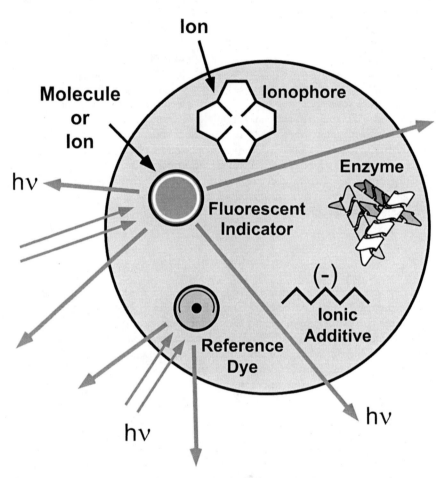

Figure 1. Schematic diagram of a PEBBLE showing many options available within this flexible platform.

the loading efficiency of the sensing components that could be fluorescent dyes or biological molecules like protein. Polyacrylamide (PAA) nanoparticle has served as a good matrix for ion sensors, due to its neutral and hydrophilic nature (*12,14,16,21-23*). It has been used especially for sensors based on biological molecules such as enzymes and proteins due to its mild synthetic condition (*16,23*). Silica or organically modified silicate (ormosil) nanoparticles has been used for dissolved gas sensors (*13, 24*).

For analytes that lack a highly selective fluorescent indicator, more sophisticated designs were adopted. The examples include Na^+, K^+ and Cl^- ion PEBBLE sensors based on the principle of ion selective optodes, where a fluorescent hydrogen selective dye plays the role of a reporter on behalf of non-fluorescent highly selective ionophore binds to the ion of interest. The degree of protonation measured from the fluorescence change of hydrogen selective dye is related to the concentration of the analyte ion by the theory developed for optical-absorption-based ion-enchange sensors (*1,25,26*). Another example is a glucose sensor based on the measurement of local oxygen change created by the enzyme activity of glucose oxidase that is coencapsulated with the fluorescent oxygen sensitive dyes and reference dyes in the same nanoplatform. There is no way in which the traditional "naked" molecular probes can be used to achieve such a synergistic task. The ion selective type PEBBLEs were made with poly (decylmethacrylate) (PDMA) nanoparticles (*27,28*) and the glucose PEBBLEs with PAA nanoparticles (*23*).

A new generation of PEBBLEs called "MOON (MOdulated Optical Nanoprobes)" has been developed for background-free intracellular measurements. With all the advantages of fluorescent nanoprobes, background fluorescence from samples and instrument optics is still a common problem. In such strong autofluorescence backgrounds, small changes in probe fluorescence tend to be washed out and become nearly undetectable (*29*). However, if the probe fluorescence is modulated with respect to background fluorescence, probe fluorescence can be distinguished and separated. This simple procedure increases the signal-to-noise (S/N) or, strictly, signal-to-background ratios by several orders of magnitude, up to 40,000 times (*30,31*). The MOONs are prepared by coating the PEBBLEs with a reflective metal on one side by vapor deposition or molecular beam epitaxy, which obscures the particle's fluorescence. The modulation can be induced both magnetically (*30,31*) (MagMOONs) and by random thermal motion (Brownian MOONs) (*32*) in nature. The MOON acts as a lock-in-amplifier by blinking at a controlled frequency. It expands the breadth of applications of PEBBLEs to include samples with highly scattering and/or fluorescent backgrounds or experiments with several fluorescent probes. The MagMOONs have also demonstrated their applicability for simultaneous measurement of chemical and physical parameters (*33*).

Design of Nanoplatform for Cancer Detection and Therapy

The nanoplatform for cancer detection and therapy is designed to carry multiple components: 1) imaging agents, 2) drugs, 3) targeting ligands and 4) "cloaking" agents if necessary to avoid interference with the immune system (See Figure 2 for a schematic diagram.) (*34*).

Unlike intracellular biosensing applications that use the nanoparticles of various matrices, only the PAA nanoparticle has been selected for the *in vivo* application, for several reasons: 1) The nanoparticles of 10-100 nm are believed to be the best option for *in vivo* application because nanoparticles that are too large would undergo renal elimination and nanoparticles too small would be recognized by phagocytes. (*19,35,36*). The typical size of our PAA nanoparticles is in the range of 30-70nm, a good fit for this size criteria.; 2) The PAA is a hydrogel, rendering a long plasma circulation time even without additional surface coating.

MRI, the current gold standard imaging method for cancer, was chosen as detection mode, thus iron oxide was incorporated as MRI contrast agent. Photo-Dynamic Therapy (PDT) was selected as therapeutic module and Photofrin® (porfimer sodium, Axcan Pharma Inc.), the FDA-approved photosensitizer, was encapsulated in the nanoparticles. For tumor-specific delivery, a targeting ligand specific to tumor vasculature was selected to be linked to the PAA nanoparticle surface: Either an arginine-glycine-aspartate (RGD) containing peptide or the F3 peptide. For additional control of the plasma lifetime, polyethylene glycol (PEG) was linked to the nanoparticle surface (*37-39*).

Production of Nanoplatform

The nanoplatform production relies on advances in nano-scale production. The production methods are based on relatively simple wet chemistry techniques, as opposed to many complicated physical and chemical nanotechnology schemes. Specific synthetic methods for nanoparticles of different matrices are optimized to produce the nanoplatforms of a specific matrix and loaded with components for a particular application. The batch-to-batch reproducibility of the nanoplatforms is good and the optimized synthetic protocol is reliable to produce the required nanoplatforms.

The production of polyacrylamide nanoparticles is based on a water-in-oil (W/O) microemulsion technique (*40*). The silica nanoparticles are prepared by a modified Stober method (*13*). The ormosil nanoplatform is prepared by a two-step method based on sol-gel process (*24,41*). The poly decylmethacrylate nanoplatform is prepared by emulsion polymerization (*27,42*). Encapsulation of drugs or nanocrystals into nanoparticles was made by adding these components to the reaction mixture at the beginning of, or during, the synthesis. Covalent

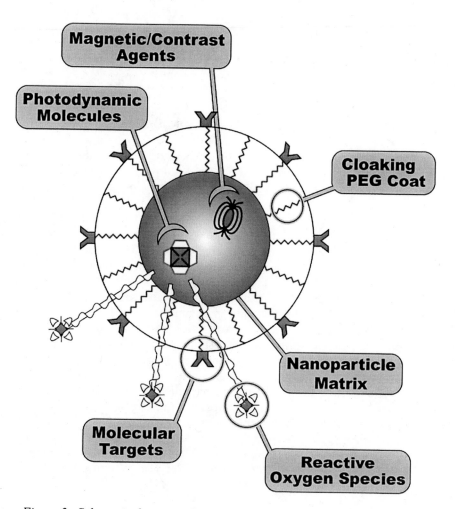

Figure 2. Schematic diagram of a multifunctional nanoplatform for cancer detection and diagnosis

linkage of drug or dye, or surface modification (PEGylation or targeting) was made by an easy coupling reaction between succinimidyl ester derivatives and amine functionalized nanoparticles. Amine-functionalized nanoparticles can be easily made by replacing a part of the original monomers with functionalized monomers (3-(aminopropyl)methacrylamide or (3-aminopropyl)triethoxysilane for PAA and sol-gel silica, respectively) in the reaction mixture. It should be noted that the nanoparticles with either single component or multiple components were produced by the same polymerization technique, with a minor modification to achieve optimal loading of components.

Characterization of Nanoplatform

The physicochemical properties of the nanoplatforms affect their functional efficiency and therefore were characterized to produce quality controlled nanoplatforms for an intended application, i.e., sensing or cancer detection and therapy.

The size and morphology of the dried nanoplatforms were determined by SEM (Scanning Electron Microscopy) while the size and extent of aggregation of nanoplatforms in aqueous solution were determined by light scattering (LS). The typical average particle diameters of the nanoplatforms are 40, 150, 150 and 600 nm for PAA, sol-gel silica, ormosil and poly decylmethacrylate particles, respectively. Modification of the surface by the attachment of targeting ligands or various size of PEG did not result in any significant change in the size (37).

The encapsulated amount of dyes or drugs was usually determined using spectrophotometric analysis, i.e., by measuring the absorbance of the prepared nanoplatform sample solution and comparing it with the calibration curve constructed from the mixture of free dye and blank nanoparticles of known concentrations. The amount of superparamagnetic iron oxide used for imaging agent was obtained from the % of iron present in the sample.

The performance of the nanoplatforms is tested in solution, the results of which are utilized, along with physicochemical characterization results shown above, as feedback to produce optimized nanoplatform for *in vitro* or *in vivo* application. Details about the performance of the nanoplatforms are shown in the application sections below.

Application of PEBBLEs for Intracellular Sensing

The PEBBLEs have been so far developed to sense important chemical and physical entities such as ions (H^+, Ca^{2+}, $Cu^{+/2+}$, Fe^{3+}, K^+, Mg^{2+}, Na^+, Zn^{2+}, Cl^- and NO_2^-), small molecules (oxygen and glucose), reactive oxygen species (OH radical and singlet oxygen), viscosity and electric field.

Performance Tests in Solution

The sensitivity, dynamic range, selectivity and stability are the key factors that determine the sensor performance. The current PEBBLE technologies have relied on fluorescence emission ratios for signal transduction (though fluorescence anisotropy and frequency modulation were also tested) and the performance of the PEBBLEs was established using either an Olympus IMT-II (Lake Success, NY, USA) inverted fluorescence microscope or a Fluoro-Max 2 spectrofluorometer (ISA Jobin Yvon-Spex, Edison, NJ, USA). The spectra and confocal images for the intracellular measurements were acquired with the same inverted fluorescent microscope and a Perkin Elmer UltraView confocal microscope system equipped with an Ar-Kr laser.

A calibration curve is constructed from the response of the sensors toward the known concentration of analytes, i.e., the ratio of the fluorescent peak intensity of sensing dye to that of reference dye for PEBBLEs, in order to determine the sensitivity and dynamic range. The sensitivity, the lowest level of analyte to be detected, is affected by the nature of sensing dye, the PEBBLE matrix and the ability of the detector. The dynamic range is the working range of the assay where signal varies in a monotonic manner with concentration of the analyte in the calibration curve. An ideal relation is a straight line over the whole range of the analytes of interest. The oxygen PEBBLE based on ormosil nanoparticles showed an ideal linear relationship over the whole range of dissolved oxygen concentrations as shown in Figure 3 (24). The sensitivity was less than 0.1 ppm. The high sensitivity and the wide dynamic range with linear response may result from the right combination of nanoparticle matrix (hydrophobic and highly permeable to oxygen), its small size, and the selection of the oxygen sensitive dye.

The selectivity is determined by monitoring the change in the response of the sensor due to the presence of possible interfering entities in the blank solution of the solution of the analyte of interest. A highly selective Mg^{2+} PEBBLE sensor has been developed using a simple laser dye, coumarine 343. The measurement of magnesium ion concentrations in biological environments has suffered from the interference from calcium ions. Coumarine 343 is much more selective (magnesium over calcium) than any commercially available probe (14) but does not pass the cell membrane by itself. The PEBBLE containing this hydrophilic dye and a commercial reference dye ("Texas Red") as hydrophilic components in a hydrophilic matrix (PAA) can, however, be used as a magnesium ion sensor and be delivered into the cell for intracellular measurement as well. The calibration curve of the Mg PEBBLE shows very little deviation, from the original calibration curve, in the presence of additional 1 mM calcium, 20 mM sodium and 120 mM potassium, demonstrating that this Mg^{2+} PEBBLE should be a reliable indicator of Mg^{2+} concentrations.

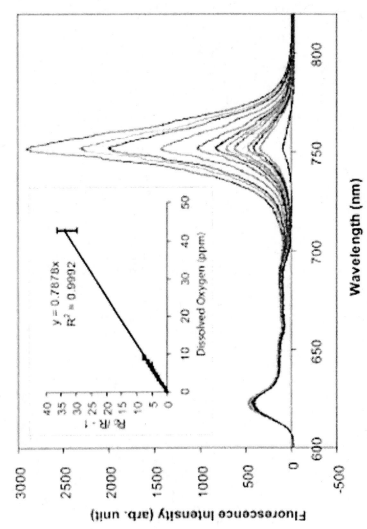

Figure 3. Response of oxygen PEBBLEs at various dissolved oxygen concentration : Fluorescent emission spectra and calibration curve(Inset). The DO concentration for each spectrum is 0.0 (nitrogen saturated), 0.5, 0.9, 1.8, 2.7, 3.6, 4.5, 5.4, 6.3, 7.2, 7.9, 9.0 (air-saturated), and 43 ppm (oxygen-saturated) from top to bottom lines. (Adapted from reference 24. Copyright 2004)

The stability and reversibility will determine the duration and frequency of the use of sensor. For any fluorescent sensor, the photostability is a very important factor, as the regeneration of unexcited dye for repeated measurements can be blocked by photobleaching (irreversible photochemical destruction of excited dye). For example, the copper ion sensitive PEBBLE based on PAA nanoparticles containing DsRed (protein) as sensing dye and AlexaFluor 488 dextran as reference shows a 17% variation in the fluorescence ratio for times longer than 70 min of constant excitation (*16*). Considering that the typical exposure time for cellular imaging is 100 ms, this decay would only occur after a significant amount of independent measurements, >10,000. Leaching of the encapsulated components out of the nanoplatform can be another factor affecting the stability of PEBBLEs during the measurement. Leaching typically depends on the size of the encapsulated components and the hydrophilicity. The hydrophobic component is less prone to leaching in aqueous solution. The leaching can be reduced using a dye linked to a polymer of high molecular weight, such as dextran. We note that the duration of an intracellular experiment is on the order of hours, and thus slow leaching is usually not a big concern for most of the PEBBLEs.

Good reversibility guarantees the repeated reliable use of sensors and also determines the response time. The typical response time is about 1 ms or shorter.

Examples of Intracellular Measurements

PEBBLE sensors were designed for intracellular use and the ability of PEBBLEs to measure intracellular analytes has been demonstrated in mouse oocytes and rat alveolar macrophages, as well as in neuroblastoma, myometrial and glioma cells (*13-15,21,24,27,43*). Two examples are given below.

PDMA-based K+ PEBBLEs

The K^+ PEBBLE is a sensor based on PDMA nanoparticles with the following components: the chromoionophore is ETH 5350, the ionophore is BME-44, and the lipophilic additive is potassium tetrakis-[3,5-bis(trifluoro-methyl)phenyl borate (KTFPB) (*27*). It operates on the principle of ion selective optodes. The K^+ PEBBLEs were gene-gun delivered into rat C6-glioma cells. After 20 seconds, and after 60 seconds, 50 µl of 0.4 mg/ml kainic acid was injected into the cells placed on an inverted fluorescent microscope. Kainic acid is known to stimulate cells by causing the opening of ion channels. Figure 4 shows the PEBBLE sensors inside the cells responding to the kainic acid addition. One can see that log (a_{K+}/a_{H+}) increases, indicating either an increase in K^+ concentration or a decrease in H^+ concentration (increase in pH). The amount

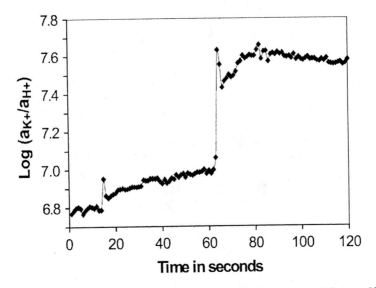

Figure 4. K+ PEBBLE response in C6-glioma cells during the addition of kanic acid at 20s and 60s. (Adapted from reference27. Copyright 2001 ACS)

of kainic acid added is not known to affect the pH of cells in culture and kainic acid by itself has no effect on the sensors. Thus the change is likely due to increasing intracellular concentration of K^+, which is the expected trend. The membrane of C6 glioma cells can initiate an inward rectifying K^+ current, induced by specific K^+ channels, a documented role in the control of extracellular potassium (*44*). Thus, when stimulated with a channel opening agonist, the K^+ concentration within the glioma cells is indeed expected to increase.

PAA-based Mg^{2+} PEBBLEs

The Mg^{2+} PEBBLE is a simple ratiometric PEBBLE based on PAA nanoparticles with encapsulated Coumarine 343 and Texas Red as indicator and reference dyes, respectively. Recently Mg^{2+} PEBBLEs have been applied to an important biological study on the role of the Mg^{2+} inside phagosomal vacuoles in the control of pathogens (Salmonella species) by macrophage (*45*). The Mg^{2+} ion has been reported as one of the potential regulation factors of PhoPQ protein (*46*). PhoPQ is a two-component regulatory system that is essential for intracellular survival of Salmonella species that cause several human diseases. Importantly, PhoPQ is also required in the first hours of invasion to prevent fusion of the Salmonella-containing vacuoles (SCV) with host cell lysosomes (*47*).

To estimate the contribution of the Mg^{2+} concentration as a modulator of PhoPQ induction within the SCV, the concentration of Mg^{2+} in the medium was varied over greater than one order of magnitude (from 0.5 to 10 mM). The concentration of Mg^{2+} within SCV and the PhoPQ induction was monitored by the Mg PEBBLEs and by the fluorescence of the GFP encoded vector introduced into the Salmonella Enterica Typhimurium, respectively, within live macrophages. Manipulating the concentration of Mg^{2+} within the SCV had no effect on the early induction of PhoPQ. Moreover, direct measurement of the concentration of Mg^{2+} within the SCV, using the Mg PEBBLE sensors, showed that, during this initial period of PhoP activation, the concentration of the divalent cation is rapidly regulated and stabilizes around 1 mM. Because the Mg^{2+} concentration had equilibrated after 30 min, a time when expression of PhoP::GFP by bacteria infecting macrophages was unambiguous, it was concluded that a decreased concentration of the divalent cation (below 1 mM) is not the primary cause of PhoP induced transcription.

Application of Nanoplatform for *In Vivo* Cancer Detection and Therapy

The nanoplatforms for PDT were made by embedding several well-known photosensitizers in nanoparticles of various matrixes used for our PEBBLE sensors (*47-49*). The PDT efficacy of the nanoplatforms was demonstrated either by chemical detection of singlet oxygen with molecular probes sensitive to singlet oxygen or by PEBBLEs containing such molecular probes or by *in vitro* cell killing tests (*49-51*). All tested nanoplatforms did show the PDT efficacy even with a large silica nanoparticle (200nm). However, further development for *in vivo* application has been focused on PAA nanoparticles due to their suitable size and hydrophilicity, as described in the Design section above. The PAA nanoparticles containing iron oxide, a superparamagnetic MRI contrast agent (*52,53*), with and without surface-attached polyethylene glycol (PEG), were prepared for cancer detection by MRI (*37-39*). Measurement of the MR relaxivity of the nanoplatforms in solution revealed an R_2 relaxivity of 620-1140 $s^{-1}mM^{-1}$, which is approximately five-fold greater than that measured with other superparamagnetic iron oxides (*54*). The *in vivo* pharmacokinetic studies were performed by MRI with the same nanoplatforms administered into rats bearing orthotopic 9L gliomas. Addition of polyethylene glycol (PEG) (0, 0.6, 2, & 10 kD) to the surface of the nanoparticles resulted in a prolonged plasma half-life and affected tumor uptake and retention, as quantitated by changes in tissue contrast using MRI.

The PAA nanoparticles containing Photofrin® as photosensitizer were prepared and i.v. injected into rats bearing intracerebral 9L tumors for the *in vivo* photodynamic therapy. Diffusion MRI was performed at various time points to monitor changes in tumor diffusion, tumor growth, and tumor load as in Figure 5. The time series of diffusion-weighted MR images show that the untreated (Figure 5A), as well as laser only treated 9L gliomas (Figure 5B), continued to grow over the lifespan of the animals. We note that the slight dark region with a central bright spot in post-treatment images of laser-treated tumor cells corresponds to a small area of damage due to insertion of the laser probe. In contrast, gliomas treated by the administration of Photofrin®-containing nanoparticles (followed by laser irradiation) produced massive regional necrosis, demonstrated by large "bright" regions in the images and a decrease in the tumor volume (Figure 5C). Re-growth occurred at 12 days post-treatment. These results demonstrated that PDT by nanoparticles containing Photofrin®, followed by light activation, is an effective and viable therapeutic approach for treating brain tumors.

Significantly improved results were obtained with targeted PDT nanoplatforms. A targeted multi-functional nanoplatform that has photosensitizers (Photofrin®), MRI contrast agents (iron oxide) and targeting moieties (F3-peptide) in the same PAA nanoparticle was prepared and applied for the synergistic operation of both PDT and diagnostic imaging in a rat 9L gliosarcoma model. Animals treated with F3-targeted nanopaticles were found to have the largest increase in MR monitored proton diffusion values and were also found to have the longest survival time among the other treatment groups, including treatment with non-targeted nanoparicles containing Photofrin® or free Photofrin®. Importantly, complete eradication of brain cancers was shown in 3 out of 5 animals treated with a single i.v. injection of F3-targeted Photofrin® nanoplatform dose, coupled with a single 7 min PDT treatment, at 60 days post-treatment.

Summary

Photonic explorers based on nanoparticle platforms have been developed for intracellular sensing and for cancer detection and photodynamic therapy. The design, employing nanoparticles as a multifunctional platform loaded with active components by encapsulation or covalent attachment, is universal and flexible. For intracellular sensing, the PEBBLE sensors have been successfully applied to measure or image important intracellular chemical species. For cancer detection and therapy, Photofrin® and iron oxide containing PAA nanoplatforms, targeted to the tumor vasculature with surface-linked F3 peptide, produced a significantly improved treatment outcome and image contrast enhancement in rat 9L glioma models.

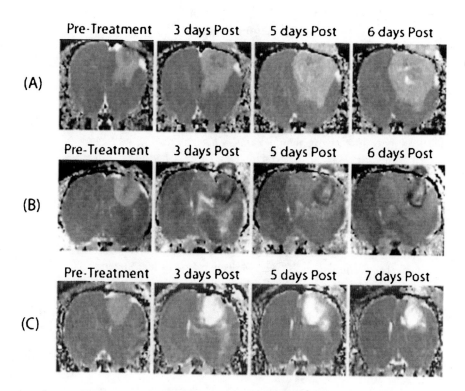

Figure 5. The time series of diffusion-weighted MR images: (A) Untreated 9L brain tumors.(B) Treated with light alone. (C) Treated with Photofrin-containing nanoparticles. Each series of images is from a representative animal from each of the three groups of animals.

Acknowledgement

This work was supported by NIH Grants 8RO1-EB00250-0909 and ES08846 as well as NCI Contract N01-CO-37123.

References

1. Bühlmann, P.; Pretsch, E.; Bakker, E. *Chem. Rev.* **1998**, *98*, 1593-1687.
2. Shi, Z. Y.; Smith, S.; Birnbaum D.; Kopelman, R. *Science* **1992**, *258*, 778-781.
3. Tan, W. H.; Kopelman, R.; Barker, S. L. R.; Miller, M. T. *Anal. Chem.* **1999**, *71*, 606A–612A.
4. Rosenzweig, Z.; Kopelman, R. *Anal. Chem.* **1996**, *68*, 1408-1413.
5. Shortreed M.; Bakker, E.; Kopelman, R. *Anal. Chem.* **1996**, *68*, 2656-2662.
6. Shortreed, M. R.; Dourado, S.; Kopelman, R. *Sens. Actuators B-Chem.* **1997**, *38*, 8-12.
7. Barker, S. L. R.; Thorsrud, B. A.; Kopelman, R. *Anal. Chem.* **1998**, *70*, 100-104.
8. Cullum, B. M.; Vo-Dinh, T. *Trends Biotechnol.* **2000**, *18*, 388-393.
9. Haugland R. P. *The Handbook: A Guide to Fluorescent Probes and Labeling Technologies 10th ed.*; Molecular Probes Inc: Eugene, 2005.
10. Buck, S. M.; Xu H.; Brasuel M.; Philbert, M. A.; Kopelman, R. *Talanta* **2004**, *63*, 41-59.
11. Graber, M. L.; Dilillo, D. C.; Friedman, B. L.; Pastorizamunoz, E. *Anal. Biochem.* **1986**, *156*, 202-212.
12. Sumner, J. P.; Monson, E.; Kopelman, R. *Analyst* **2002**, *127*, 11-16.
13. Xu, H.; Aylott, J. W.; Kopelman, R.; Miller, T. J.; Philbert, M. A. *Anal. Chem.* **2001**, *73*, 4124-4133.
14. Park, E. J.; Brasuel, M.; Behrend, C.; Philbert, M. A.; Kopelman, R. *Anal. Chem.* **2003**, *75*, 3784-3791.
15. Clark, H. A.; Hoyer, M.; Parus, S.; Philbert, M. A.; Kopelman, R. *Mikrochim. Acta* **1999**, *131*, 121-128.
16. Sumner, J. P.; Westerberg, N.; Stoddard, A. K.; Fierke, C. A.; Kopelman, R. *Sensors and Actuators B* **2005**, *113*, 760-767.
17. Maeda, H.; Matsumura, Y. *Crit. Rev. Ther. Drug. Carrier Sys.* **1989**, *6*, 193-210.
18. Muggia, F. M. *Clin, Cancer Res.* **1999**, *5*, 7-8.
19. Moghimi, S. M.; Hunter, A. C.; Murray, J. C. *Pharmacol. Rev.* **2001**, *52*, 283-318.
20. Maeda H. *Advan. Enzyme Regul.* **2001**, *41*, 189-207.
21. Clark, H. A.; Kopelman, R.; Tjalkens, R.; Philbert, M. A. *Anal Chem* **1999**, *71*, 4837-4843.

22. Sumner, J.; Kopelman, R. *Analyst* **2005**, *130*, 528-533.
23. Xu, H.; Aylott, J. W.; Kopelman, R. *Analyst* **2002**, *127*, 1471-1477.
24. Koo, Y.-E. L.; Cao, Y.; Kopelman, R.; Koo, S. M.; Brasuel, M.; Philbert, M. A. *Anal. Chem.* **2004**, *76*, 2498-2505.
25. Bakker, E.; Simon, W. *Anal Chem* **1992**, *64*, 1805-1812.
26. Morf, W. E.; Seiler, K.; Lehmann, B.; Behringer, C.; Hartman, K.; Simon, W. *Pure Appl. Chem.* **1989**, *61*, 1613-1618.
27. Brasuel, M.; Kopelman, R.; Miller, T. J.; Tjalkens R.; Philbert, M. A. *Anal Chem* **2001**, *73*, 2221-2228.
28. Brasuel, M. G.; Miller, T. J.; Kopelman, R.; Philbert, M. A. *Analyst* **2003**, *128*, 1262-1267.
29. Xu, H.; Buck, S. M.; Kopelman, R.; Philbert, M. A.; Brasuel, M.; Monson, E.; Behrend, C.; Ross, B.; Rehemtulla, A.; Koo, Y.-E. L. In *Topics in Fluorescence Spectroscopy*; Geddes, C.D.; Lakowicz, J. R., Eds.; Kluwer Academic/Plenum Press: New York, 2005; Vol. 10, pp 69-126.
30. Anker, J. N.; Behrend, C.; Kopelman, R. *J. Appl. Phys.* **2003**, *93*, 6698-6700.
31. Anker, J. N.; Kopelman, R. *Appl. Phys. Lett.* **2003**, *82*, 1102-1104.
32. Behrend, C. J.; Anker, J. N.; Kopelman, R. *Appl. Phys. Lett.* **2004**, *84*, 154-156.
33. McNaughton, B. H.; Kehbein, K. A.; Anker, J. N.; Kopelman, R. *J. of Phy. Chem. B*, **2006**, *In Press*.
34. Harrell, J. A.; Kopelman, R. *Biophotonics International* **2000**, *7*, 22-24.
35. Moghimi, S. M.; Bonnemain, B. *Adv. Drug Delivery Rev.* **1999**, *37*, 295-312.
36. Wiessleder, R.; Heautot, J. F.; Schaffer, B. K.; Nossiff, N.; Papisov, M. I.; Bogdanov, A. A.; Brady, T. J. *Radiology* **1994**, *191*, 225-230.
37. Moffat, B. A.; Reddy, G. R.; McConville, P.; Hall, D. E.; Chenevert, T. L.; Kopelman, R.; Philbert, M.; Weissleder, R.; Rehemtulla, A.; Ross, B. D. *Mol. Imaging* **2003**, *2*, 324-332.
38. Ross, B.; Rehemtulla, A.; Koo, Y.-E. L.; Reddy, R.; Kim, G.; Behrend, C.; Buck, S.; Schneider II, R. J.; Philbert, M. A.; Weissleder, R.; Kopelman, R. In *Nanobiophotonics and Biomedical Applications*, BiOS 2004: International Biomedical Optics Symposium, San Jose, CA, USA, 26 January, 2004; Cartwright, A. N., Ed. SPIE (International Society of Photonic Engineering): San Jose, CA, USA, 2004; pp 76-83.
39. Kopelman, R; Philbert, M.; Koo, Y.-E. L.; Moffat, B. A.; Reddy, G. R.; McConville, P.; Hall, D. E.; *J. of Magnetism and Magnetic Materials* **2005**, *293*, 404-410.
40. Daubresse, C.; Grandfils, C.; Jerome, R.; Teyssie, P. *J. Colloid Interf. Sci.* **1994**, *168*, 222-229.
41. Hah, H. J.; Kim, J. S.; Jeon, B. J.; Koo, S. M.; Lee, Y. E. *ChemComm*, **2003**, *14*, 1712-1713.

218

42. Cao, Y.; Koo, Y.-E. L.; Kopelman, R. *Analyst*, **2004**, *129*, 745-750.
43. Clark, H. A.; Barker, S. L. R.; Brasuel, M.; Miller, M. T.; Monson, E.; Parus, S.; Shi, Z. Y.; Song, A.; Thorsrud, B.; Kopelman, R.; Ade, A.; Meixner, W.; Athey, B.; Hoyer, M.; Hill, D.; Lightle, R.; Philbert, M.A. *Sens. Actuators B-Chem.* **1998**, *51*, 12-16.
44. Emmi, A.; Wenzel, H. J.; Schwartzkroin, P. A.; Taglialatela, M.; Castaldo, P.; Bianchi, L.; Nerbonne, J.; Robertson, G. A.; Janigro D. *J. Neurosci.* **2000**, *20*, 3915-3925.
45. Orozco, N. M.; Touret, N.; Zaharik, M.; Park, E.; Kopelman, R.; Miller, S. ; Finlay, B.; Gros, P.; Grinstein, S. *Mol. Biol. Cell* **2006**, *17*, 498-510.
46. Garcia Vescovi, E.; Soncini, F. C.; Groisman, E. A. *Cell* **1996**, *84*, 165-174.
47. Garvis, S. G.; Beuzon, C. R.; Holden, D. W. *Cell Microbiol.* **2001**, *3*, 731-744.
48. Yan, F.; Kopelman, R. *Photochem. Photobio.* **2003**, *78*, 587-591.
49. Moreno, M. J.; Monson, E.; Reddy, R. G.; Rehemtulla, A.; Ross, B. D.; Philbert, M.; Schneider, R. J.; Kopelman, R. *Sens. Actuator B-Chem.* **2003**, 90, 82-89.
50. Tang W.; Xu, H.; Kopelman, R.; Philbert, M. A. *Photochem. Photobio.* **2005**, *81*, 242-249.
51. Cao, Y.; Koo, Y.-E. L.; Koo, S. M.; Kopelman, R. *Photochem. Photobio.* **2005**, *81*, 1489-1498.
52. Weissleder, R.; Elizondo, G.; Wittenberg, J.; Rabito, C. A.; Begele, H. H.; Josephson, L.; *Radiology* **1990**, *175*, 489-493.
53. Weissleder, R.; Elizondo, G.; Wittenberg, J.; Lee, A. S.; Josephson, L.; Brady, T. J. *Radiology* **1990**, *175*, 494-498.
54. Wang, Y. X. J.; Hussain, S. M.; Krestin, G. P. *Eur. Radiol.* **2001**, *11*, 2319-2331.

Chapter 15

Intrinsic Optical Signals in Neural Tissues: Measurements, Mechanisms, and Applications

Christopher Fang-Yen and Michael S. Feld

G. R. Harrison Spectroscopy Laboratory, Massachusetts Institute of Technology, Cambridge, MA 02139

Signaling phenomena in nerve tissues are accompanied by a number of intrinsic optical signals, including changes in absorption, scattering, birefringence, refractive index, and mechanical motions. We review these signals, various techniques for their measurement, possible mechanisms for their origin, and discuss their present and potential applications in functional neural imaging.

Introduction

In 1949 Hille and Keynes (*1*) demonstrated that electrical stimulation modulates the light-scattering properties of an excised crab leg nerve. The nerve's opacity exhibited a transient change of about 1 part in 10^5 during the action potential. These initial studies were followed by reports of changes in scattering, birefringence (*2*), and optical activity (*3*) during the action potential in numerous invertebrate and vertebrate nerve preparations, including lobster leg nerve (*4*), squid fin nerve (*4*), squid giant axon (*2, 5, 6*), rabbit vagus nerve (*7*), pike olfactory nerve (*7*), and garfish olfactory nerve (*8*). The measurement of such intrinsic changes in almost every nerve preparation tested suggested that the optical effects could be used to monitor neuronal activity without invasive electrophysiological probes or contrast agents (9). More recently, the repertoire of optical phenomena has expanded to include scattering and refractive index changes in dissociated neuron cultures (10, 11), fast and slow scattering changes in brain slices (12), and scattering in intact brain (13). Activity-dependent optical changes are often called intrinsic optical signals, to distinguish them from *extrinsic* optical effects such as fluorescence changes of externally applied dyes.

In parallel with optical recordings, researchers have also observed small, rapid mechanical motions in nerves during the action potential, initially in crab nerve bundles (14-16). Motions are typically on the order of 1 nanometer and correspond to a "swelling", i.e. a transient increase in diameter. The mechanical and optical phenomena are likely to be related, since scattering properties depend sensitively on scatterer geometry.

Optical studies of the intact brain surface have revealed a different type of intrinsic optical signal: near-infrared absorption changes caused by changes in local blood volume and/or oxygenation. Activity-dependent modulation of brain reflectance by ~0.5% has been observed (17) and attributed primarily due to the neurovascular response. Similar hemodynamic changes in relatively deep brain tissue have been measured transcranially using diffuse optical tomography (18).

Interest in intrinsic optical signals has generally concerned their potential applications as functional neural imaging techniques. Of the intrinsic optical signals reviewed here, only the imaging of slow changes in brain surface reflectance has so far become firmly established as such a technique, having been used very successfully in studies of functional architecture of the mammalian cortex (19). In fact, the term "intrinsic optical signals" is often used to refer specifically to near-infrared imaging of hemodynamic changes on the brain surface. What then are the prospects for application of the "other" intrinsic signals: the birefringence, scattering, and refractive index changes, and mechanical motions?

In this paper we review the various types of intrinsic optical signals and the technologies for measuring them, with emphasis on non-hemodynamic signals. We discuss interpretations and proposed mechanisms for the intrinsic signals and evaluate the status of efforts to use them as a basis for functional neuroimaging techniques.

Intrinsic signals in nerves and cultured neurons

Initial studies

Early experiments for measuring scattering and birefringence changes (1, 2, 5, 6) used variations on a simple experimental setup. A nerve was dissected from a specimen and placed in a nerve chamber containing a physiological bath solution and stimulation and recording electrodes. Extracellular or intracellular electrodes were used for nerve bundle and giant axon experiments, respectively. Light from an incandescent lamp was focused onto the nerve and scattered light imaged onto a photodiode or photomultiplier tube. Most experiments considered the angular dependence of scattered light, typically via measurements in the near-forward (0 degree) and near-right angle (90 degree) directions, via a movable detector assembly. Photodetector outputs were amplified and averaged by a signal averager. Several thousand traces were averaged to obtain a reasonable signal to noise ratio.

Typical data from nerve bundles (e.g. Figure 1 from Tasaki et al (4)) generally showed a fast $\sim 10^{-5}$ relative increase in scattering at 45-90 degrees during the action potential, followed by an decrease in scattering over longer time scales (4). Results from single axons of squid were roughly an order of magnitude smaller (2). Considerable variability (over a factor ~ 10) in responses between specimens of the same type was found.

For birefringence measurements, the sample was placed between crossed polarizers and scattering measured in the forward direction. Largest effects were observed with the axis of the nerve bundle oriented 45 degrees from the direction of either polarizer (2). Transmitted intensity changes varied from $\sim 10^{-6}$ to $\sim 10^{-4}$, with a sign corresponding to an increase in birefringence (4). Nerves exhibit an activity-independent intrinsic birefringence due to longitudinally aligned microtubules in axoplasm and may also have significant form birefringence.

Several explanations were suggested to account for the activity-dependent modulation of scattering and birefringence. Scattering changes may result from transient increases in diameter or changes in ionic composition near the axon membrane (8). Birefringence changes could be caused by polarization and alignment of axon membrane protein molecules due to changes in membrane potential (1, 2, 10). Although the membrane is only a few nanometers thick, order-of-magnitude estimates for polar molecules suggest that such an electro-optic mechanism would be sufficiently large to account for the birefringence observations. Another explanation for birefringence changes involves the mechanical compression of the membrane due to membrane charge (electrostriction effect) (2). These mechanisms will be revisited in the discussion of other intrinsic optical experiments.

Scattering in cultured neurons

A study by Stepnoski et al (10) of scattering changes in cultured neurons from Aplysia californica (a marine invertebrate) represents the most detailed attempt so far to interpret intrinsic optical changes in terms of a specific biophysical model.

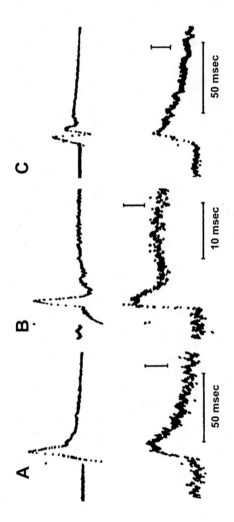

Figure 1. Top: electrophysiological recordings of action potentials of (a) spider crab nerve, (B) squid fin nerve, (C) lobster nerve. Bottom: associated changes in light scattering at 90 degrees. Vertical bars: 10^{-5}, 2×10^{-5}, and 3×10^{-5} relative increases in light intensity for A, B, and C, respectively. From Tasaki et al (4).
Copyright (c) 1968 National Academy of Sciences.

Cultured neurons containing an extensive array of processes (nerve fibers) were placed in a microscope illuminated with a darkfield configuration such that only light which scatters by 3 degrees or more is detected. Scattered light intensity from a field of view containing many processes was found to be modulated nearly linearly with changes in membrane potential (Figure 2), with a proportionality constant on the order of 10^{-5} relative scattering change per mV. Changes were large enough to be detected without signal averaging. Individual action potentials could be resolved in a spike train induced by injecting a constant level current pulse into the cell body.

To further investigate the intrinsic changes, the authors performed angle-resolved light scattering measurements. A 633 nm laser was focused onto single axons and the scattering intensity and voltage-dependence of scattering were measured for as a function of azimuthal scattering angle. The baseline scattering pattern fell off from 0 degrees in a series of peaks and troughs, as expected from scattering from a cylindrically symmetric dielectric. The change in scattering during a voltage spike was observed to alternate in sign as a function of angle. The magnitude of small-angle scattering was shown to be in agreement with the darkfield scattering results.

To explain the angular dependence results quantitatively, the authors developed a model based on dipoles in the axon membrane with voltage-dependent reorientation in the transmembrane field. The dipole polarization leads to a difference in refractive index between light polarized normal and tangential to the membrane. A linear change in refractive index with voltage was assumed; this is the linear electro-optic or Pockels effect. This model based on dipole reorientation produced a reasonably good agreement with experimental data. By contrast, the data excluded models based on (i) an isotropic change in membrane index, (ii) modulation of radial refractive index only, and (iii) voltage-dependent modulation of the bulk refractive index in the axon. Surprisingly, the authors do not discuss the possibility that scattering changes may be due at least in part to transient diameter changes in the axons, observations of which are discussed later in this review.

Voltage-induced refractive index shifts will lead to shifts in optical path length (OPL) of transmitted light; in a related study, Laporta and Kleinfeld (*11*) use a Mach-Zehnder laser interferometer resembling a differential interference contrast (DIC) microscope to probe OPL shifts through a few axons of a lobster leg nerve during the action potential. With an averaging of 100 trials, a peak OPL change of about 0.3 angstroms was observed. Recently, efforts to measure similar signals in single cultured neurons using interferometric quantitative phase microscopy techniques have been reported (*20*).

Nerve changes measured with Optical Coherence Tomography

Optical Coherence Tomography (OCT) (*21*) is a relatively new imaging technique in which a probe beam of broadband light is scanned across a sample

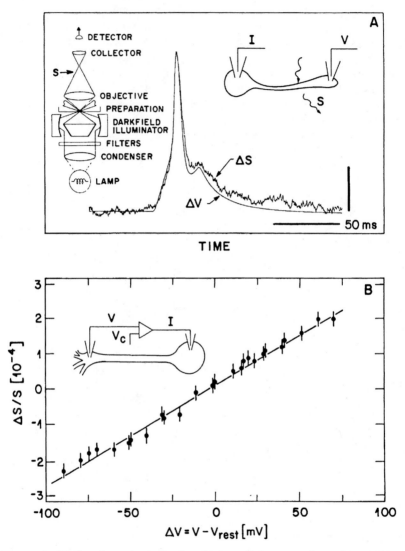

Figure 2. (A) Simultaneous optical and intracellular recordings from Aplysia neuron. ΔS: scattering change; ΔV: voltage change. Inset: schematic of darkfield light scattering setup. (B) Demonstration of linear relationship between scattered change and potential change. From Stepnoski et al (10).Copyright (c) 1991 National Academy of Sciences.

and the backscattered light is mixed with a reference field to create depth-resolved, 2D or 3D scattering profiles in biological tissues. OCT can be thought of as an optical analogue of ultrasound imaging, with 2-5 micron spatial resolution and 1-2 mm penetration depths. Although it is primarily known for applications in retinal, skin, and intravascular imaging (22), recently OCT has begun to be applied to measurements of intrinsic signals in nerve tissues (23). The depth gating property of OCT allows imaging scattering changes in three dimensions and helps to reject the multiply-scattered light contributions in reflectance imaging.

Maheswari et al (24) used OCT to measure slow intrinsic scattering signals in exposed cat visual cortex during horizontal and vertical grating stimuli. In order to take advantage of OCT's coherence gating properties, the authors focused on depth-dependent rather than lateral-dependent changes in reflectance, and found preliminary evidence that the modular organization of visual cortex also extends in the depth direction. Lazebnik et al (25) measured scattering changes in dissociated *Aplysia* nerve fibers in vitro and demonstrated localized reversible scattering increases when electrical stimulation was applied.

Two recent OCT studies measured slow scattering changes in the dissociated vertebrate retina, in which intrinsic signals had been measured by other techniques (26). In one study, functional changes from light-activated frog retina in vitro (27) were recorded. In another, increases in scattering at particular depths were reported in rabbit retina (28).

Mechanical displacements: optical and non-optical approaches

Observations of nerve mechanical motions during the action potential have been performed with various contact and combined contact/optical techniques, including piezoelectric sensors (29), volume change measurements (30), measurements of scattering of attached beads (15) and optical lever recordings (31).

Optical interferometry is capable of measuring changes many orders of magnitude smaller than the wavelength of light, and is well-suited for measuring nanometer-scale changes in nerve tissues. Applications of optical interferometry to measure motions in excitable tissues were described in Sandlin et al (32) and Hill et al (33), who measured a 1.8 nm, ~1 ms contraction displacement followed by a slow swelling in a crayfish giant axon coated with gold particles to increase reflectivity. Recently two optical interferometric techniques (34, 35) been developed for measuring action potential-induced displacements in nerves without scattering contrast agents to the tissues. Two keys to these approaches have been the use of low coherence (broadband) light to achieve depth-selective measurements, and phase-referencing (36), in which the phase of light reflected from a sample is measured relative to a fixed reference reflection, to reduce interferometer noise.

In Akkin *et al*, a polarization-sensitive low coherence fiber interferometer was used to measure ~0.5 nm, 1 ms displacements in crayfish leg nerves. Calcite prisms are used to compensate for optical path delay between the nerve reflection and a reference cover glass. The authors suggested that the technology may allow noninvasive detection of various neuropathies.

Fang-Yen *et al* used a heterodyne dual-beam low coherence interferometer to measure displacements in a lobster leg nerve (Figure 3). A free space Michelson interferometer containing acousto-optic modulators was used to compensate for the path length difference between sample and reference surfaces. An upward displacement of the upper nerve surface of ~5 nm was measured without signal averaging. The threshold and saturation stimulus current amplitudes for the optical and electrical signals were shown to be nearly identical, strongly suggesting that displacements were directly related to the action potentials.

Initial explanations for the surface displacements were concerned with hydration due to imbalances of sodium and potassium ions involved in the generation of the action potential (*37*). However, estimates of the scale of such an effect give swelling magnitudes at least 2 orders of magnitude smaller than observed (*38*). It was suggested that intrinsic motions may arise from a gel phase transition in axoplasm due to a sodium/calcium ion binding exchange (*38*), but direct evidence for this model has not yet been found.

More recently, it has been suggested that voltage-dependent motions may arise from a fundamental electromechanical coupling between membrane potential and membrane mechanical properties. A study combining patch clamp electrophysiology and atomic force microscopy (AFM) measured voltage-dependent modulations in membrane potential (*39*). A theoretical study (*40*) considered the effect of two electrostatic properties of dielectrics, electrostriction and piezoelectricity, on the nerve membrane to predict changes in axon dimensions during the action potential.

Intrinsic signals in brain slices

The recording of optical changes in brain tissue slices was pioneered by Lipton (*12*), who used a photodiode to measure reflectance at 176 degrees from the surface of hippocampal slices submerged in a bath medium. Electrical stimulation led to decrease in scattering, which was attributed to cell swelling. The association between transparency and cell swelling had been established on the basis of experiments with cell suspensions (*41*), and protein solutions (*42*).

Current experiments use near-infrared illumination and a charged-coupled device (CCD) camera to image the slice in reflection mode, transmission mode, or both. Scattering changes may be very large, more than 20% over several minutes, and may be readily observed using relatively simple sources and detectors.

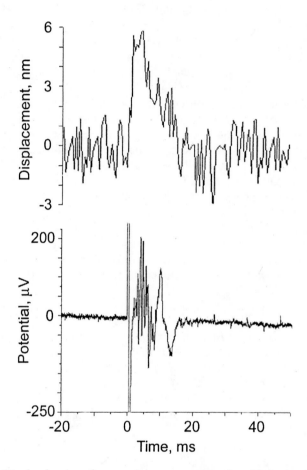

Figure 3. Single-shot interferometric measurement of rapid displacements in a lobster leg nerve during the action potential. Top, displacement signal; bottom: simultaneous recording from extracellular electrodes showing compound action potential. Spike at t=0 is stimulus artifact From Fang-Yen et al (35). Copyright (c) 2004 Optical Society of America.

A great deal of the literature on optical changes in brain slices concerns the phenomenon of spreading depression (SD) (*43*), a self-propagating wave of depolarization associated with depression of neuronal activity for several minutes, dramatic changes in extracellular and intracellular ionic concentrations, and cellular swelling. SD is implicated in a number of pathological conditions, although its mechanisms and significance remain unclear. Optically, SD generally creates an increase in hippocampal slice reflectance despite the cell swelling effect, although this depends on the conditions of the experiments and the method of SD induction (*44*). One study (*45*) attempts to clarify the picture by suggesting that at least two different mechanisms underlie intrinsic signals in hippocampal slices. First, light scattering decreases due to cell swelling and increase of interstitial volume. Second, during SD and strong hypotonic conditions with severely decreased interstitial volume, light scattering increases due to swelling of cellular organelles, especially mitochondria.

MacVicar and Hochman (*46*) examined intrinsic optical changes in hippocampal slices induced by synaptic transmission. Based on results from perturbing extracellular ionic concentrations and administering various ion transport inhibitors, the authors suggested that intrinsic signals may arise from glial swelling due to potassium released by neurons during excitation.

Although most brain slice studies have measured slow (~1 sec) optical changes, optical changes on time scales comparable to the electrical signals have been reported in some preparations.

Large and rapid changes in light scattering was seen in nerve terminals of the mouse neurohypophysis, the posterior lobe of the pituitary gland (*47*). By detecting transmitted light of a 600-850 nm source through a microscope, large ~0.3% intrinsic peaks were observed, without averaging, and with time course similar to the action potential. The origin of the transient decreases in light scattering (increases in transmission) was unclear. Suggestions included loss of scattering from secretory vesicles during exocytosis, changes in contents of secretory vesicles, or structural changes associated with rapid changes in calcium concentrations.

Intrinsic signals in the intact brain

Brain surface imaging

Functional imaging of the exposed and intact brain surface, the most well-known type of intrinsic signal imaging, was developed by Grinvald and colleagues (*19, 48, 49*). The brain surface is illuminated with light at one or multiple wavelengths, and reflectance images are captured with a CCD camera during stimulus presentation (e.g. Figure 4). Since typical activity-dependent changes in reflectivity are in the range 0.1% – 0.5%, reflectance images are typically measured relative to a baseline "blank" image.

Intrinsic signals in the brain are much more complex than nerve or brain slice preparations, in part due to vascular responses to neural activity. Brain reflectance is modulated by at least three factors: (i) chromophore oxidation states, notably of hemoglobin and cytochrome oxidase, (ii) changes in blood volume, and (iii) light scattering changes (50). Metabolic demands associated with neuronal activity create a local depletion in hemoglobin oxygenation in microcapillaries, which increases tissue absorption at wavelengths 600-650 nm. The vascular system responds to this change by increasing blood flow to the region, leading to a net *increase* in oxy-hemoglobin within 1-2 seconds. The vascular response is less well-localized than the original oxygen depletion response. Similar dynamics including an "initial dip" in oxygenation have been reported in fMRI (51). Scattering changes are thought to be similar in nature to those observed in brain slices.

Different components in the brain reflectance images can be distinguished by their different time courses and wavelength dependence. Wavelength-selective optical filters can be used to emphasize different signal components. At 605 nm, the oxymetry component is dominant; at 630 nm blood volume and hemoglobin saturation; at wavelengths greater than 700 nm absorption changes are small and the scattering signal dominates (50).

Intrinsic imaging has become an important tool for studying functional cortical maps – the spatial patterning of activation during a certain type of stimulus or behavior – in visual cortex, motor cortex, somatosensory cortex, olfactory bulb, and auditory cortex. For a detailed review of this large field see (52). A particular advantage of optical imaging, compared with techniques such as electrode recording and radiolabel tracing, is the ability to record multiple maps in succession in the same subject. This has led to important findings about relationships between different cortical columnar systems (53-55).

Intrinsic signal imaging of the brain surface has been applied to humans intra-operatively. One area of study concerns the mapping of functional boundaries to minimize damage during operations to remove tumors, epileptic foci, or other tissues (56-58). Studies have delineated human somatosensory maps (58) and areas involved in language (57).

While most brain surface imaging studies are concerned with slow absorption and scattering changes, a few measurements of fast scattering changes have been reported. Rector et al (59) used a deep-brain fiber optic illuminator and fiber imaging system to record ~0.1% stimulus-dependent light scattering changes in cat dorsal hippocampus. Light scattering peaked about 20 ms after stimulation and coincided with neuronal population spiking. A longer-lasting scattering component peaked 100-500 ms post-stimulus, probably reflecting postsynaptic potentials. Recordings of similar changes in rat brain stem (13) and whisker barrel cortex (60) have also been reported.

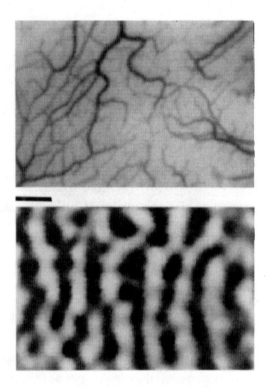

Figure 4. Imaging of ocular dominance columns in monkey visual cortex obtained by monocular presentation of a video movie. Top: vasculature of imaged area, illuminated with green light. Bottom: Ocular dominance map obtained by dividing average of 48 cortical images during right eye stimulus with average of 48 cortical images during left eye stimulus. Scale bar = 1 mm. From Grinvald et al (49). Copyright (c) 1991 National Academy of Sciences.

Transcranial measurements: diffuse optical methods

For intrinsic optical effects in the human brain to be detected noninvasively, light must pass through the skin and skull, interact with the brain, then pass again through skull and skin to detectors (*61*). In diffuse optical imaging, an array of light sources (typically light emitting diodes) and an array of photodetectors are arranged on the head. Sources used are typically in the wavelength range 650nm-1000nm, at which tissue absorption is negligible compared to scattering. The density of photons in tissue is considered to obey a diffusion equation with absorption (*62*). Two wavelengths are often used to perform spectroscopy of the oxygenation states.

Three types of diffuse optical imaging have been developed: (i) time domain imaging, (ii) frequency domain image, and (iii) continuous-wave imaging. In time domain imaging, extremely short (~ps) pulses of light are introduced into the head and the temporal distribution of photons measured by the detectors gives information about absorption and scattering in the brain. Frequency domain techniques use sources with high frequency (~10-100 MHz or more) modulation in amplitude; detectors record amplitude and phase shift of the field. Continuous-wave systems use unmodulated or slowly modulated sources and measure only the intensity of the returning light. (*18*). As in brain surface imaging, diffuse imaging signals are primarily dominated by absorptive hemodynamic changes.

Imaging of intrinsic changes by diffuse techniques is performed via inversion algorithms (*61*). Because of the strong scattering, spatial resolution is typically limited to several centimeters, and penetration depths to about 3 cm.

Diffuse optical imaging has been used to investigate cerebral responses to visual, auditory, and other sensory stimuli, motor activity, and language, among others. Although diffuse imaging has a lower spatial resolution than fMRI and cannot generally used to probe deep brain regions, diffuse optical imaging technology has the advantage of being portable, relatively inexpensive, and can be applied for the study of states and behaviors incompatible with MRI imaging.

Several authors have reported fast scattering measurements in transcranial diffuse optical experiments (*63, 64*), although the robustness of these measurements has been questioned (*65*).

Discussion and conclusions

The wide range of intrinsic effects reviewed here can be seen as falling into two classes. On one hand are the absorption-based changes, primarily caused by blood oxygenation and flow changes in response to neural activity. These effects are slow and have modest spatial resolution which can resolve cortical columns but not single neurons. However, they arise from relatively robust mechanisms and are closely correlated with other measures of activity. As such, brain surface reflectance imaging is widely used to measure spatial distributions of cortical activity, and applications of noninvasive diffuse optical techniques are growing steadily. On the other hand are the non-hemodynamic phenomena such as scattering, birefringence, and mechanical changes, which can have time resolution comparable to the underlying electrical signals, and spatial resolution to the level of single nerve fibers. However, they are generally difficult to measure and interpret, and their mechanisms are relatively poorly understood. As a result, applications of these measurements remain quite limited.

What will be required to develop these non-hemodynamic measurements into useful functional neuroimaging tools? Three related components are needed: (i) continued development of optical techniques for measuring these signals robustly and with high sensitivity, (ii) improved understanding of the mechanisms underlying intrinsic signals, and (iii) a clear interpretation of the measurements, which may involve elucidating and separating contributions from different processes.

Where will measurement of non-hemodynamic intrinsic changes first have important applications? We suggest that in the near future the most likely useful role will be for imaging activity in cultured neurons and neural networks (10, 66). Cultured neurons are widely used to study nervous systems at the molecular, cellular, and synaptic levels. Collections of connected neurons are emerging as a model system for biological neural circuits; such networks have been shown to exhibit basic forms of network plasticity (67). In these relatively simple, controlled preparations, very small optical changes can be measured without artifacts from animal motions or cross-talk from large numbers of neurons. Interpretations of the intrinsic signals may be easier due to a reduced complexity of the overall system. Manipulations of the preparations can be more readily performed, for investigation of signal mechanisms or optimization of conditions for signal measurement. As a technique for monitoring signaling in many neurons simultaneously, a fast intrinsic optical method would offer better spatial localization compared with multi-electrode array (MEA) techniques (68), and avoid phototoxicity and photobleaching problems of calcium-sensitive and voltage-sensitive fluorescent dye imaging (69).

In summary, nerve tissues display a wide variety of intrinsic optical phenomena, most of which are not yet well understood. Intrinsic signals which are hemodynamic in origin are now being commonly applied in functional imaging. Others such as fast scattering and nerve geometry changes have made inroads more slowly, due to challenges in measuring, understanding, and interpreting the signals. Novel optical imaging techniques have the potential to overcome these challenges, particularly in studies with cultured neurons.

233

References

1. Hill, D.; Keynes, R. *J Physiol* **1949**, *108*, 278-281.
2. Cohen, L. B.; Keynes, R. D.; Hille, B. *Nature* **1968**, *218*, 438-441.
3. Watanabe, A. *J Physiol* **1987**, *389*, 223-253.
4. Tasaki, I.; Watanabe, A.; Sandlin, R.; Carnay, L. *Proc Natl Acad Sci U S A* **1968**, *61*, 883-888.
5. Cohen, L. B.; Keynes, R. D.; Landowne, D. *J Physiol* **1972**, *224*, 701-725.
6. Cohen, L. B.; Keynes, R. D.; Landowne, D. *J Physiol* **1972**, *224*, 727-752.
7. von Muralt, A. *Philos Trans R Soc Lond B Biol Sci* **1975**, *270*, 411-423.
8. Tasaki, I.; Byrne, P. M. *Jpn J Physiol* **1993**, *43 Suppl 1*, S67-S75.
9. Cohen, L. *Annual Review of Physiology* **1989**, *51*, 487-490.
10. Stepnoski, R. A.; Laporta, A.; Raccuia-Behling, F.; Blonder, G. E.; Slusher, R. E.; Kleinfeld, D. *Proceedings of the National Academy of Sciences of the United States of America* **1991**, *88*, 9382-9386.
11. Laporta, A.; Kleinfeld, D. *Interferometric detection of action potentials in vitro*; Imaging in Neuroscience and Development: A Laboratory Manual Cold Spring Harbor Laboratory Press: New York, 2005, pp 539-543.
12. Lipton, P. *J Physiol* **1973**, *231*, 365-383.
13. Rector, D. M.; Rogers, R. F.; Schwaber, J. S.; Harper, R. M.; George, J. S. *Neuroimage* **2001**, *14*, 977-994.
14. Iwasa, K.; Tasaki, I.; Gibbons, R. C. *Science* **1980**, *210*, 338-339.
15. Iwasa, K.; Tasaki, I. *Biochem Biophys Res Commun* **1980**, *95*, 1328-1331.
16. Tasaki, I.; Iwasa, K.; Gibbons, R. C. *Jpn J Physiol* **1980**, *30*, 897-905.
17. Grinvald, A.; Lieke, E.; Frostig, R. D.; Gilbert, C. D.; Wiesel, T. N. *Nature* **1986**, *324*, 361-364.
18. Villringer, A.; Chance, B. *Trends Neurosci* **1997**, *20*, 435-442.
19. Boenhoffer, T.; A., G. *Optical imaging based on intrinsic signals. The methodology.*; Brain Mapping: the methods; Academic Press: London, 1996; Vol. pp 55-97.
20. Fang-Yen, C.; Oh, S.; Song, S.; Seung, H. S.; Dasari, R. R.; Feld, M. S., *Differential Heterodyne Mach-Zehnder Interferometer for Measurement of Nanometer-Scale Motions in Living Cells. In *Optical Society of America Topical Meeting in Biomedical Optics*, ed.; 'Ed.'^'Eds.' Optical Society of America: Ft. Lauderdale, FL, 2006
21. Huang, D.; Swanson, E. A.; Lin, C. P.; Schuman, J. S.; Stinson, W. G.; Chang, W.; Hee, M. R.; Flotte, T.; Gregory, K.; Puliafito, C. A.; Fujimoto, J. G. *Science* **1991**, *254*, 1178-1181.
22. Fercher, A. F.; Drexler, W.; Hitzenberger, C. K.; Lasser, T. *Reports on Progress in Physics* **2003**, *66*, 239-303.
23. Boppart, S. A. *Psychophysiology* **2003**, *40*, 529-541.
24. Maheswari, R. U.; Takaoka, H.; Kadono, H.; Homma, R.; Tanifuji, M. *J Neurosci Methods* **2003**, *124*, 83-92.

234

25. Lazebnik, M.; Marks, D. L.; Potgieter, K.; Gillette, R.; Boppart, S. A. *Opt Lett* **2003**, *28*, 1218-1220.
26. Harary, H. H.; Brown, J. E.; Pinto, L. H. *Science* **1978**, *202*, 1083-1085.
27. Yao, X. C.; Yamauchi, A.; Perry, B.; George, J. S. *Appl Opt* **2005**, *44*, 2019-2023.
28. Bizheva, K.; Pflug, R.; Hermann, B.; Povazay, B.; Sattmann, H.; Qiu, P.; Anger, E.; Reitsamer, H.; Popov, S.; Taylor, J. R.; Unterhuber, A.; Ahnelt, P.; Drexler, W. *Proc Natl Acad Sci U S A* **2006**, *103*, 5066-5071.
29. Tasaki, I.; Kusano, K.; Byrne, P. M. *Biophys J* **1989**, *55*, 1033-1040.
30. Tasaki, I.; Byrne, P. M. *Biophys J* **1990**, *57*, 633-635.
31. Yao, X. C.; Rector, D. M.; George, J. S. *Appl Opt* **2003**, *42*, 2972-2978.
32. Sandlin, R.; Lerman, L.; Barry, W.; Tasaki, I. *Nature* **1968**, *217*, 575-576.
33. Hill, B. C.; Schubert, E. D.; Nokes, M. A.; Michelson, R. P. *Science* **1977**, *196*, 426-428.
34. Akkin, T.; Davé, D.; Milner, T.; Rylander, H. *Opt. Express* **2004**, *12*, 2377-2386.
35. Fang-Yen, C.; Chu, M. C.; Seung, H. S.; Dasari, R. R.; Feld, M. S. *Opt Lett* **2004**, *29*, 2028-2030.
36. Yang, C.; Wax, A.; Hahn, M. S.; Badizadegan, K.; Dasari, R. R.; Feld, M. S. *Optics Letters* **2001**, *26*, 1271-1273.
37. Hill, D. K. *J Physiol* **1950**, *11*, 304-327.
38. Tasaki, I. *Physiol Chem Phys Med NMR* **1988**, *20*, 251-268.
39. Zhang, P. C.; Keleshian, A. M.; Sachs, F. *Nature* **2001**, *413*, 428-432.
40. Gross, D.; Williams, W. S.; Connor, J. A. *Cell Mol Neurobiol* **1983**, *3*, 89-111.
41. Shapiro, H.; Parpart, A. *J Cell Comp Physiol* **1937**, *10*, 147-163.
42. Barer, R.; Ross, K. F.; Tkaczyk, S. *Nature* **1953**, *171*, 720-724.
43. Leão, A. *J Neurophysiol* **1944**, *7*, 359-390.
44. Aitken, P. G.; Fayuk, D.; Somjen, G. G.; Turner, D. A. *Methods* **1999**, *18*, 91-103.
45. Fayuk, D.; Aitken, P. G.; Somjen, G. G.; Turner, D. A. *J Neurophysiol* **2002**, *87*, 1924-1937.
46. MacVicar, B. A.; Hochman, D. *J Neurosci* **1991**, *11*, 1458-1469.
47. Salzberg, B. M.; Obaid, A. L.; Gainer, H. *J Gen Physiol* **1985**, *86*, 395-411.
48. Frostig, R. D.; Lieke, E. E.; Ts'o, D. Y.; Grinvald, A. *Proc Natl Acad Sci U S A* **1990**, *87*, 6082-6086.
49. Grinvald, A.; Frostig, R. D.; Siegel, R. M.; Bartfeld, E. *Proc Natl Acad Sci U S A* **1991**, *88*, 11559-11563.
50. Zepeda, A.; Arias, C.; Sengpiel, F. *J Neurosci Methods* **2004**, *136*, 1-21.
51. Kim, D. S.; Duong, T. Q.; Kim, S. G. *Nat Neurosci* **2000**, *3*, 164-169.
52. Pouratian, N.; Sheth, S. A.; Martin, N. A.; Toga, A. W. *Trends Neurosci* **2003**, *26*, 277-282.
53. Bartfeld, E.; Grinvald, A. *Proc Natl Acad Sci U S A* **1992**, *89*, 11905-9.

54. Bosking, W. H.; Crowley, J. C.; Fitzpatrick, D. *Nat Neurosci* **2002**, *5*, 874-882.

55. Hubener, M.; Shoham, D.; Grinvald, A.; Bonhoeffer, T. *J Neurosci* **1997**, *17*, 9270-9284.

56. Cannestra, A. F.; Black, K. L.; Martin, N. A.; Cloughesy, T.; Burton, J. S.; Rubinstein, E.; Woods, R. P.; Toga, A. W. *Neuroreport* **1998**, *9*, 2557-2563.

57. Haglund, M. M.; Ojemann, G. A.; Hochman, D. W. *Nature* **1992**, *358*, 668-671.

58. Toga, A. W.; Cannestra, A. F.; Black, K. L. *Cereb Cortex* **1995**, *5*, 561-565.

59. Rector, D. M.; Poe, G. R.; Kristensen, M. P.; Harper, R. M. *J Neurophysiol* **1997**, *78*, 1707-1713.

60. Rector, D. M.; Carter, K. M.; Volegov, P. L.; George, J. S. *Neuroimage* **2005**, *26*, 619-627.

61. Strangman, G.; Boas, D. A.; Sutton, J. P. *Biol Psychiatry* **2002**, *52*, 679-693.

62. Boas, D. A.; Dale, A. M.; Franceschini, M. A. *Neuroimage* **2004**, *23 Suppl 1*, S275-88.

63. Gratton, G.; Corballis, P. M. *Psychophysiology* **1995**, *32*, 292-299.

64. Franceschini, M. A.; Boas, D. A. *Neuroimage* **2004**, *21*, 372-386.

65. Steinbrink, J.; Kempf, F. C.; Villringer, A.; Obrig, H. *Neuroimage* **2005**, *26*, 996-1008.

66. Banker, G.; Goslin, K. *Culturing Nerve Cells*; MIT Press: Cambridge, 1998

67. Bi, G.; Poo, M. *Nature* **1999**, *401*, 792-796.

68. Novak, J. L.; Wheeler, B. C. *J Neurosci Methods* **1988**, *23*, 149-159.

69. Jin, W.; Zhang, R. J.; Wu, J. Y. *J Neurosci Methods* **2002**, *115*, 13-27.

Chapter 16

Nonlinear Interferometric Vibrational Imaging

A Method for Distinguishing Coherent Anti-Stokes Raman Scattering from Nonresonant Four-Wave-Mixing Processes and Retrieving Raman Spectra Using Broadband Pulses

Daniel L. Marks and Stephen A. Boppart

Biophotonics Imaging Laboratory, Beckman Institute for Advanced Science and Technology, Department of Electrical and Computer Engineering, University of Illinois at Urbana-Champaign, 405 North Mathews, Urbana, IL 61801

When utilizing broadband, short pulses to stimulate Coherent Anti-Stokes Raman Scattering (CARS), frequently the peak power is sufficient to excite other nonlinear, nonresonant processes in the material. These processes produce a four-wave-mixing component in the same frequency band as the CARS signal, so that the two signal types cannot be distinguished on the basis of frequency band alone. Typically in biological materials, the nonresonant component produced by the bulk medium can overwhelm the CARS signals produced by the usually much lower concentration target molecular species. Resonant processes can be distinguished from nonresonant processes in that molecular vibrations and rotations typically last longer than a picosecond, while nonresonant processes are not persistent and last shorter than 10 fs. Interferometry allows the arrival time of the signal to be determined to extremely high accuracy, limited by the bandwidth of the reference pulse. Because resonances are persistent, they can produce anti-Stokes radiation that persists after the nonresonant excitation is produced. Interferometry can distinguish the later arrival of the anti-Stokes radiation and therefore distinguish the resonant and nonresonant signals.

More complicated pulse-shaping and interferometry schemes can allow additional flexibility to allow simultaneous sampling of the Raman spectrum while rejecting the nonresonant component. This technique enables an entire region of the Raman spectrum of a sample to be measured with a single brief broadband pulse.

Theory of CARS Interferometry

Nonlinear Interferometric Vibrational Imaging (NIVI) is a previously described method used to measure the three-dimensional distribution of molecular species in various samples (biological or otherwise) (*1-3*). Its basic operation is to stimulate the excitation of molecular bonds with particular resonance frequencies, and then use these excitations to produce radiation distinct from the excitation that can be measured. The physical process of excitation and stimulation of radiation is called Coherent Anti-Stokes Raman Scattering (CARS). Unlike previous methods that use CARS in microscopy to probe for the presence of molecular species (*4-6*), NIVI utilizes a heterodyne approach where a reference signal is separately generated and interferomerically compared to the signal received from the sample. In this way, additional information can be inferred from the emitted radiation such as the distance to the sample and phase relationships containing additional details about the sample molecular composition (*7,8*). It also has other advantages in sensitivity and the ability to screen out background radiation that is not produced by the sample. This technique also can allow more flexibility in the choice of laser illumination source, because the coherent detection process does not rely on photon frequency alone to discriminate emitted radiation.

In this chapter we describe how NIVI can be used to estimate the Raman spectrum of a sample, which can be specifically applied to distinguishing resonant CARS processes and nonresonant four-wave-mixing processes (*3*). We are concerned with how to measure the concentration of several vibrational states simultaneously, and how to distinguish these signals from the nonresonant background. At each point in a three-dimensional sample, a further dimension of Raman frequency specificity can be extracted. This can achieve further contrast between the constituents of biological tissues enabling a more thorough determination of molecular contents.

We consider a focused optical pulse, with an electric field at the focus given in the frequency domain by $\tilde{E}_i(\omega)$. At the focus there is a medium with a complex Raman spectrum given by $\chi^{(3)}(\Omega)$ where Ω is the Raman frequency. The anti-Stokes/Stokes electric field produced by the CARS/CSRS (Coherent Stokes Raman Scattering) process at the focus is given by:

$$P^{(3)}(\Omega) = \chi^{(3)}(\Omega) \int_0^\infty \widetilde{E}_i(\omega + \Omega)\widetilde{E}_i(\omega)^* d\omega \tag{1}$$

$$\widetilde{E}_A(\omega) = \int_0^\omega P^{(3)}(\Omega)\widetilde{E}_i(\omega - \Omega)d\Omega \quad \text{(CARS)} \tag{2}$$

$$\widetilde{E}_S(\omega) = \int_0^\omega P^{(3)}(\Omega)^* \widetilde{E}_i(\omega + \Omega)d\Omega \quad \text{(CSRS)} \tag{3}$$

These equations assume that the optical field is strong and can be treated classically, a perturbative interaction with the sample that begins and ends with the vibrational ground state, and that there are no levels directly resonant with any of the individual frequencies in the pulse (resonant one-photon interactions do not occur). Near-infrared light (800-1400 nm), which is only weakly absorbed by biological tissues, contains frequencies typically well above the vibrational frequencies of molecular bonds, but below the electronic excitation frequencies, and so is suitable for NIVI. These equations give the time evolution of a CARS/CSRS process involving many possible simultaneous Raman-active vibrations. In CARS, the molecule is excited by two frequencies ω_1 and ω_2 that are separated by the Raman vibrational frequency $\omega_1 - \omega_2 = \Omega$. The upper frequency ω_1 is called the pump, and the lower frequency ω_2 the Stokes. In general, the signal consists of not just two discrete frequencies but a continuous spectrum. In this case, Eq. 1 represents the excitation of the molecule through stimulated Raman scattering (SRS). A vibration at frequency Ω is excited by every pair of optical frequencies with a frequency difference of Ω. The excitation at any given Raman frequency is given by the polarization $P^{(3)}(\Omega)$. This excitation is converted to anti-Stokes radiation $\widetilde{E}_A(\omega)$ by a second SRS process whereby the pump signal at frequency ω_1 is partially converted to the anti-Stokes signal at frequency $\omega_A = \omega_1 + \Omega = 2\omega_1 - \omega_2$. For broadband signals, the anti-Stokes signal must be summed over all pump frequencies, and so Eq. 2 results. The CSRS process is likewise modeled by Eq. 1 and Eq. 3. In general, both CARS and CSRS are generated simultaneously for any given pulse.

The information about the sample is contained in its Raman spectrum $\chi^{(3)}(\Omega)$. By sending in a pulse with a known electric field $\widetilde{E}_i(\omega)$, and measuring the returned anti-Stokes signal $\widetilde{E}_A(\omega)$, we would like to infer $\chi^{(3)}(\Omega)$. Unfortunately, at optical frequencies, the electric field $\widetilde{E}_A(\omega)$ is not directly measurable because it oscillates on a time scale too fast to be electronically demodulated, requiring interferometric demodulation. A schematic of the setup to do this is given in Figure 1.

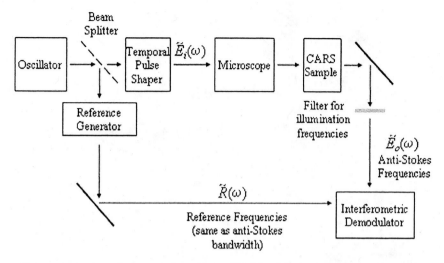

Figure 1. Basic schematic of Nonlinear Interferometric Vibrational Imaging.

To demodulate the anti-Stokes field, we generate a reference field $\tilde{R}(\omega)$ from the oscillator field that contains frequencies same as the anti-Stokes radiation. The interferometric demodulator measures the intensity of the cross-correlation of the anti-Stokes and reference fields, which is given by the frequency spectrum $\tilde{I}(\omega) = \tilde{R}(\omega)^* \tilde{E}_A(\omega)$. A good method of interferometric demodulation for these techniques is spectral interferometry (7,9-10). Spectral interferometry may be especially attractive because it allows the signal to be sampled in one shot to minimize transient effects, and can achieve higher signal-to-noise ratio than conventional temporal interferometry. Putting the steps of generating the anti-Stokes radiation and interferometry together, the cross-correlation spectrum is given by:

$$\tilde{I}(\omega) = \tilde{R}(\omega)^* \int_0^\omega \chi^{(3)}(\Omega)\tilde{E}_i(\omega-\Omega)\int_0^\infty \tilde{E}_i(\omega'+\Omega)\tilde{E}_i(\omega')^* d\omega' d\Omega \quad \text{(CARS)} \quad (4)$$

$$\tilde{I}(\omega) = \tilde{R}(\omega)^* \int_0^\omega \chi^{(3)}(\Omega)^* \tilde{E}_i(\omega+\Omega)\int_0^\infty \tilde{E}_i(\omega'+\Omega)^* \tilde{E}_i(\omega') d\omega' d\Omega \quad \text{(CSRS)} \quad (5)$$

An important consequence of Eqs. 4 and 5 is that the cross-correlation spectrum $\tilde{I}(\omega)$ is a linear function of the Raman spectrum $\chi^{(3)}(\Omega)$. Because of this, linear estimation can be used to infer $\chi^{(3)}(\Omega)$ from $\tilde{I}(\omega)$. To better see this, we reform Eqs. 4 and 5 as linear operators $\hat{A}$ and $\hat{S}$ with kernels $A(\omega,\Omega)$ and $S(\omega,\Omega)$ respectively:

$$\tilde{I}(\omega) = \hat{A}\chi = \int_0^\omega \chi^{(3)}(\Omega) A(\omega,\Omega) d\Omega \quad \text{where}$$

$$A(\omega,\Omega) = \tilde{R}(\omega)^* \tilde{E}_i(\omega'-\Omega) \int_0^\infty \tilde{E}_i(\omega'+\Omega)\tilde{E}_i(\omega')^* d\omega' \quad \text{(CARS)} \qquad (6)$$

$$\tilde{I}(\omega) = \hat{S}\chi^* = \int_0^\omega \chi^{(3)}(\Omega)^* S(\omega,\Omega) d\Omega \quad \text{where}$$

$$S(\omega,\Omega) = \tilde{R}(\omega)^* \tilde{E}_i(\omega'+\Omega) \int_0^\infty \tilde{E}_i(\omega'+\Omega)^* \tilde{E}_i(\omega') d\omega' \quad \text{(CSRS)} \qquad (7)$$

With the CARS interferometry process now expressed as a linear operator $\hat{A}$ mapping $\chi^{(3)}(\Omega)$ to $\tilde{I}(\omega)$, an inverse operator can be found to estimate $\chi^{(3)}(\Omega)$ from $\tilde{I}(\omega)$. Many inverse operators are possible and should be chosen on the basis of stability, ability to model the physical system and noise, and ability to incorporate confidence measures of the data. We present an inverse operator based on the weighted Tikhonov regularized least-squares solution. The least-squares solution by itself attempts to find a solution for $\chi^{(3)}(\Omega)$ that minimizes the Euclidean distance $\|I - \hat{A}\chi\|$ for the linear system $I = \hat{A}\chi$. Unfortunately, this in practice tends to produce poor results due to poor conditioning and the inability to incorporate confidence information about the data. A solution is to augment the linear system with a weighting operator

$$\hat{W}I = \tilde{W}(\omega)\tilde{I}(\omega) \qquad (8)$$

where $\tilde{W}(\omega)$ is a weighting function that is of unit value inside the measured anti-Stokes bandwidth and zero outside. The weighting operator ensures that only data in the measured bandwidth contributes to the estimate of the Raman spectrum. The weighted linear system becomes $\hat{W}I = \hat{W}\hat{A}\chi$. Formally, the Tikhonov regularized solution to this system is $\chi = \left(\hat{A}^H \hat{W}^H \hat{W}\hat{A} + \varepsilon\hat{I}\right)^{-1} \hat{A}^H \hat{W}^H \hat{W}I$ where the operator $\hat{I}$ is the identity operator (for CSRS it is $\chi = \left(\hat{S}^H \hat{W}^H \hat{W}\hat{S} + \varepsilon\hat{I}\right)^{-1} \hat{S}^H \hat{W}^H \hat{W}I$). The constant $\varepsilon > 0$ is set to account for the magnitude of noise in the measurement, e.g. from thermal or photon noise. In practice, this operator can be inverted numerically using iterative linear solution methods such as the preconditioned conjugate gradient method including the Fast Fourier Transform to implement the numerical cross-correlations.

Resonant CARS versus Nonresonant Four-Wave-Mixing

While the above formalism specifies how the measured interferometric cross-correlation is related to the Raman spectrum, and an inverse operator for this relation, it does not tell us what input signal $\tilde{E}_i(\omega)$ or reference signal $\tilde{R}(\omega)$ will enable reconstruction of the desired features of the Raman spectrum $\chi^{(3)}(\Omega)$.

Two important features of the Raman spectrum in practice that need to be mutually distinguished are the resonant and nonresonant components. The resonant components are specific to features of a molecule, which include vibrational frequencies, rotational frequencies, and electronic resonances. Nonresonant features are not specific to a particular molecule, and are weakly dependent on frequency. A typical Raman spectrum can be decomposed into a sum of resonance and nonresonant components:

$$\chi^{(3)}(\Omega) = \chi_{NR}^{(3)}(\Omega) + \sum_n \chi_n^{(3)}(\Omega) \tag{9}$$

The nonresonant component $\chi_{NR}^{(3)}(\Omega)$ is a slowly changing function of Ω, and is usually approximated by a real-valued constant $\chi_{NR}^{(3)}(\Omega) = \chi_{NR}$. The resonant components $\chi_n^{(3)}(\Omega)$ are sharply peaked at vibrational frequencies, and are typically described by a homogeneously broadened Lorentzian spectrum:

$$\chi_n^{(3)}(\Omega) = \frac{2\Omega_n}{\Omega^2 - 2i\Gamma_n\Omega - (\Omega_n^2 - \Gamma_n^2)} \tag{10}$$

where $\Omega = \sqrt{\Omega_n^2 - 2\Gamma_n^2}$ is the center frequency of the resonance, and Γ_n is the line width. A resonance has a Raman magnitude that is sharply peaked around the center frequency. Interferometric detection can distinguish the real-valued nearly constant nonresonant spectra and the sharply peaked resonant CARS spectra. In addition, the imaginary part of $\chi^{(3)}(\Omega)$ is determined by only the resonant Raman component, while the real part is determined by both resonant and nonresonant components. Another feature distinguishing resonant components is the phase reversal of π that occurs through the center frequency.

Noninterferometric instruments cannot distinguish the phase, while the interferometric method measures the complex $\chi_n^{(3)}(\Omega)$. As well as being a signature of CARS, this phase reversal may be used to separate several resonances close together in frequency.

Distinguishing CARS resonance from nonresonant four-wave-mixing has become problematic with the advent of ultrafast laser sources. These sources produce pulses on the order of 5-200 fs, much shorter than the lifetime of the resonance, which is $2\pi/\Gamma_n$. If transform-limited ultrafast pulses are used, only a small polarization $P^{(3)}(\Omega)$ can be produced in the molecule, while the nonresonant components are enhanced. Since the lifetime of the resonance is typically 1-100 ps, transform-limited ultrafast pulses excite resonant transitions inefficiently and nonresonant transitions efficiently. Many current noninterferometric CARS instruments utilize narrowband pump and Stokes pulses of picosecond length for this reason (*11,12*).

There are compelling reasons to use broadband sources to excite CARS (*1, 13-17*). An ultrafast pulse can be shaped into a longer picosecond pulse that can excite CARS more efficiently. In addition, it can be used to excite many resonances simultaneously. Unfortunately, unlike narrowband pulses, the anti-Stokes radiation produced will not be narrowband. If many resonances are excited simultaneously, then the anti-Stokes radiation they produce have overlapping spectra. Because noninterferometric detection can only measure the spectrum of the anti-Stokes radiation, the contributions of each resonance to the anti-Stokes radiation are difficult to separate. Interferometric detection allows the demodulation of the anti-Stokes field so the Raman spectrum can be inferred when exciting multiple simultaneous resonances.

In addition, broadband sources should allow for more Raman-frequency agile imaging instruments. When utilizing narrowband pump and Stokes pulses, the frequency difference between them must be tuned to the Raman frequency of interest. Retuning lasers or amplifiers is often difficult to make reliable and automatic. Pulse shaping, however, is achieved by movable gratings, prisms, or mirrors to adjust dispersion and delay, or by acousto-optic or liquid-crystal Fourier-plane pulse shapers without moving parts (*14-20*). Because pulse shapers do not typically involve feedback, oscillation, or overly sensitive alignment, pulse shapes can be changed much more easily. In addition, a computer can control these mechanisms automatically, so that changing the pulse shape should be much easier than retuning a laser source. One of the aims of this work is to manipulate the pulse shape of a broadband laser to stimulate specific Raman frequency bands, and use the resulting anti-Stokes radiation can be decoded to estimate the Raman frequency spectrum.

Distinguishing resonant CARS and nonresonant four-wave-mixing using interferometry

To see how interferometry can detect the difference between resonant and nonresonant processes, consider the experimental set-up shown in Figure 2 (a) and the shape of the anti-Stokes pulses produced by the long pump and short Stokes pulses previously mentioned. This combination of pulses is illustrated in Figure 2 (b). When the pump and Stokes pulses overlap, the molecule will be excited by SRS. This excitation will remain after the Stokes pulse passes. At the moment of the overlap, nonresonant four-wave-mixing processes can also be excited. However, because there is no persistent state associated with nonresonant processes, the nonresonant emission will end quickly after the Stokes pulse passes. With a Raman-active resonance, however, the pump can produce anti-Stokes radiation via SRS even after the Stokes has passed, because the resonance persists. Therefore, the nonresonant component can be discarded by rejecting any anti-Stokes radiation that occurs coincident with the Stokes pulse.

The benefit of interferometry is that the time of arrival of the anti-Stokes radiation can be found very precisely by cross-correlating a broad bandwidth reference pulse with the anti-Stokes radiation. Incoherent detection can detect the interference between the resonant and nonresonant components of the anti-Stokes radiation, but does not directly detect the time of arrival of the anti-Stokes light. Figure 2 shows the actual measured cross-correlations between a short reference pulse and the anti-Stokes radiation for a cuvette of acetone with a resonance at 2925 cm^{-1} as the frequency difference between the pump and Stokes is tuned (3).

The apparatus to acquire these interferogram is shown in Figure 2 (a) and consists of several components. A regenerative amplifier produces a pulse at 808 nm and 30 nm bandwidth that is used as a seed pulse for a second-harmonic-generation optical parametric amplifier, and also as the pump for the CARS sample. The pump is lengthened to approximately 200 fs by passing it through a dispersive Dove prism made of BK7 glass of 105 mm length. The idler of the optical parametric amplifier produces a Stokes pulse at 1056 nm and 70 fs in length, which is combined with the pump at a dichroic mirror. The pump is delayed so the Stokes arrives on the leading edge of the pump pulse. The signal of the optical parametric amplifier is at 653 nm and serves as the reference pulse because it matches the frequency of the anti-Stokes produced in the CARS sample when the pump and Stokes are focused into it. The anti-Stokes and reference are cross-correlated by delaying them relative to each other and detecting their interference power on a silicon photodetector. Figure 2(c) shows interferograms acquired with various frequency differences between the pump and Stokes pulses. When this difference matches a Raman resonance, a resonant "tail" is generated, otherwise only the non-resonant impulse remains.

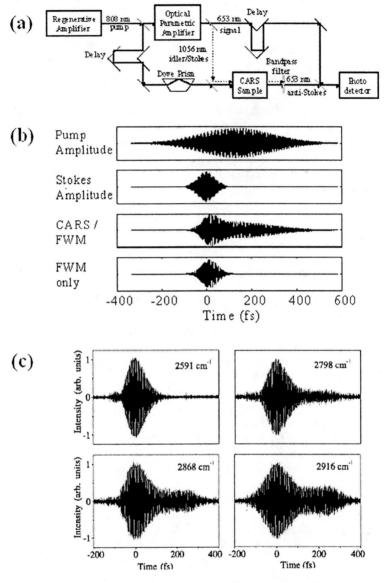

Figure 2. (a) Schematic of interferometer used to cross-correlate the reference and anti-Stokes signals. (b) Illustration of the anti-Stokes and four-wave-mixing signals generated by resonant and nonresonant media illuminated by a long pump and short Stokes pulse. (c) Cross-correlation interferograms between anti-Stokes generated from acetone and a reference pulse with various pump/Stokes frequency separations.

Note that to detect the difference between resonant and nonresonant processes did not require that the pump be lengthened to the lifetime of the resonance of the acetone. This is because the nonresonant component stops abruptly at the end of the Stokes pulse. Unlike the incoherent detection case, one can actually reject the nonresonant component based on its time of arrival, and not just depend on minimizing the amount of nonresonant four-wave-mixing that is generated.

Raman spectra retrieval with a chirped pump pulse

We now present a method using a pulse sequence to recover the Raman spectrum $\chi^{(3)}(\Omega)$ in a particular frequency range. The method uses a pump that is broadband but dispersed to be lengthened in time, while using a short, nearly transform-limited Stokes pulse. These two pulses will be overlapped so that the Stokes pulse is on the leading edge of the pump pulse.

The idea is to stimulate a broad bandwidth of Raman transitions using a short Stokes pulse overlapped with the leading edge of a pump pulse. After the Stokes pulse is over, the pump pulse continues to produce stimulated Raman scattering, which is demodulated with the reference pulse. The portion of the cross-correlation that overlaps the Stokes pulse arrival can be discarded to remove the nonresonant four-wave-mixing component of the anti-Stokes signal. Because the field used to produce the resonant CARS is only due to the pump pulse, the cross-correlation amplitude and the pump pulse can be used to estimate the Raman spectrum $\chi^{(3)}(\Omega)$, perhaps using an inverse of Eq. 6. An alternate method of measuring the Raman spectrum with chirped pulses without interferometry is has also been described (21).

To understand in better detail how this method can recover the Raman spectrum, consider a dispersed pump pulse and a Stokes pulse with an electric field

$$\widetilde{E}_p(\omega) = A_p(\omega)\exp(i\phi(\omega)) \text{ for } \omega_0 \leq \omega \leq \omega_0 + \Delta\omega_P \tag{11}$$
$$\widetilde{E}_p(\omega) = 0 \text{ otherwise}$$
$$\widetilde{E}_s(\omega) = A_s \text{ for } \omega_0 - \Omega_0 - \Delta\omega_S/2 \leq \omega \leq \omega_0 - \Omega_0 + \Delta\omega_S/2 \tag{12}$$
$$\widetilde{E}_s(\omega) = 0 \text{ otherwise}$$

The frequency Ω_0 corresponds to the center of the resonant Raman frequency bandwidth of interest, while the bandwidths $\Delta\omega_P$ and $\Delta\omega_S$ give the bandwidths of the pump and Stokes signals, respectively. The amplitudes $A_p(\omega)$ and A_s correspond to the real, positive amplitudes of the frequency components of the pump and Stokes signals, respectively. To find these pulses in the time domain, we use the stationary phase approximation (which should apply well to dispersed pulses). We define $d\phi/d\omega = t'(\omega)$, $\omega'(t) = t'^{-1}(\omega)$ (the inverse function of $t'(\omega)$), and $d\Phi/dt = \omega'(t)$. Note that $t'(\omega)$ must be a strictly increasing or decreasing function of ω. The time-domain versions of these signals are:

$$E_p(t) = A_p(\omega'(t))\left|\frac{d^2\Phi}{dt^2}\right|^{-1/2} \exp(i\Phi(t)) \text{ for } t'(\omega_0) \le t \le t'(\omega_0 + \Delta\omega_P) \quad (13)$$

$$E_p(t) = 0 \text{ otherwise}$$

$$E_s(t) = A_s \exp(i(\omega_0 - \Omega_0)t)\sin(t\Delta\omega_S)/(t\Delta\omega_S) \quad (14)$$

To place the Stokes pulse on the leading edge of the pump pulse, we assume that $d\phi/d\omega\big|_{\omega=\omega_0} = 0$. At time t=0, the Raman frequencies $\Omega_0 - \Delta\omega_S/2 < \Omega < \Omega_0 + \Delta\omega_S/2$ are excited. Because the Stokes is nearly transform-limited, the phases of Eq. 1 cancel and the polarization is proportional to the Raman spectrum:

$$P^{(3)}(\Omega) \propto \tilde{E}_p(\omega_0)\tilde{E}_s(\omega_0 - \Omega)^* \chi^{(3)}(\Omega) \text{ for } \Omega_0 - \Delta\omega_S/2 < \Omega < \Omega_0 + \Delta\omega_S/2 \quad (15)$$

$$P^{(3)}(\Omega) = 0 \text{ otherwise}$$

The anti-Stokes generated from this polarization is:

$$\tilde{E}_A(\omega) = \int_{\Omega_0 - \Delta\omega_S/2}^{\Omega_0 + \Delta\omega_S/2} P^{(3)}(\Omega)A_p(\omega - \Omega)\exp(i\phi(\omega - \Omega))d\Omega \quad (16)$$

If we define $P^{(3)}(t) = \int_{\Omega_0 - \Delta\omega/2}^{\Omega_0 + \Delta\omega/2} P^{(3)}(\Omega)\exp(-i\Omega t)dt$ then this becomes a product:

$$E_A(t) = P^{(3)}(t)A_p(\omega'(t))\left|\frac{d^2\Phi}{dt^2}\right|^{-1/2} \exp(i\Phi(t)) \text{ for } t'(\omega_0) < t < t'(\omega_0 + \Delta\omega_P) \quad (17)$$

where $E_A(t)$ is the complex analytic continuation (22) with the positive frequency content of $\tilde{E}_A(\omega)$. This suggests that $P^{(3)}(t)$ can be recovered by multiplying $E_A(t)$ by the conjugate of the phase of the probe field

$\exp(-i\Phi(t))\left|\dfrac{d^2\Phi}{dt^2}\right|^{1/2}$ in the time domain. The $P^{(3)}(\Omega)$ and therefore $\chi^{(3)}(\Omega)$ can be

recovered from $P^{(3)}(t)$ by means of a Fourier transform.

To demonstrate the measurement of high-resolution Raman spectra with broadband pulses, we acquired the complex Raman spectra of isopropanol using the set-up in Figure 3(a) (*10*). A regenerative amplifier produced pulses at 250 kHz repetition-rate, 805 nm center wavelength, and 25 nm full-width-half-maximum (FWHM) bandwidth. The pulses were divided into two, with 90% of the light used as the pump a second-harmonic-generation optical parameter amplifier (SHG-OPA). The other 10% was dispersed to 6-10 ps pulse length by propagation through an 85 cm bar of BK7 optical glass. The dispersed pulse acted as the CARS pump and probe and was 45 mW average power at the sample. The SHG-OPA generated nearly transformed-limited signal pulses at 655 nm center wavelength and 25 nm FWHM, which were used to interferometrically demodulate the anti-Stokes signal. The idler was 2 mW average power, centered at 1050 nm, with a bandwidth of 30 nm FWHM, which served as the Stokes pulses.

The pump pulse and Stokes pulse were combined using a dichroic beamsplitter so that the Stokes pulse overlapped the leading edge of the pump pulse. The combined pulses were focused into the sample liquid by a 30-mm focal length near-infrared achromatic objective lens. A gold mirror at the bottom of the liquid sample dish reflected the generated anti-Stokes light backwards to be recollected by the objective. The anti-Stokes light was separated from the remaining pump and Stokes radiation using a dichroic beam splitter, and then was superimposed onto the reference SHG-OPA signal pulse with a broadband 50/50 beam splitter. A spectral interferogram was sampled on the photodetector linear array, which corresponded to the real part of the Fourier transform of the temporal cross-correlation between the anti-Stokes and SHG-OPA signal pulses. The interferogram was numerically resampled so that the spacing of the cross-correlation samples was uniform in frequency. From the resampled interferogram, the complex analytic signal was computed by assuming the cross-correlation signal was zero for negative time delays, so that the Hilbert transform could be used. Effectively this was accomplished by taking the inverse Fourier transform of the interferogram and setting the signal to zero for negative delays. (*10,22*).

To calibrate the dispersion of the interferometer and the spacing of frequencies on the linear array, a nonresonant sample (sapphire) was used. It was found that the OPA signal and the anti-Stokes are almost transform limited and nearly identical, and the interferometer had negligible dispersion. Therefore, only the angular dispersion of the wavelengths on the array needed to be characterized. To calibrate the chirp introduced by the 85 cm BK7 bar, the interferogram was sampled from a sample of acetone. Acetone has a narrowly peaked resonance at 2925 cm^{-1}. The chirp was calibrated by searching for the chirp value that reconstructed the computed Raman spectrum as a sharp peak. Finally, we experimentally demonstrated our method by measuring the Raman spectrum of isopropanol, which was chosen because the fine detail in the spectrum would test the spectral resolution of the instrument. Figure 3 (b) shows the spectral interferogram obtained for the anti-Stokes signal, from which the Raman spectrum was inferred. The primary noise source was photon noise because the number of generated photoelectrons was much greater than the thermally generated charge. The spectral interferogram signal was integrated over a 25 ms interval by averaging together 100 line scans captured at 4000 Hz.

The time-resolved anti-Stokes decay signal magnitude of Figure 3 (c) was computed from the inverse Fourier transform of the spectral interferogram. One can see the envelope of the signal shaped by the beats between various Raman resonances. The chirp of the probe pulse was numerically conjugated out of the time-resolved complex-valued decay signal, from which the Raman spectrum was computed by using a Fourier transform. Figure 3 (d) is a plot of the real part of the nonlinear susceptibility, which contains the nonresonant component. Figure 3 (e) contains the imaginary component of the nonlinear susceptibility, which is compared to the spontaneous Raman spectrum obtained by a commercial spectrometer given in Figure 3 (f). There is a close resemblance between the imaginary component of the nonlinear susceptibility and the spontaneous Raman spectrum. In fact, the magnitude of the imaginary part of the Raman susceptibility is proportional to the spontaneous Raman spectrum (*23*), so these should be similar. The differences in heights of spectral features are due to the nonuniform Stokes pulse spectrum. These results show that broadband pump pulses can be used to measure spectra with high resolution. This resolution is manifest in the two peaks at 2928 cm^{-1} and 2942 cm^{-1} that are clearly discerned by this method. Even though the pump bandwidth is almost 400 cm^{-1}, a high-resolution Raman spectrum can be obtained because the chirp of the pump can be interferometrically measured and numerically removed.

Raman Spectra Retrieval using Broadband Pulses

One of the primary benefits of these methods is that they enable the use of ultrabroadband lasers with thousands of wavenumbers of bandwidth to be used to measure either narrow (less than 10 cm^{-1}) or a very wide range of Raman frequencies. Therefore, the same laser can be used to produce light that both acts as the pump and Stokes frequencies (rather than requiring a separate Stokes to be derived), and the entire power spectral bandwidth of the laser can be used to contribute to the SRS excitation of the sample. There has been increasing interest in using these ultrabroadband sources in CARS imaging and microscopy (*24-30*). When using narrowband pulses, the power spectral density of the pump and Stokes must be very high to ensure sufficient excitation. With ultrabroadband excitation one can use its relatively low power spectral density to excite many vibrations simultaneously, but a one-to-one correspondence between anti-Stokes and Raman frequencies is not preserved.

The essence behind these methods is a reinterpretation of Eq. 1. In the time domain, Eq. 1 becomes:

$$P^{(3)}(t) = \int_0^\infty \chi^{(3)}(\tau)|E_i(t-\tau)|^2 d\tau \tag{18}$$

Eq. 18 expresses the causal convolution of the impulse Raman response $\chi^{(3)}(t) = \int_{-\infty}^{\infty} \chi^{(3)}(\Omega)\exp(-i\Omega t)dt$ with the instantaneous intensity of the incoming pulse. If there is a modulation of the intensity at the same frequency as a Raman resonance, the Raman resonance will become excited. By using a pulse that contains a modulation of the intensity of the excitation wave that spans a range of frequencies, an entire range of the same Raman frequencies can be excited and studied. For example, pulses have been shaped with a periodic phase modulation in the Fourier domain (*14-18,31*), and by interfering chirped pulses together (*32*), have created the intensity modulation needed for SRS. Previously, narrowband lasers were the primary source available for stimulating CARS, so that the intensity modulation was created by the beats between two narrowband pulses. Broadband sources have the benefit of providing a much more easily tunable source of beats of a particular frequency, because the laser need not be retuned. Only the pulse need be reshaped.

An ultrabroadband source of pulsed radiation can be used to both stimulate CARS and provide the reference pulse. The pulses from the source will be divided by a frequency-selective element such as a dichroic beam splitter into lower and higher frequency pulses. The higher frequency pulse bandwidth will correspond to the anti-Stokes frequencies emitted by the sample, and will act as the reference pulse to demodulate the anti-Stokes signal. The lower frequency pulse will be shaped to stimulate the Raman frequencies of the sample in a particular bandwidth.

The strategy used here is to split the lower frequency pulse into two copies. Each copy will propagate through separate dispersive elements, and then recombined afterwards with a time delay between them. When the two pulses are overlapped, a beat frequency is produced at the difference between the instantaneous frequencies of the two pulses at a given instant. By causing the frequency difference of the two pulses to vary between a lower and higher difference frequency during the time interval they are overlapped, the beats can stimulate the Raman frequencies in the same range. By varying the instantaneous frequency of the two pulses, and their relative delay, the range of Raman frequencies that the overlapped pulses stimulate can be varied. A similar scheme was proposed to stimulate a single Raman-active vibration (*32*).

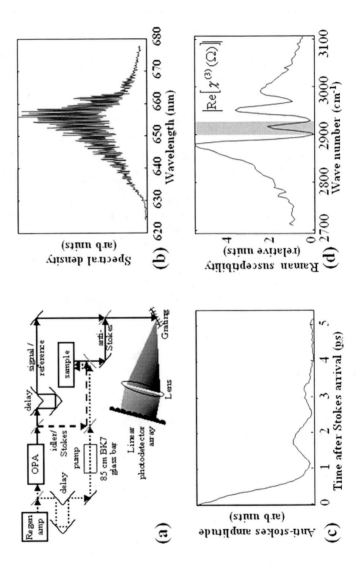

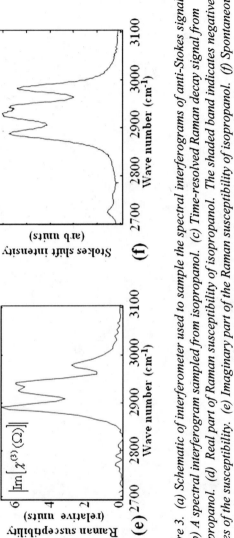

Figure 3. (a) Schematic of interferometer used to sample the spectral interferograms of anti-Stokes signals. (b) A spectral interferogram sampled from isopropanol. (c) Time-resolved Raman decay signal from isopropanol. (d) Real part of Raman susceptibility of isopropanol. The shaded band indicates negative values of the susceptibility. (e) Imaginary part of the Raman susceptibility of isopropanol. (f) Spontaneous Raman spectrum of isopropanol sampled with a commercial spectrometer.

To see how two dispersed pulses can produce a beat frequency spectrum that ranges from $\Omega_L < \Omega < \Omega_H$, consider two bandlimited pulses with different dispersion phases imposed on them :

$$\tilde{E}_1(\omega) = E_1 \exp(i\phi_1(\omega)) \text{ for } \omega_0 - \Delta\omega/2 < \omega < \omega_0 + \Delta\omega/2 \qquad (19)$$
$$\tilde{E}_1(\omega) = 0 \text{ otherwise}$$

$$\tilde{E}_2(\omega) = E_2 \exp(i\phi_2(\omega)) \text{ for } \omega_0 - \Delta\omega/2 < \omega < \omega_0 + \Delta\omega/2 \qquad (20)$$
$$\tilde{E}_2(\omega) = 0 \text{ otherwise}$$

The dispersion phase $\phi_1(\omega)$ and $\phi_2(\omega)$ correspond to the total phase a frequency ω of its respective pulse accumulates in its respective dispersive pulse shaper. For example, if pulse 1 travels through a medium with dispersion relation $k(\omega)$ and thickness d, then the dispersion phase for that pulse $\phi_1(\omega) = k(\omega)d$. In the stationary phase approximation, the time-domain signals for these pulses are:

$$E_1(t) = E_1 \left| \frac{d^2\Phi_1}{dt^2} \right|^{-1/2} \exp(i\Phi_1(t)) \text{ for } t_1{'}(\omega_0 - \Delta\omega/2) < t < t_1{'}(\omega_0 + \Delta\omega/2) \quad (21)$$

$$E_1(t) = 0 \text{ otherwise}$$

where $\frac{d\phi_1}{d\omega} = t_1{'}(\omega)$, $\omega_1{'}(t) = t_1{'}^{-1}(\omega)$, and $\frac{d\Phi_1}{dt} = \omega_1{'}(t)$.

$$E_2(t) = E_2 \left| \frac{d^2\Phi_2}{dt^2} \right|^{-1/2} \exp(i\Phi_2(t)) \text{ for } t_2{'}(\omega_0 - \Delta\omega/2) < t < t_2{'}(\omega_0 + \Delta\omega/2) \quad (22)$$

$$E_2(t) = 0 \text{ otherwise}$$

where $\frac{d\phi_2}{d\omega} = t_2{'}(\omega)$, $\omega_2{'}(t) = t_2{'}^{-1}(\omega)$, and $\frac{d\Phi_2}{dt} = \omega_2{'}(t)$.

We wish to further restrict the time interval of the overlap to be confined from $-T/2 < t < T/2$. To do this and utilize the full bandwidth of the signal, we force pulse 1 to end at time $T/2$, and force pulse 2 to begin at time $-T/2$:

$$\left. \frac{d\phi_1}{d\omega} \right|_{\omega=\omega_0+\Delta\omega/2} = \frac{T}{2} \text{ and } \left. \frac{d\phi_2}{d\omega} \right|_{\omega=\omega_0-\Delta\omega/2} = \frac{-T}{2} \qquad (23)$$

We would also like the frequency difference between the two pulses to start at Ω_H and end at Ω_L. This best places the anti-Stokes frequencies outside of the bandwidth of the pump and Stokes frequencies:

$$\left.\frac{d\Phi_1}{dt}\right|_{t=-T/2} - \left.\frac{d\Phi_2}{dt}\right|_{t=-T/2} = \Omega_H \text{ and } \left.\frac{d\Phi_1}{dt}\right|_{t=T/2} - \left.\frac{d\Phi_2}{dt}\right|_{t=T/2} = \Omega_L \qquad (24)$$

$$\Omega_L < \left[\frac{d\Phi_1}{dt} - \frac{d\Phi_2}{dt}\right] < \Omega_H \text{ for } \frac{-T}{2} < t < \frac{T}{2} \qquad (25)$$

$$\frac{d^2\Phi_2}{dt^2} > \frac{d^2\Phi_1}{dt^2} > 0 \qquad (26)$$

The inequalities of Eq. 25 ensure that the beat frequency is between Ω_L and Ω_H.

Based on the pulses of Eqs. 19 and 20, we can determine a method of finding the CARS signal. Based on Eqs. 1 and 2, these will be:

$$P^{(3)}(\Omega) = \chi^{(3)}(\Omega) \int_{\omega_0-\Delta\omega/2}^{\omega_0+\Delta\omega/2-\Omega} E_1 E_2^* \exp(i\phi_1(\omega+\Omega) - i\phi_2(\omega))d\omega \qquad (27)$$

$$\tilde{E}_A(\omega) = \int_0^\omega P^{(3)}(\Omega)[E_1\exp(i\phi_1(\omega-\Omega)) + E_2\exp(i\phi_2(\omega-\Omega))]d\Omega \qquad (28)$$

The last equation can be recast as:

$$E_A(t) = P^{(3)}(t)\left[E_1\left|\frac{d^2\Phi_1}{dt^2}\right|^{-1/2}\exp(i\Phi_1(t)) + E_2\left|\frac{d^2\Phi_2}{dt^2}\right|^{-1/2}\exp(i\Phi_2(t))\right] \qquad (29)$$

The inversion in all cases including that of Eqs. 27 and 28 can be implemented by the regularized least-squares solution already described. In general, because the anti-Stokes radiation generated by pulse 1 and pulse 2 overlap in frequency, a solution like that implemented for Eq. 17 will not suffice for this case.

We note that if the lowest Raman frequency Ω_L is stimulated exactly at the end of the overlap of the two pulses at time $t = T/2$, then there is no time to accumulate anti-Stokes signal from that Raman frequency. Therefore, in practice, one should choose an Ω_L slightly lower than the minimum desired Raman frequency, perhaps decreased by 20% of $\Omega_H - \Omega_L$. This ensures that the upper frequency pulse continues long enough to read out the minimum desired Raman frequency.

It is also possible to reverse the beat frequency sweep in time so that the frequency is swept from Ω_L to Ω_H. To do this, one can pose the following constraint on the signals:

$$\frac{d\Phi_1}{dt}\bigg|_{t=-T/2} - \frac{d\Phi_2}{dt}\bigg|_{t=-T/2} = \Omega_L \text{ and } \frac{d\Phi_1}{dt}\bigg|_{t=T/2} - \frac{d\Phi_2}{dt}\bigg|_{t=T/2} = \Omega_H \quad (30)$$

$$\frac{d^2\Phi_1}{dt} > \frac{d^2\Phi_2}{dt} > 0$$

with the added constraint of Eq. 25. In addition, one can filter the output signal from the sample for frequencies less than the minimum frequency of the illumination $\omega_0 - \Delta\omega/2$. In this case, one captures the CSRS output of the sample. A reference pulse of this same frequency band will then demodulate this signal, and a regularized least-squares inversion operator for Eq. 7 can be employed to find the Raman spectrum.

As an example of a pulse combination that can stimulate a band of Raman frequencies, we design a pulse that sweeps the beat frequency linearly from Ω_L to Ω_H. To do this, we will split a transform-limited pulse of bandwidth $\omega_0 - \Delta\omega/2 < \omega < \omega_0 + \Delta\omega/2$ into two components. Each component will have a linear chirp (quadratic phase) applied to it in the frequency domain. This chirp is imparted by a dispersive pulse shaper such as optical glass or a prism pair.

The spectrum of the pulse combination that produces beats from Ω_L to Ω_H over an interval T is given by:

$$\tilde{E}_i(\omega) = E_0 \cos\left(\frac{\pi(\omega-\omega_0)}{\Delta\omega}\right)\left[\left(\frac{1+\kappa}{2}\right)\exp\left(\frac{-i(\omega-\omega_0)\tau}{2} - \frac{i(\omega-\omega_0)^2}{2(\alpha-\beta)}\right)\right. \quad (31)$$

$$\left. + \left(\frac{1-\kappa}{2}\right)\exp\left(\frac{i(\omega-\omega_0)\tau}{2} - \frac{i(\omega-\omega_0)^2}{2(\alpha+\beta)}\right)\right] \text{ for } \omega_0 - \frac{\Delta\omega}{2} < \omega < \omega_0 + \frac{\Delta\omega}{2}$$

$$\tilde{E}_i(\omega) = 0 \text{ otherwise}$$

where $\alpha = (2\Delta\omega - \Omega_H - \Omega_L)/2T$, $\beta = (\Omega_H - \Omega_L)/2T$, and $\tau = (T/2)\left[\Omega_H/(\Delta\omega - \Omega_H) + \Omega_L/(\Delta\omega - \Omega_L)\right]$. A cosine apodization window has been added to improve the stability to the inverse. The constant α is the common chirp to both pulses, β is the difference chirp between both pulses, τ is the time delay between the pulses, and κ is the difference in field magnitude between the pulses. To be able to form beats at all Raman frequencies, $\Delta\omega > \Omega_H$.

As an example of this method, we present a simulation of this technique using experimentally realistic values (*1*). We assume that the laser source is an ultrabroadband titanium-sapphire laser producing transform-limited pulses of uniform power spectral density between 700-1000 nm. The bandwidth from 800-1000 nm is reserved for stimulating CARS, while the bandwidth from 700-800 nm is used as the reference pulse. In this simulation, we desire to simulate the measurement of the relative amounts of DNA (deoxyribonucleic acid) which has a resonance at 1094 cm^{-1}, and RNA (ribonucleic acid) which has a resonance at 1101 cm^{-1}. A hypothetical Raman spectrum was created which has Lorentzian resonances at both frequencies. A pulse is designed such that Ω_L =1070 cm^{-1}, Ω_H = 1130 cm^{-1}, and T =5 ps. The simulation of the anti-Stokes signal and the reconstruction of the Raman spectrum are shown in Figure 4. Because of the relatively short pulse interval, there is a noticeable resolution loss to the spectrum, but the resonance lines are distinct. The short dephasing time in liquids, especially water, is the practical limit on the instrument resolution *in vivo*.

In this chapter, we have shown the potential of Nonlinear Interferometric Vibrational Imaging (NIVI), an interferometric technique that is capable of differentiating CARS from non-resonant four-wave-mixing processes, and for retrieving large regions of the Raman spectrum of a sample by using ultrashort broadband pulses. NIVI offers many advantages over current photon-counting methods commonly employed in CARS microscopy. In the future, we endeavor to improve the accuracy of this Raman spectroscopy technique and extend it to ultimately become a versatile and configurable Raman spectroscopy method for probing biological specimens.

Acknowledgements

We thank Prof. Martin Gruebele, Prof. Dana Dlott, Dr. Claudio Vinegoni, Gareth Jones, and Jeremy Bredfeldt for their many technical contributions to this work. This research was supported in part by grants from NASA (NAS2-02057) and the National Institutes of Health (1 R01 EB001777). Additional information can be found at http://biophotonics.uiuc.edu.

References

1. Marks, D. L.; Boppart, S. A. *Phys. Rev. Lett.* **2004**, *92*, 123905.
2. Bredfeldt, J. S.;Vinegoni, C.; Marks, D. L.; Boppart, S. A.. *Opt. Lett.* **2005**, *30*, 495.
3. Marks D. L.; Vinegoni, C.; Bredfeldt, J. S.; Boppart, S. A. *Appl. Phys. Lett.* **2004**, *85*, 5787.

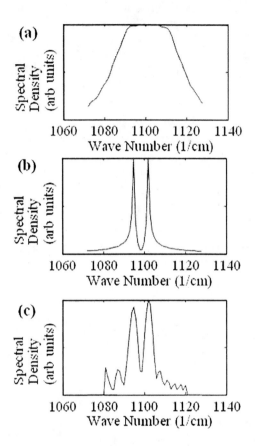

Figure 4. Simulated experiment sampling the Raman spectrum of a combination of DNA and RNA. (a) Spectrum of beat frequencies in excitation pulse, which indicates the possible Raman frequencies that can be excited. (b) Ideal simulated Raman spectrum containing two Lorentzian peaks. (c) Computed reconstruction of Raman spectrum.

4. Duncan, M. D.; Reintjes, J.; Manuccia, T. J. *Opt. Lett.* **1982**, *7*, 350.
5. Zumbusch, A.; Holtom, G. R.; Xie, X. S. *Phys. Rev. Lett.* **1999**, *82*, 4142.
6. Wurpel, G. W. H.; Schins, J. M.; Muller, M. *Opt. Lett.* **2002**, *27*, 1093.
7. Evans, C. L.; Potma, E. O.; Xie, X. S. *Opt. Lett.* **2004**, *29*, 2923.
8. Potma, E. O.; Evans, C. L.; Xie, X. S. *Opt. Lett.* **2006**, *31*, 241.
9. Lepetit, L.; Cheriaux, G.; Joffre, M. *J. Opt. Soc. Am. B* **1995**, *12*, 2467.
10. Jones, G. W.; Marks, D. L.; Vinegoni, C.; Boppart, S. A. *Opt. Lett.* **2006**, *31*, 1543.
11. Potma, E. O.; Jones, D. J.; Cheng, J. X.; Xie, X. S.; Ye, J. *Opt. Lett.* **2002**, *27*, 1168.
12. Cheng, J. X.; Volkmer, A.; Book, L. D.; Xie, X. S. *J. Phys. Chem. B*, **2001**, *105*, 1277.
13. Vinegoni, C.; Bredfeldt, J. S.; Marks, D. L.; Boppart, S. A. *Opt. Expr.* **2004**, *12*, 331.
14. Weiner, A. M.; Leaird, D. E.; Weiderrech, G. P.; Nelson, K. A. *Science* **1990**, *247*, 1317.
15. Oron, D.; Dudovich, N.; Yelin, D.; Silberberg, Y. *Phys. Rev. Lett* **2002**, *88*, 063004.
16. Dudovich, N.; Oron, D.; Silberberg, Y. *Nature* **2002**, *418*, 512.
17. Dudovich, N.; Oron, D.; Silberberg, Y. *J. Chem. Phys.* **2003**, *118*, 9208.
18. Weiner, A. M.; Leaird, D. E.; Patel, J. S.; Wuller, J. R. *Opt. Lett* **1990**, *15*, 326.
19. Wefers, M. M.; Nelson, K. A. *Opt. Lett.* **1995**, *20*, 1047.
20. Fetterman, M. R.; Goswami, D.; Keusters, D.; Yang, W.; Rhee, J.-K.; Warren, W. S. *Opt. Expr.* **1998**, *3*, 366.
21. Knutsen, K. P.; Johnson, J. C.; Miller, A. E.; Petersen, P. B.; Saykally, R. J. *Chem. Phys. Lett.* **2004**, *387*, 436.
22. Mandel, L.; Wolf, E. *Optical Coherence and Quantum Optics;* Cambridge University Press: Cambridge, UK, 1995.
23. Hellwarth, R. W. *Prog. Quant. Electr.* **1977**, *5*, 1.
24. Chin, S. L.; Petit, S.; Borne, F.; Miyazaki, K. *Jpn. J. Appl. Phys.* **1999**, *38*, L126.
25. Nagura, C.; Suda, A.; Kawano, H.; Obara, M.; Midorikawa, K. *Appl. Opt.* **2002**, *41*, 3735.
26. Dudley, J. M.; Provino, L.; Grossard, N.; Mailliote, H.; Windeler, R. S.; Eggleton, B. J.; Coen, S. *J. Opt. Soc. Am. B* **2002**, *19*, 765.
27. Alfano, R. R. *The Supercontinuum Laser Source*; Springer-Verlag: New York, 1989.
28. Paulsen, H. N.; Hilligsoe, K. M.; Thogersen, J.; Keiding, S. R.; Larsen, J. J. *Opt. Lett.* **2003**, *28*, 1123.
29. Andresen, E. R.; Paulsen, H. N.; Birkedal, V.; Thøgersen, J.; Keiding, S. R. *J. Opt. Soc. Am. B* **2005**, *22*, 1934.

30. Wadsworth, W. J.; Ortigosa-Blanch, A.; Knight, J .C.; Birks, T. A.; Martin-Man, T. P.; Russell, P. St. J. *J. Opt. Soc. Am. B* **2002**, *19*, 2148.
31. Oron, D.; Dudovich, N.; Silberberg, Y. *Phys. Rev. Lett.* **2002**, *21*, 213902.
32. Gershgoren, E.; Bartels, R. A.; Fourkas, J. T.; Tobey, R.; Murnane, M. M.; Kapteyn, H. C. *Opt. Lett.* **2003**, *28*, 361.

Chapter 17

Single Molecule Detection at Surfaces: Dual-Color Fluorescence Fluctuation Spectroscopy with Total Internal Reflection Excitation

M. Leutenegger[1], K. Hassler[1,2], P. Rigler[3], A. Bilenca[4], and T. Lasser[1]

[1]Laboratoire d'Optique Biomédicale, École Polytechnique Fédérale de Lausanne, Lausanne, Switzerland
[2]Biomolekylär Fysik, Kungliga Tekniska Högskolan, Stockholm, Sweden
[3]Department of Chemistry, Universität Basel, Basel, Switzerland
[4]Harvard Medical School and Wellman Center for Photomedicine, Massachusetts General Hospital, Boston, MA 02114

Fluorescence fluctuation spectroscopy based on an evanescent field excitation scheme leads to an alternative concept for single molecule detection with interesting experimental features. This concept is described and analyzed regarding the total fluorescence process including excitation, fluorescence emission and collection close to a dielectric interface. The realized experimental scheme showed superior performance of this concept with substantially enhanced molecular brightness when compared to the classical confocal setup based on water immersion objectives. Selected applications in life sciences including dual-color surface fluorescence correlation spectroscopy, enzyme kinetics and receptor–ligand binding are underlining the interest of this experimental approach.

Introduction

Fluorescence fluctuation spectroscopy (FFS) is a general designation for the observation of the fluorescence process at the single molecule level. A well-known technique for single molecule detection is fluorescence correlation spectroscopy (FCS), which was conceived to study kinetic processes through the statistical analysis of fluctuations at thermodynamic equilibrium (1). The temporal correlation of the photon emission trace of a molecular system (generally a biomolecule specifically labeled with a fluorophore) allows extracting the characteristic time scales and the relative weights of different transitions of this molecular system. Therefore, with an appropriate model of the system dynamics, different characteristic kinetic rates can be accessed. For example, fluctuations in the number of few fluorescent particles unravel the diffusion dynamics within the highly confined sampling volume.

The principal concept was already described in the early 70's by the first paper on this topic by Magde et al. (2). A real renaissance arrived in 1993 with the introduction of the confocal illumination scheme for FCS by Rigler et al. (3). Since then FCS has been developed towards probably the most important technique for single molecule detection (SMD). FCS allows measuring rates of binding/unbinding reactions (4, 5), coefficients of translational and rotational diffusion (6, 7, 8, 9), conformational states and a manifold of photophysical parameters (10, 11, 12) of the induced fluorescent process.

For FFS the confocal illumination scheme led to a highly confined excitation and detection volume. Alternative optical schemes for confining the excitation volume, for instance evanescent field excitation, were demonstrated already in the 80's (13, 14). However, they did not show a performance comparable to confocal FFS for single molecule detection.

In this contribution, we summarize and present numerous recent results of total internal reflection FCS, i.e. single molecule detection in the proximity of dielectric surfaces, emphasizing interesting features which overcome the past limitations of evanescent field FCS (15, 16). First, we outline the general aspects of FFS for SMD and describe the essential steps of a FFS measurement. In the following, a novel dual-color setup for SMD at surfaces is presented (17). This setup enables high photon count rates per molecule well beyond count rates known from classical excitation schemes. Finally, we demonstrate this concept on selected biological applications, such as enzymatic catalysis and monitoring of membrane embedded proteins.

The fluorescence fluctuation process

FFS applied to SMD involves recording and analyzing traces of emitted fluorescence photons emanating from a tiny probe volume. Figure 1 outlines the

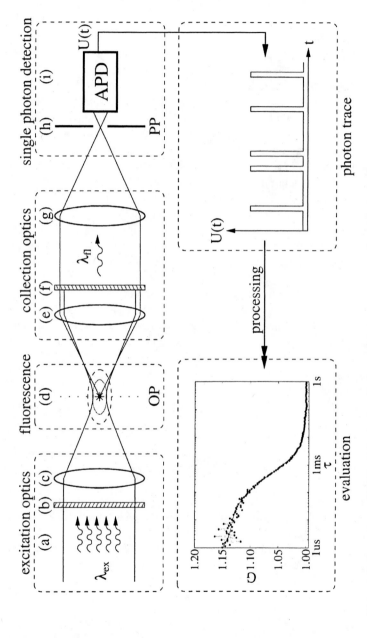

Figure 1. Overview of FFS methods for single molecule detection. OP: object plane, PP: conjugated pinhole plane. (a) Excitation light, (b) excitation band-pass filter, (c) focusing optics, (d) excitation (solid line) and detection (dashed line) volumes, (e) collection optics, (f) emission band-pass filter, (g) tube lens, (h) pinhole, (i) single photon detector.

instrumental chain including the molecular fluorescence process as happening in a typical SMD experiment. The excitation light at wavelength λ_{ex} is tightly focused into the sample and generates a highly confined excitation volume as indicated in Figure 1d in solid-line ellipse. Fluorescently labeled biomolecules diffusing through this confined light field are excited and emit photons at a wavelength $\lambda_{fl} > \lambda_{ex}$. These fluorescence photons are partly gathered by the collection optics. A dielectric band-pass filter rejects residual excitation light and suppresses Raman scattered light contributions.[1]

The fluorescence photons are focused onto the pinhole rejecting stray and out-of-focus light. The pinhole and the collection optics determine the detection volume as indicated in Figure 1d in dashed-line ellipse. For optimum FFS measurements, the excitation and detection volume must be perfectly matched, i.e. the pinhole must be in a conjugated position to the excitation volume. The diameter of the pinhole determines the lateral extent of the detection volume and is chosen according to the criteria of diffraction-limited imaging.

The single photon detector – typically an avalanche photo diode (APD) driven in counting mode – detects the arrival of fluorescence photons. The recorded photon trace is finally evaluated according to the chosen FFS method, i.e. the sequence of detection events is numerically processed for yielding information about the investigated sample.

FFS retrieves essential information about biomolecules by analyzing the fluctuations of the fluorescence intensity $I(t)$ during a short time interval T. Typical sources for these fluctuations are particle motions through the sampling volume via diffusion and flow, the stochastic nature of the photophysical response of the fluorophores, for instance singlet or triplet transitions, and variations of the quantum yield with the molecular environment or binding/unbinding reactions.

A low particle number causes high relative fluctuations of the fluorescence intensity rendering FFS measurements more robust amplitudes. Therefore, a very small excitation volume is required, which is normally achieved with a high numerical aperture (NA) objective. Excitation volumes as low as $V_{ex} \approx 0.3\mathrm{fl}$ are not uncommon for measuring particle concentrations in the range up to $\approx 100\mathrm{nM}$.

The fluorescence emission of individual fluorophores is very weak. Consequently, the collection optics should capture as much fluorescence photons as possible, which demands the use of high NA optics. Furthermore, since the fluorescence intensity is typically 8 to 10 orders of magnitude weaker than the excitation intensity, dichroic filters must block the excitation light by at

---

[1] Raman scattering of individual particles is many orders of magnitude weaker than fluorescence. However, the high solvent concentration (55M for water) largely compensates this low efficiency, making Raman scattering one of the most important background sources.

least 9 to 11 orders of magnitude for obtaining a good signal to noise ratio (SNR). An excitation band-pass filter is used in front of the focusing optics for achieving a spectrally pure excitation, which is subsequently blocked efficiently by a complementary emission band-pass filter.

In summary, FFS relies on instrumental features, such as the generation of a confined sampling volume, the high collection efficiency of the observation system, the detection of single photons, the processing of these photon events as well as the stochastic characteristics of the investigated molecular system.

In the following sub-section, we describe, compare and analyze in detail the sampling volume, in particular for an evanescent field excitation. The dipole response and its interaction with a dielectric substrate within the framework of the molecule detection efficiency will be considered in detail as a relevant model for the overall FFS process. Finally, fluorescence correlation spectroscopy (FCS) as a prominent member of FFS techniques will be described in the context of SMD.

Molecule detection efficiency and confined sampling volume

The analysis and interpretation of the fluctuation statistics obtained with FFS measurements requires knowledge of the excitation and detection volume and the detected fluorescence brightness profile, i.e. the molecule detection efficiency of the fluorescent particle at position $\vec{r}$. The brightness $Q(\vec{r})$ of a single fluorophore is determined by its photophysical properties, the excitation intensity $\Phi(\vec{r})\, hc/\lambda_{ex}$ (where $\Phi(\vec{r})$ is the excitation photon flux, h Planck's constant, c the speed of light and λ_{ex} the excitation wavelength), the collection efficiency $CEF(\vec{r})$ and the pinhole transmission efficiency $T(\vec{r})$.[2]

The fluorophore's brightness is given by

$$Q(\vec{r}) = q_d\, R_{fl}(\vec{r})\, CEF(\vec{r})\, T(\vec{r}) \tag{1}$$

with q_d being the detection quantum yield. $R_{fl}(\vec{r})$ describes the average fluorescence emission rate and is expressed as

$$R_{fl}(\vec{r}) = \frac{q_{fl}(\vec{r})}{\tau_{ex}(\vec{r}) + \tau_S(\vec{r}) + q_{isc}(\vec{r})\, \tau_T(\vec{r})} \tag{2}$$

---

[2] The pinhole transmission efficiency is frequently approximated by convolving the projected pinhole with the point-spread function of the collection optics. However, a complete wave-optical calculation has to account for the anisotropic emission of fluorophores close to a surface.

where q_{fl} is the fluorescence quantum yield, τ_{ex} the average excitation time, τ_S the excited state lifetime, q_{isc} the transition probability of the singlet-triplet intersystem crossing, and τ_T the triplet state lifetime. Assuming fast rotation of the fluorophore, the absorption at λ_{ex} is isotropic and the excitation rate reads $\tau_{ex}^{-1} = \sigma_{abs}\Phi(\vec{r})$, with σ_{abs} representing the absorption cross-section at wavelength λ_{ex}. It is worth mentioning that the fluorophore's brightness $Q(\vec{r})$ saturates for high excitation intensities. Usually, the excitation intensity is kept in the linear regime (well below saturation); otherwise, saturation effects would introduce artifacts in the recorded photon traces.

In general, the fluorophore's brightness $Q(\vec{r})$ varies with position. Particularly, the fluorophore's photophysical parameters become distance-dependent near a surface, as the presence of the surface modifies the local density of states, and consequently also the transition rates between the fluorophore's electronic states *(18, 19)*. Additionally, the proximity of the fluorophore to the dielectric surface, modeled as a dipole emitter close to a dielectric substrate *(20, 21, 22)*, results in a deformation of the angular power density of the dipole emission, thus altering essentially the collection efficiency $CEF(\vec{r})$.

Figure 2 depicts the calculated radiation profile of a fluorophore at the coverslip–sample interface $z = 0$. The fluorophore is located at the origin of the coordinate system with a dipole moment $\vec{\mu}_0 \,//\,(1,0,1)$. The polar plot indicates the anisotropic and asymmetric angular power density. Thin straight lines represent the critical angle of refraction in the coverslip ($z < 0$) and thick lines show the radiated power density in the $x = 0$ and $y = 0$ planes. For comparison, the power density of an identical dipole $\vec{\mu}_0$ in an isotropic environment is shown in thin lines.

In close vicinity to the coverslip, the dipole "senses" the higher refraction index of the coverslip and radiates power into the coverslip not only at sub-critical angles but also at super-critical angles. This effect stems from the evanescent coupling of the dipole's near field to propagating waves in the coverslip. The power radiated at super-critical angles significantly enhances the total power radiated into the coverslip. However, wave propagation in the sample space is diminished due to destructive interference between the direct radiation and the back-reflection at the interface.

Figure 3 compares the power collection efficiency of two optical collection schemes, system I, a trans-illumination scheme and system II, an epi-illumination scheme. The power collection efficiency is computed with respect to system 0 gathering the power P_0 from a randomly oriented fluorophore located at a distance $z > 0$ from a glass–water interface using a 1.20NA water immersion objective positioned at $z < 0$. The dotted line describes the fluorescence power P_1 collected by system I, consisting of a 1.20NA water

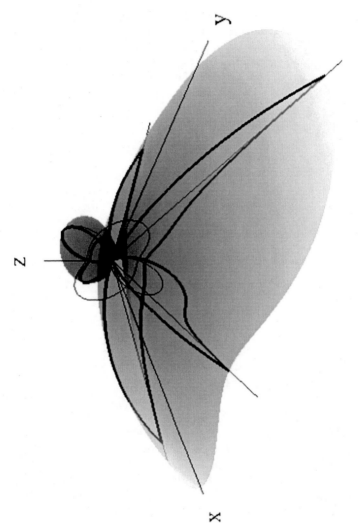

Figure 2. Anisotropic dipole radiation profile modeling a fluorophore at a glass–water interface. The critical angle is indicated by straight lines. Depending on distance and dipole orientation, this radiation profile shows a pronounced anisotropy and asymmetry.

immersion objective positioned at $z > 0$. The observed undulation stems from interference of direct and reflected fluorescence radiation at the interface. Note that this optical collection configuration is commonly employed in prism-based evanescent field excitation setups *(23)*. The solid curve shows the fluorescence power P_2 collected by system II comprising a 1.45NA oil immersion objective positioned at $z < 0$. The exponential decrease (up to $z \approx \lambda_{fl}$) results from the reduced coupling of the fluorophore emission located far from the interface. Finally, the enhancement factor P_2 / P_1 is plotted in dashed line. It is notable that for $z < \lambda_{ex}/3$, the epi-illumination setup II achieves a two-fold increase in power collection efficiency compared to the trans-illumination setup I. This difference is caused by the increased radiation towards the coverslip and the high collection efficiency of the high NA objective collecting the substantial light contribution beyond the critical angle.

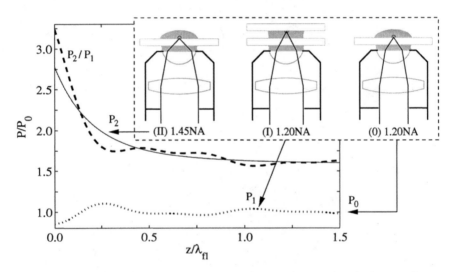

Figure 3. Collected power versus fluorophore position. The glass coverslip ($z < 0$) has a refraction index of 1.520 and the water sample 1.335, respectively. In the region of an evanescent excitation field ($z < \lambda_{ex}/3$), the trans-illumination setup I with a NA of 1.20 collects less power (dotted curve) than the epi-illumination reference system ($P_0 \equiv 1$), whereas the epi-illumination setup II with a NA of 1.45 collects two times more power (solid curve). The enhanced power collection of system II compared to system I is represented by the dashed curve.

Figure 4 depicts brightness profiles $Q(\vec{r})$ of an evanescent field epi-illumination scheme employing a 1.45NA oil immersion objective focused at the glass–water interface at $z = 0$ (system II). Using Equation (1), the brightness calculations were performed assuming excitation and average emission wavelengths of $\lambda_{ex} = 488\text{nm}$ and $\overline{\lambda}_{fl} = 525\text{nm}$, respectively, and a projected pinhole diameter of $0.5\mu\text{m}$ at the glass–water interface. Figure 4a is computed for an evanescent field excitation on a circular area of $16\mu\text{m}$ diameter (c.f. Figure 4a). This results in a cylindrical sampling volume of approximately $\pi (0.3\mu\text{m})^2 0.1\mu\text{m}$. Figure 4b shows the calculations for a linearly polarized field focused to a diffraction-limited spot at the glass–water interface (c.f. Figure 4b), resulting in a hemi-ellipsoidal sampling volume of $0.4\mu\text{m}$ and $0.3\mu\text{m}$ semi-axes and a z-axis dimension of $0.2\mu\text{m}$, respectively.

The effective sampling volume V_{eff} is determined by integrating $Q(\vec{r})$ over the entire volume *(24, 25)*, that is

$$V_{eff} = \left(\iiint Q(\vec{r}) \, d\vec{r} \right)^2 / \iiint \left(Q(\vec{r}) \right)^2 d\vec{r} \tag{3}$$

yielding 39al for total internal reflection (TIR) excitation and 22al for confocal excitation (c.f. 26,27,28). For comparison, confocal excitation with a 1.20NA water immersion objective yields a sampling volume $V_{eff} \approx 250\text{al}$ in liquid. Therefore, FFS at the surface measures particle concentrations in the range up to $\approx 1\mu\text{M}$ instead of $\approx 100\text{nM}$ in liquid with state-of-the-art instrumentation.

Fluorescence correlation spectroscopy

In this sub-section, we introduce an important member of FFS, namely fluorescence correlation spectroscopy (FCS). FCS is an evaluation method based on the temporal intensity auto- or cross-correlation analyses of the photon traces *(1)*. Whereas auto-correlations obtained from single photon traces yield information about particle mobility, particle concentration, as well as kinetics of other fluorescence fluctuation sources, cross-correlations computed using two photon traces provide information mainly about correlated particle motions i.e. binding kinetics. In coincidence studies, the auto- and cross-correlations are simultaneously measured, enabling, for example, the evaluation of the fraction of bound particles in binding/unbinding experiments. The temporal correlation function $G_{m \times n}(\tau)$ is given by

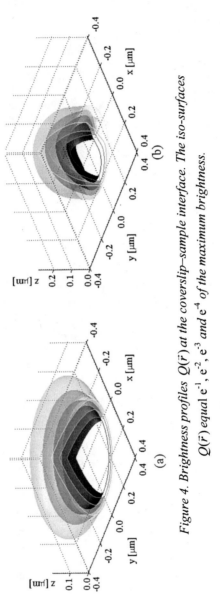

Figure 4. Brightness profiles $Q(\bar{r})$ at the coverslip–sample interface. The iso-surfaces $Q(\bar{r})$ equal e^{-1}, e^{-2}, e^{-3} and e^{-4} of the maximum brightness.

$$G_{m \times n}(\tau) = \frac{\langle I_m(t)I_n(t+\tau)\rangle}{\langle I_m(t)\rangle\langle I_n(t)\rangle} = (T-\tau)\frac{\int\limits_0^{T-\tau} I_m(t)I_n(t+\tau)dt}{\int\limits_0^{T-\tau} I_m(t)dt \int\limits_\tau^T I_n(t)dt} \qquad (4)$$

with τ being the lag time, T the measurement interval, and I_m and I_n the count rates (intensities) in the detection channels m and n, respectively. For example, $G_{A \times A}$ is the temporal auto-correlation of a single channel A, and $G_{A \times B}$ is the temporal cross-correlation of channels A and B.

It is important to point out that FFS methods assume that the observed process is stationary in time and position, i.e. at steady state.[3] However, FFS experiments are usually carried out assuming that the process is *sufficiently* stationary in time, i.e., the sample does not alter significantly during the measurement. Position stationarity is requested primarily if the observation point is scanned during the measurement.

Typically, parameterized model equations are used for describing the temporal correlations $G_{m \times n}(\tau)$. For instance, model equations of temporal auto-correlations are usually given by

$$G(\tau) = G_\infty + G_0 D(\tau)K(\tau)\left(1 + \frac{P_t}{1-P_t}\exp\left(-\frac{\tau}{\tau_t}\right)\right) \qquad (5)$$

where $G_\infty \approx 1$ is the correlation amplitude at infinite lag time and G_0 the amplitude of the particle motion. $D(\tau)$ represents diffusion/flow processes and $K(\tau)$ reads for kinetic processes (e.g. chemical reactions). Finally, P_t and τ_t are the triplet probability and the correlation time of the triplet state, respectively.

Dual-color total internal reflection FCS setup

Total internal reflection FCS proved to be a versatile tool for studying particle motion and photophysics on a single molecule level near a substrate. Its outstanding performance due to the fluorophore–substrate interaction makes epi-illumination TIR-FCS a method of choice for biologically driven applications at the surface, for instance, membrane studies and enzymatic reactions. Very recently, we proposed a new setup for dual-color TIR-FCS *(16, 17)*.

Figure 5 outlines the dual-color epi-fluorescence setup allowing both TIR and confocal excitation. A 18mW HeNe laser (633nm: LHRP-1701, Laser 2000,

---

[3] It is assumed that the investigated sample is ergodic, i.e. the temporal average equals the ensemble (spatial) average.

Weßling, Germany) and a 22mW solid-state laser (488nm: Protera™ 488-15, Novalux, Sunnyvale, CA) provided two linearly polarized beams with 633nm and 488nm wavelength, respectively. These beams were expanded to an e^{-2} diameter of $\approx$ 2mm and collinearly aligned to the microscope objective (BE & BS: 2.5× beam expanders & periscope beam steerers). The laser powers were controlled by neutral density filters. Laser-line clean-up filters (CF: Chroma[4] Z488/10x (blue); Chroma Z633/10x (red)) assured spectrally pure excitations, which were combined by a dichroic mirror (BC: Chroma Z488bcm). An achromatic lens (FC: f = 130mm) focused the beams into the back-focal plane (BFP) of the high NA oil immersion objective (α-Plan-Fluar 100×1.45 with Immersoil™ 518F, Carl Zeiss Jena, Jena, Germany), which resulted in circular areas with e^{-2} diameters of $\approx$ 16μm (blue) and $\approx$ 20μm (red) at the coverslip–sample interface, respectively. In the BFP, a lateral beam focus offset of $\approx$ 2.3mm resulted in a super-critical angle illumination, i.e. in an evanescent field excitation. The sample was a droplet (containing biomolecules in low concentration) on a 150μm thick glass coverslip mounted on a xyz-translation stage (ULTRALign 561D with μDrive Controller ESA-C, Newport Corp., Darmstadt, Germany).

In this epi-illumination setup, the fluorescent light was collected with the same high NA objective, focused onto the pinholes and detected via the single photon detectors. The pinholes were realized by two multimode fibers with a core diameter of 50μm (ASY50/105 silica fibers, Thorlabs Inc., Grünberg, Germany). A dichroic mirror (BS: Omega[5] DML625) separated the green and red fluorescence light, whereas the combination of the main dichroic mirror and band-pass filters (BF: Chroma HQ540/80m and Omega 520DF40 (green); Chroma HQ690/80m and Omega 685DF70 (red)) blocked the back-reflected laser light by more than 10 orders of magnitude. The fluorescence light was detected by fiber-coupled single photon counting modules (APD: SPCM-AQR-14-FC, PerkinElmer Optoelectronics, Wiesbaden, Germany), whose signals were recorded and correlated with a USB hardware correlator (Flex02-08D, Correlator.com, Bridgewater, NJ) linked to a standard PC. Within this setup, the focusing lens, the λ/4 plate (PL: OWIS, Staufen, Germany) and the main dichroic mirror (DM: Omega DM488/633) combination was laterally shifted by a linear translator, thereby positioning the beam foci off-axis in the BFP of the objective. This allows an independent excitation angle adjustment while keeping the beams focused on the BFP of the objective. As already mentioned, the configuration can easily be changed to a confocal epi-illumination by removing the focusing lens and by centering the collimated beams in the BFP.

---

[4] Chroma Technology Corp., Brattleboro, VT
[5] Omega Optical Inc., Brattleboro, VT

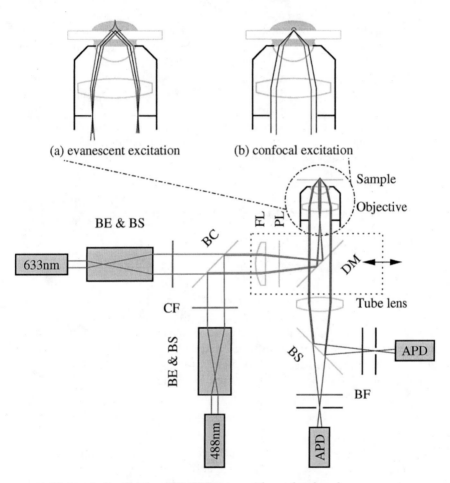

(a) evanescent excitation   (b) confocal excitation

Sample

Objective

BE & BS

BC

FL PL

633nm

DM

Tube lens

CF

BE & BS

BS

APD

488nm

BF

APD

Figure 5. Dual-color TIR-FFS setup with confocal and evanescent epi-illumination at the coverslip–sample surface.

Applications from photophysics to life sciences

Single molecule coincidence assay measured with dual-color TIR-FCS

A synthetic binding assay based on free Rhodamine Green (RhG), Cyanine[5] (Cy5) fluorophores and a 40mer double-stranded desoxyribonucleic acid (dsDNA) labeled with Alexa488 and Cy5 (Zeiss cross-correlation standard) was investigated using the dual-color TIR-FCS system *(17)*. The laser settings were optimized with the dsDNA sample, whereas the background was measured using a NaCl/EDTA/TRIS pH 8.0 buffer. Different mixtures of the double-labeled dsDNA solution with a solution of 9nM RhG / 50nM Cy5 were investigated and measured during 20s. Plasma cleaning of the coverslips was found to be a necessary processing step and improved the SNR due to a strong suppression of unspecific binding at the glass surface.

The amplitudes of the experimental auto-correlation curves were corrected for afterpulsing and for background to avoid systematic biases *(28)*. The following model equation was used for the FCS analysis of these three diffusing species:

$$G_{mn}(\tau) = G_{mn\infty} + \frac{1}{2}\left(1 - \frac{B_m}{I_m}\right)\left(1 - \frac{B_n}{I_n}\right)\frac{\sum Q_{mi}Q_{ni}N_i D_{mni}(\tau)}{\sum Q_{mi}N_i \sum Q_{ni}N_i}$$
$$+ G_{mnt}\exp\left(-\frac{\tau}{\tau_{mnt}}\right) \tag{6}$$

Here, indices m and n represent the green and red detection channels; hence, G_{gg} and G_{rr} are the auto-correlations and G_{gr} the cross-correlation. B is the average background count rate and I the average total count rate. B and I were used for background correction of the diffusion amplitude. The index i represents the diffusing species: g for RhG, r for Cy5 and c for the dsDNA. Q_{mi} is the average brightness in channel m of species i, N_i are the average number of molecules in the sampling volumes. $Q_{mi}Q_{ni}N_i$ is the joint count rate of species i in both detection channels, whereas $Q_{mi}N_i$ and $Q_{ni}N_i$ are the count rates in either channel. G_{mnt} and τ_{mnt} are the amplitudes and the correlation times of photophysical relaxations, such as triplet state population or isomerization. It was assumed that τ_{mnt} is much shorter than the diffusion times, which allows writing the triplet contribution as an additional term, thus simplifying the model. D_{mni} describes the diffusion and is given by

$$D_{mni}(\tau) = \left(1 + \frac{\tau}{\tau_{ixy}}\right)^{-1}\left(\sqrt{\frac{\tau}{\pi\tau_{iz}}} + \left(1 - \frac{\tau}{2\tau_{iz}}\right)\mathrm{erfcx}\left(\sqrt{\frac{\tau}{4\tau_{iz}}}\right)\right) \tag{7}$$

where τ_{iz} and τ_{ixy} are the axial and lateral diffusion times of species i, respectively, and the scaled complementary error function reads as $\mathrm{erfcx}(x) = \exp(x^2) \times \mathrm{erfc}(x)$. A multidimensional least-squares Gauss-Newton algorithm was used to fit the experimental data to the above model equations.

Figure 6 demonstrates the auto- and cross-correlation curves obtained for the dsDNA sample. With a mixture of free fluorophores, the cross-correlation amplitude ($G_{br} \approx 1$) was very small compared to the auto-correlation amplitudes. Thin solid lines show the fits with the model equations (6). For $\tau > 10\mu s$, the fit residuals were lower than 10^{-2}. For smaller lag times, shot noise and afterpulsing reduced the signal to noise ratio. The measured fraction of dsDNA scaled linearly with the mixed fraction from $\approx 1\%$ (no dsDNA) to $\approx 28\%$ (only dsDNA) with a relative scatter of $\approx \pm 15\%$. The measured fraction was at best one third of the mixed fraction due to an excess of molecules with a single green label, possibly further enhanced by photobleaching of the red label during the two-color excitation (29). The overlap of the sampling volumes was estimated to be $\approx 60\%$, which is close to the theoretical maximum. To reduce the influence of photobleaching, we used excitation intensities of $\approx 10\mu W/\mu m^2$. The molecular brightness was about two times higher compared to a confocal epi-illumination employing a 1.20NA water immersion objective at identical excitation intensities (16). The diffusion times of the dsDNA were $\tau_z \approx 51\mu s$ axially and $\tau_{xy} \approx 2.3ms$ laterally. With an evanescent field depth of 160nm and a waist radius of 370nm, the diffusion constant was calculated to be $D \approx 1.5 \times 10^{-7} cm^2/s$, which is about 22% of the estimated diffusion constant $D_{DNA} \approx 6.8 \times 10^{-7} cm^2/s$ for a rod-like molecule with 24Å diameter and 140Å length. We attribute the differences mainly to an increased hydrodynamic drag near the interface and unspecific binding (ionic interaction between DNA and coverslip).

This study demonstrated dual-color single molecule FCCS measurements based on epi-illumination TIR. This TIR-FCCS concept offers distinct advantages to confocal FCCS for coincidence assays at solid/liquid surfaces, in particular by virtue of the much higher fluorescence collection efficiency and the high confinement of the excitation field at the surface.

Single enzyme reaction kinetics measured with TIR-FFS

We investigated the kinetics of the catalytic cycle of single horseradish peroxidase enzymes when hydrogen peroxide (H_2O_2) as an electron donor is processed for oxidizing (dihydro)Rhodamine 123, thereby generating Rhodamine 123 (Rh123) (30 and references therein).

Horseradish peroxidase is a 44kDa heme protein, which efficiently catalyses the decomposition of H_2O_2 in the presence of hydrogen donors. For these experiments, we used the fluorogenic substrate (dihydro)Rh123 as hydrogen donor. After oxidation, it yields the highly fluorescing Rh123.

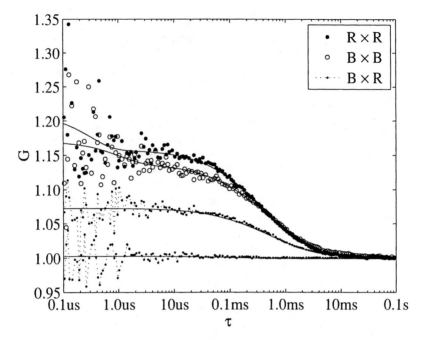

Figure 6. 2C-TIR-FCCS correlation curves and fits. G_{rr}, G_{bb} and $G_{br} > 1$ were measured with doubly labeled dsDNA. $G_{br} \approx 1$ was measured with a solution of RhG and Cy5 fluorophores.

The enzyme, the substrate and the enzyme–substrate complex are non-fluorescent. However, the product and the enzyme–product complex are fluorescent. For each catalytic cycle, two (dihydro)Rh123 are bound to the horseradish peroxidase and turned over into Rh123, which finally dissociate from the enzyme. Edman and Rigler *(30)* suggested that the enzyme retains some conformation memory resulting in a fluctuating enzyme activity, which is non-Markovian by nature. In a simplified model, the enzyme processes the substrate at a very high rate if it runs along a preconditioned reaction pathway, which is supposed to correspond to an "active conformation". In contrast, the enzyme processes the substrate at a very low rate if following a sub-optimal reaction pathway i.e. the "inactive conformation". It is supposed that once a pathway is adopted it favors the substrate processing due to some persistent structural information (conformation memory) retained between consecutive catalytic cycles. Overall, this leads to a fluctuating processing rate whenever the enzyme is changing the pathway i.e. the catalytic cycle.

The production rate of a single enzyme can be observed by detecting the Rh123 emission at the single enzyme–single molecule level. As the evanescent excitation confines the excitation to the surface-immobilized enzyme, the background is efficiently reduced, which translates in an increased SNR when measuring the enzyme activity. For illustration, Figure 7 shows a typical photon trace during a short interval. Fluorescence bursts indicate periods of high enzyme activity. Interruptions are due to inactive periods *(31, 32)*.

Membrane protein detection by image correlation microscopy

Sinner et al. recently published a novel method for *in vitro* synthesis of complex mammalian membrane proteins into artificial planar lipid membrane structures. The cellular extract of rabbit reticulocytes contains the protein synthesis machinery for *de novo* synthesis of an olfactory receptor species starting from the mere DNA from the receptor. We are investigating the density of the inserted receptor proteins by image correlation microscopy *(33)* using evanescent excitation at the surface. For instance, Figure 8 shows the fluorescently labeled antibodies tagging affinity labels of the individual membrane proteins. The excitation area has a diameter of about $20\mu m$. High-resolution image correlation microscopy, i.e. the *spatio-temporal* auto-correlation of the membrane proteins, yields information about the spatial protein distribution (incorporation density) as well as the protein mobility (diffusion in the membrane) *(34)*.

Conclusions

Dual-color total internal reflection fluorescence fluctuation spectroscopy provides substantial improvements compared to other existing confocal or evanescent illumination FFS setups used for single molecule studies at surfaces. Evanescent field excitation by total internal reflection at the coverslip–sample interface and the enhanced fluorescence detection efficiency within the evanescent field are major benefits for investigating biological processes and materials immobilized on glass slides. Fluorescent labeling of molecules in combination with very efficient fluorescence detection features a high signal to noise ratio for single molecule detection and imaging. In addition, dual-color excitation and detection improve the selectivity in coincidence measurements.

Acknowledgements

We would like to thank Eva-Kathrin Sinner and Rudolf Robelek for generously contributing the membrane protein application example, Hans Blom for assisting the cross-correlation measurements, and Jerker Widengren for numerous discussions and improvements of this work.

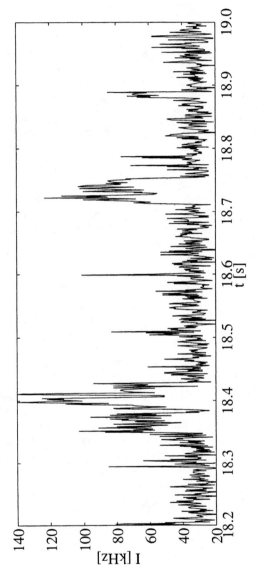

Figure 7. Photon trace of a single horseradish peroxidase producing Rh123.

Figure 8. Image of labeled membrane proteins inserted into an artificial planar membrane surface.

References

1. *Fluorescence Correlation Spectroscopy: Theory and Applications;* Rigler, R.; Elson, E. S., Eds.; Springer Series in Chemical Physics: Berlin Heidelberg, Germany, 2001; Vol. 65.
2. Magde, D.; Webb, W. W.; Elson, E. Thermodynamic Fluctuations in a Reacting System – Measurement by Fluorescence Correlation Spectroscopy. *Phys. Rev. Lett.* **1972**, *29*, 705–708.
3. Rigler, R.; Mets, Ü.; Widengren, J.; et al. Fluorescence Correlation Spectroscopy with high Count Rate and low Background – Analysis of Translational Diffusion. *Eur. Biophys. J. Biophys. Lett.* **1993**, *22*, 169–175.
4. Bacia, K.; Majoul, I. V.; Schwille, P. Probing the endocytic pathway in live cells using dual-color fluorescence cross-correlation analysis. *Biophys. J.* **2002**, *83*, 1184–1193.
5. Weidemann, T.; Wachsmuth, M.; Tewes, M.; Rippe, K.; Langowski, J. Analysis of Ligand Binding by Two-Colour Fluorescence Cross-Correlation Spectroscopy. *Single Mol.* **2002**, *3*, 49–61.
6. Schwille, P.; Korlach, J.; Webb, W. W. Fluorescence Correlation Spectroscopy With Single-Molecule Sensitivity on Cell and Model Membranes. *Cytometry* **1999**, *36*, 176–182.

278

7. Delon, A.; Usson, Y.; Derouard, J.; Biben, T.; Souchier C. Photobleaching, Mobility, and Compartmentalisation: Inferences in Fluorescence Correlation Spectroscopy. *J. Fluorescence* **2004**, *14*, 255–267.

8. Aragón, S. R.; Pecora, R. Fluorescence correlations spectroscopy as a probe of molecular dynamics. *J. Chem. Phys.* **1976**, *64*, 1791–1803.

9. Ehrenberg, M.; Rigler, R. Rotational Brownian motion and fluorescence intensify fluctuations. *Chem. Phys.* **1974**, *4*, 390–401.

10. Widengren, J.; Schwille, P. Characterization of Photoinduced Isomerization and Back-Isomerization of the Cyanine Dye Cy5 by Fluorescence Correlation Spectroscopy. *J. Phys. Chem. A* **2000**, *104*, 6416–6428.

11. Widengren, J.; Rigler, R. Mechanisms of photobleaching investigated by fluorescence correlation spectroscopy. *Bioimaging* **1996**, *4*, 149–157.

12. Widengren, J.; Mets, Ü.; Rigler, R. Fluorescence correlation spectroscopy of triplet states in solution: A theoretical and experimental study. *J. Phys. Chem.* **1995**, *99*, 13368–13379.

13. Thompson, N. L.; Burghardt, T. P.; Axelrod, D. Measuring Surface Dynamics of Biomolecules by Total Internal-Reflection Fluorescence with Photobleaching Recovery or Correlation Spectroscopy. *Biophys. J.* **1981**, *33*, 435–454.

14. Thompson, N. L.; Axelrod, D. Immunoglobulin Surface-Binding Kinetics studied by Total Internal-Reflection with Fluorescence Correlation Spectroscopy. *Biophys. J.* **1983**, *43*, 103–114.

15. Hassler, K.; Anhut, T.; Rigler, R.; Gösch, M.; Lasser, T. High Count Rates with Total Internal Reflection Fluorescence Correlation Spectroscopy. *Biophys. J.* **2005**, *88*, L1–L3.

16. Hassler, K.; Leutenegger, M.; Rigler, P.; Rao, R.; Rigler, R.; Gösch, M.; Lasser, T. Total internal reflection fluorescence correlation spectroscopy (TIR-FCS) with low background and high count-rate per molecule. *Opt. Express* **2005**, *13*, 7415–7423.

17. Leutenegger, M.; Blom, H.; Widengren, J.; Eggeling, C.; Gösch, M.; Leitgeb, R. A.; Lasser T. Dual-color Total Internal Reflection Fluorescence cross-Correlation Spectroscopy. *J. Biomed. Opt.* **2006**, *11*, accepted for publication.

18. Paulus, M.; Martin, O. J. F. Light propagation and scattering in stratified media: a Green's tensor approach. *J. Opt. Soc. Am. A* **2001**, *18*, 854–861.

19. Ruppin, R.; Martin, O. J. F. Lifetime of an emitting dipole near various types of interfaces including magnetic and negative refractive materials. *J. Chem. Phys.* **2004**, *121*, 11358–11361.

20. Hellen, E. H.; Axelrod, D. Fluorescence emission at dielectric and metal-film interfaces. *J. Opt. Soc. Am. B* **1987**, *4*, 337–350.

21. Novotny, L. Allowed and forbidden light in near-field optics. I. A single dipolar light source. *J. Opt. Soc. Am. A* **1997**, *14*, 91–104.

22. Mertz, J. Radiative absorption, fluorescence, and scattering of a classical dipole near a lossless interface: a unified description. *J. Opt. Soc. Am. B* **2000**, *17*, 1906–1913.

23. Pero, J. K.; Haas, E. M.; Thompson, N. L. Size Dependence of Protein Diffusion Very Close to Membrane Surfaces: Measurement by Total Internal Reflection with Fluorescence Correlation Spectroscopy. *J. Phys. Chem. B* **2006**, *110*, 10910–10918.

24. Schwille, P. In *Fluorescence Correlation Spectroscopy: Theory and Applications;* Rigler, R.; Elson, E. S., Eds.; Springer Series in Chemical Physics: Berlin Heidelberg, Germany, 2001; Vol. 65, pp 364–366.

25. Wohland, T.; Rigler, R.; Vogel, H. The standard deviation in fluorescence correlation spectroscopy. *Biophys. J.* **2001**, *80*, 2987–2999.

26. Ruckstuhl, T.; Seeger, S. Attoliter detection volumes by confocal total-internal-reflection fluorescence microscopy. *Opt. Lett.* **2004**, *29*, 569–571.

27. Levene, M. J.; Korlach, J.; Turner, S. W.; Foquet, M.; Craighead, H. G.; Webb, W. W. Zero-Mode Waveguides for Single-Molecule Analysis at High Concentrations. *Science* **2003**, *299*, 682–686.

28. Leutenegger, M.; Gösch, M.; Perentes, A.; Hoffmann, P.; Martin, O. J. F.; Lasser, T. Confining the sampling volume for Fluorescence Correlation Spectroscopy using a sub-wavelength sized aperture. *Opt. Express* **2006**, *14*, 956–969.

29. Eggeling, C.; Widengren, J.; Brand, L.; Schaffer, J.; Felekyan, S.; Seidel, C. A. M. Analysis of Photobleaching in Single-Molecule Multicolor Excitation and Förster Resonance Energy Transfer Measurements. *J. Phys. Chem. A* **2006**, *110*, 2979–2995.

30. Edman, L.; Rigler, R. Memory landscapes of single-enzyme molecules. *PNAS* **2000**, *97*, 8266–8271.

31. Hassler, K. Ph.D. thesis, École Polytechnique Fédérale de Lausanne, Lausanne, Switzerland, 2005.

32. Hassler, K.; Rigler, P.; et al. *to be published.*

33. Petersen, N. O.; Brown, C.; Kaminski, A.; Rocheleau, J.; Srivastava, M.; Wiseman, P. W. Analysis of membrane protein cluster densities and sizes in situ by image correlation spectroscopy. *Faraday Discuss.* **1998**, *111*, 289–305.

34. Leutenegger, M.; Sinner, E. K.; Robelek, R.; Lasser, T.; et al. *to be published.*

Chapter 18

High-Resolution THz Spectroscopy of Crystalline Peptides: Exploring Hydrogen-Bonding Networks

K. Siegrist, C. R. Bucher, C. Pfefferkorn, A. Schwarzkopf, and D. F. Plusquellic

Optical Technology Division, National Institute of Standards and Technology, Gaithersburg, MD 20899-8443

The sensitivity of CW THz spectroscopy to hydrogen bonding networks of peptide crystals is explored through investigations of the lowest frequency vibrational modes including three forms of trialanine, and two isostructural dipeptides which form hydrophobic nanotubes, alanyl isoleucine and isoleucyl alanine. THz spectra were obtained in the range 0.6 cm^{-1} to 100 cm^{-1} at 4.2 K for the peptides under investigation. Three issues related to the THz absorption features have been addressed: i) the impact of the intermolecular hydrogen bonding network, ii) effects arising from weak hydrophobic interactions with water and iii) line broadening due to crystal defects. First, large variations in the THz spectra of antiparallel β-sheet trialanine are observed due to changes in the degree of hydration. Second, subtle frequency shifts of ~ 1 cm^{-1} due to the weak hydrophobic interactions of dipeptide molecules with intracrystalline water are reported for the hydrated dipeptide alanyl isoleucine. In contrast, little or no shift was observed for the unhydrated but isostructural retroanalogue, isoleucyl alanine. Finally, line-broadening effects which can be attributed to the relative abundance of crystal defects are reported for the parallel-β-sheet form of trialanine. The small crystalline peptides studied here provide benchmark systems for validating computational models at the level of semi-empirical force fields and, because of the small molecular size, at the *ab initio* levels of theory. The experimental results give insight into the nature and scope of computational models necessary to accurately capture the structure and spectra of these simple peptides. Studies of simple systems are a necessary first step towards delineating the model requirements imposed by the very low energy modes of this regime, where long range and many body interactions can have a significant impact. Comparison with computational results will aid in advancing the current state-of-the-art in computational software to accurately model more complex biological systems.

Introduction

The vibrational modes in the THz region of the spectrum involve large-amplitude motions, often of an entire molecule. Motions involving backbone bending, tail-wagging, and torsional motions of methyl groups typify these very low energy modes. Such large amplitude motions have an entirely different character than the local modes probed in the mid-IR region, which are typified by bond stretches and bends. The global motions probed in the THz region are directly relevant to the dynamics of biological molecules, and THz spectroscopy of biomolecules has consequently generated much current interest. Previous studies of polypeptides (*1,2*), proteins (*3-7*), DNA (*8-10*), sugars (*11-13*), and other biomolecules (*14-16*) have examined the dependence of THz absorption on such factors as the degree of hydration, and amorphous versus crystalline forms. However, the THz vibrational spectra reported for these systems are often unresolved, due to the conformational flexibility and the variety of environmental interactions that exist in solutions and amorphous solids. The potential energy surfaces associated with these bending and torsional modes are "soft", having shallow minima so that mechanical anharmonicity is expected to be quite important. Multiple minima may be separated only by very small barriers and many conformational modes can exist for unrestricted molecular motions. The resulting unresolved spectra do not lend themselves to the definitive interpretation of individual resonances in the THz region. In contrast, conformational flexibility is constrained in a crystalline network such that the bending and torsional motions of individual molecules must be collective, giving rise to oscillations characteristic of the entire crystalline structure rather than eigenmodes of individual molecules. Since conformational flexibility is constrained in these systems, THz spectra of low-defect crystalline solids can be very sharp, with line widths <1 cm^{-1} (*11-16*).

In addition, the shallow potential energy surfaces are intrinsically more sensitive than the surfaces associated with local modes in the mid-IR region. Experimentally, THz vibrational frequencies are indeed found to be sensitive to small perturbations of the molecular environment. In computations, they are likewise sensitive to the accuracy of detail of the model, as well as the numerical accuracy of the method. These factors increase the difficulty of making a descriptive vibrational assignment of the nuclear motions, and underline the advantages of studying well resolved spectra of small crystalline systems, for which accurate models may be devised.

In previous studies of trialanine (*17*), it was found that the spectral features in the THz region are extremely sensitive to the different crystalline forms of a peptide, including the presence of intra-crystalline water molecules. The intracrystalline water in trialanine is an integral part of the hydrogen bonding network, interlinking the sheets of anti-parallel β-sheet trialanine (*ap*-trialanine). This single result helps to clarify the impact of intermolecular hydrogen bonding on THz spectra. In sharp contrast, for spectra in the mid-IR region, the spectral patterns of hydrated and unhydrated forms of *ap*-trialanine are relatively insensitive to changes in the intermolecular hydrogen bonding networks that stabilize peptide crystals.

Changes in the well-defined hydrogen bonding networks that exist in high quality peptide crystals can have quite a dramatic impact on the THz spectra, but smaller effects can provide insight as well. In the present work, spectral changes caused by different degrees of hydration of *ap*-trialanine are first investigated. Next, in a second series of experiments, we have begun studying a group of structurally well-characterized dipeptides that are known to have very similar three dimensional hydrogen bonding networks. The members of this group, the Val-Ala dipeptide family (*18-20*), all belong to the space group P6₁ and form hydrophobic-core nanotubes with pore diameters between 3.3 and 5.2 Å, depending upon the size of the residues that line the pore. All these dipeptides share the same hydrogen bonding pattern, although the strength of an individual bond in the network naturally varies between different peptides. Like *ap*-trialanine, some of the crystal structures of these dipeptides are known from x-ray studies to contain co-crystallized water molecules (*18-21*). Unlike trialanine, in which water molecules are an integral part of the hydrogen bonding network, the intracrystalline water molecules which reside inside the pores of these dipeptide nanotubes can have only weak hydrophobic interactions with the peptide molecules. In light of the changes in *ap*-trialanine spectra on removal of water, it is natural to ask if removal of non-bonded intracrystalline water can affect the THz spectra of these dipeptides. Of the two dipeptides studied here, the crystal structure of alanyl isoleucine (AI) is hydrated, while its retroanalogue, isoleucyl alanine (IA), is unhydrated.

Finally, in the last series of experiments, the more subtle effects of hydrogen bond disruption are investigated. Crystalline structures can have varying degrees of disorder depending on crystal quality, or some disorder can be inherent in the structure. Here, the effects of disorder in the crystalline structure arising from crystal defects are investigated in the spectra of parallel β-sheet trialanine.

Experimental

A full description of the continuous-wave THz spectrometer and its performance for high resolution THz laser studies is given elsewhere (5,14,22). Briefly, the system consists of a low-temperature-grown GaAs photomixer (23-26) driven at the difference frequency of two near-infrared lasers. The first, a fixed frequency diode laser, operates near 850 nm ($\Delta v_{FWHM} \approx 0.0001$ cm[-1]) while the second, a newly constructed standing-wave Ti:Sapphire (Ti:Sapp) laser having a resolution of $\Delta v_{FWHM} \approx 0.04$ cm[-1], is broadly tunable and gives more than an order-of-magnitude improvement in resolution over that of the ring laser used in previous work (5,13). This second laser, similar to grating tuned lasers reported elsewhere (27,28) is seeded by feedback from an external grating-tuned cavity through a 4 % output coupler. This configuration improves the scan repeatability of the spectrometer to better than ±0.03 cm[-1], as determined from repeated scans of water vapor in the THz region. The scan repeatability is important to minimize "noise" induced by small shifts in the standing wave interference patterns (with modulation depths that exceed 50 % of the transmitted power) when background ratios are performed.

The tunable Ti:Sapp and the fixed frequency diode laser outputs are combined, chopped at 400 Hz, and focused by an aspherical lens onto the photomixer in vacuum. A maximum power of ≈ 1 μW of THz radiation is obtained from 0.06 THz to 3 THz, with the THz output power decreasing as ω^{-4} beyond the peak value at 0.6 THz. The focused THz beam passes through the cryogenically cooled sample at 4.2 K and is detected by a liquid-helium-cooled silicon-composite bolometer. Power detection sensitivity of the bolometer is <1 nW up to 3 THz in a 400 Hz bandpass (NEP of the bolometer is 1 pW/Hz$^{1/2}$).

The alanine tripeptide, alanyl isoleucine, and isoleucyl alanine were obtained at 98 % purity. Tri-L-alanine was recrystallized in different ways to produce two different crystalline forms. Following the recipes of Hempel, Camerman and Camerman (29), platelet-type crystals of the parallel β-sheet form were obtained and needle-shaped crystals of the anti-parallel β-sheet form were produced. The crystalline dipeptides were used as received.

The peptides were mixed with high purity polyethylene (PE) powder at concentrations of 5% to 35%, either by grinding with mortar and pestle or by shaking in a container with stainless steel bearings. The latter method ensured fine pulverization of recrystallized samples, where this was desired. Peptide samples were prepared from either 100 mg or 200 mg of the mixture, pressed in a vacuum die to make 1.25 cm diameter disk-shaped samples. All 200 mg samples used a large grain (≈150 μm grain diameter) PE for dilution of the peptide, were pressed at 6.89×10^7 Pa (10,000 psi), and had a thickness of ≈ 2 mm. Samples of ≈ 500 μm thickness were produced for all the 100 mg samples, which all used small grain (4 μm mean grain diameter) PE for dilution, and were pressed at a pressure of 2.07×10^7 Pa (3000 psi). The small grain polyethylene was found to minimize Mie scattering losses which were problematic at the higher frequency end of the spectra. Pure polyethylene disks were similarly pressed for background scans.

One mixed sample and one pure PE disk were secured by screw-in rings in the cryogenically cooled brass sample holder. Background and sample spectra were obtained by raising or lowering the sample holder about 1 cm without appreciably altering the incident angle of the THz beam. THz power transmitted is then detected by the bolometer. Both the amplified bolometer signal and a voltage proportional to the photomixer current were oversampled using lock-in bolometer, to eliminate lineshape distortion. The scanning Ti:Sapp frequency is amplifiers, at an interval 5 times smaller than the 30 ms time constant of the measured with a wavemeter every 0.5 second to an accuracy of 0.02 cm^{-1}. Typically, six to nine spectral scans were performed consecutively.

Each scan was first scaled for changes in the lock-in input sensitivity, normalized to changes in the AC component of the photocurrent and linearized according to the wavemeter scale. Finally, the data were averaged, normalized to the transmitted power through the PE blank, and the absorption spectra were obtained from the normalized transmission spectra using a base 10 log scale.

Results and Discussion

Dehydration and Rehydration of *ap*-trialanine.

In previous work on the three crystalline forms of trialanine (parallel β-sheet, antiparallel β-sheet, and dehydrated antiparallel β-sheet), three dramatically different THz spectra were reported (*17*). In particular, it was found that the two anti-parallel β-sheet forms which differ only in the presence or absence of water interlinking the sheets give vastly different spectra, in contrast to spectra in the mid-IR (*30*). A series of THz spectra for *ap*-trialanine taken for various states of hydration are shown in Figure 1. The samples were prepared from a single recrystallization of trialanine, grown over a four week period by slow evaporation of 200 mg trialanine from 150 ml aqueous solution of 15% dimethyl formamide (DMF). For sample characterization, the FTIR spectra (from 400 cm^{-1} to 4000 cm^{-1}) were taken of samples composed of 1%

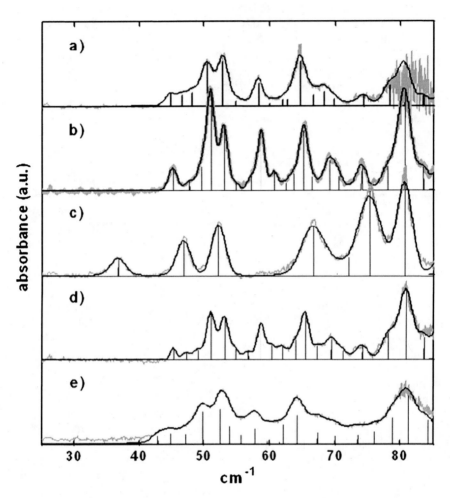

Figure 1. Absorbance spectra of anti-parallel β-sheet trialanine for a) freshly recrystallized trialanine b) partially hydrated, 30-50% water lost c) fully, or almost fully, dehydrated trialanine d) partially rehydrated e) fully rehydrated, or overhydrated, trialanine.

peptide (by weight) in KBr pressed at 6.89×10^7 Pa (10,000 psi). The FTIR spectra were nearly identical with previous work (30) and therefore served to verify the anti-parallel β-sheet structure. For all data of Figure 1, 200 mg of the sample was used and spectra were taken at 4.2K over the range from 25 cm^{-1} to 85 cm^{-1}. No features were observed below 25 cm^{-1} and the regions above 85 cm^{-1} are precluded from discussion because of the limited S/N ratio. To aid in the comparisons between spectra, the individual resonances were identified using a non-linear least squares lineshape fitting procedure. For each line, the center frequency, ω_e, width, $\Delta\omega_e$, and peak absorption intensity, k_e were iteratively varied. The best-fit Gaussian lineshapes are superimposed on the experimental data. The same number of features was used to account for the observed THz absorption intensity of the hydrated spectra of trialanine shown in Figures 1a, 1b, and 1c, 1d. The change in the spectrum of the fully dehydrated sheet is too large to be characterized by the same resonances.

Shown in Figure 1a is the spectrum from freshly recrystallized *ap*-trialanine, prior to any dehydration by exposure to vacuum. Figure 1b shows a spectrum which is only slightly different. The absorption peaks are narrower but positions and relative intensities of the main features are similar, excepting the increased intensity of the peak near 51 cm^{-1}. This sharper spectrum was found for the sample, prepared from vacuum-exposed crystals, which had lost 2% of its weight under vacuum. It should be noted that this is the spectrum obtained from a trialanine sample received from the manufacturer.

Another recrystallized sample of *ap*-trialanine was found to lose a maximum of 4.8% of its weight under vacuum, slightly more than the 4.2% water content expected from the crystal structure. This may indicate excess water in the freshly recrystallized sample, but even so, implies that at least 30% of the structurally integral intracrystalline water is lost under vacuum, in going from the spectrum of Figure 1a to the similar one in Figure 1b. Further vacuum dehydration of the sample of Figure 1b at some point elicits a dramatic change in the spectrum, as demonstrated in Figure 1c. Except for some small peak shifts, this change was also observed for the sample used in Figure 1a after 4.8% weight loss. The dehydrated sample of Figure 1c was then sealed in an empty desiccator with a beaker of water for four days, in an attempt to rehydrate the *ap*-trialanine. The spectrum from this rehydrated sample appears in Figure 1d and clearly shows that a hydrated *ap*-sheet structure has been recovered. The features are broader than those of Figure 1b, but the spectral signature is nearly the same. When this rehydrated sample was allowed to return to room temperature and then immediately cooled to 4.2K before evaporation of intracrystalline water could take place, the final spectrum of Figure 1e resulted. This final spectrum resembles the original spectrum of Figure 1a, having the lower peak intensity at 51 cm^{-1} and broadened features. The broadening may be attributed to further rehydration, since water visibly condenses on the sample during warming from 4.2K to room temperature. In these results, subtle variations in the spectra of *ap*-trialanine are seen with variations in the amount of water in the crystals, as well as the obvious striking change which occurs at some point after 30% water loss.

Closer Examination of Dehydrated *ap*-trialanine Spectra

A second series of spectra of the *ap*-sheet of trialanine are shown in Figure 2 over the interval from 30 cm^{-1} to 60 cm^{-1}. All three spectra are characteristic of the dehydrated *ap*-sheet structure. However, small frequency shifts of the three prominent features are apparent in these spectra, which are arranged by increasing shift. Although the spectra in Figures 2a and 2b were taken from the same sample, the spectrum 2b was first taken and then the sample was warmed quickly to room temperature where it remained under vacuum at 1.33 × 10^{-2} Pa (10^{-4} Torr) for 24 hours, when spectrum 2a was acquired. The last spectrum of Figure 2c was taken from a different, less dehydrated sample.

The slight frequency shifts in the peaks of Figure 2a and 2b can be attributed to loss of remaining traces of water in an already dehydrated sample, after further vacuum exposure. The crystals used in Figure 1c were only coarsely ground, rather than well pulverized and therefore, were expected to have a larger grain size. The larger crystals were found to lose water at a comparatively slower rate. These samples required two days under vacuum to lose 2% of their weight, at which point they produced the spectrum shown in Figure 1b, and finally after four more days under vacuum, they produced the characteristic dehydrated spectrum of Figure 2c. By comparison, finely pulverized crystals (as well as trialanine powder, as received) produced the characteristic spectrum of well-dehydrated trialanine after 8 to 12 hours of under vacuum.

The crystals used for Figure 2c, therefore, were probably less dehydrated than those of Figures 2a and 2b. Finally, it should be noted that in several other cases, extended vacuum exposure always resulted in the most red-shifted peak positions (to within 0.1 cm^{-1}). The small variations in the center frequencies of the three peaks can therefore be attributed to the presence of varying amounts of trace water resulting in the peak positions steadily red-shifting as the crystalline matrix approaches a completely dehydrated state.

The largest shift is seen in the lowest frequency mode located at 35.23 cm^{-1}, 35.44 cm^{-1} and 36.38 cm^{-1} in Figures 2a, 2b and 2c, respectively, showing that trace water can cause shifts on the order of 1 cm^{-1}. The next frequency peak is located at 46.22 cm^{-1}, 46.45 cm^{-1} and 46.86 cm^{-1} in Figure 2a, 2b and 2c, respectively while the third peak is shifted by less than 0.2 cm^{-1} from Figure 1a to 1c. These small shifts are not as dramatic as the change in the spectrum of the *ap*-sheet of trialanine that occurs when a large percentage of the hydrogen bonds cross linking the sheets are broken (i.e., on removal of more than 30% of the intra-crystalline water) but these results give yet another example of the extreme sensitivity of THz spectra to these types of intermolecular interactions.

288

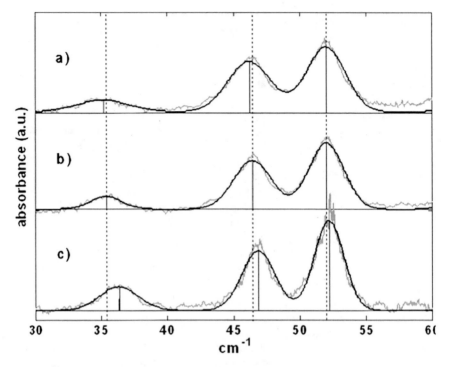

Figure 2. Absorbance spectra of anti-parallel β-sheet trialanine, fully dehydrated or nearly fully dehydrated, showing slight shifts due to small differences in trace water content. Continued dehydration always results in the most down-shifted peak positions.

Dehydration and Rehydration of Dipeptides AI and IA.

The isostructural dipeptides, Ala-Ile and Ile-Ala (or AI and IA), have been well characterized by Gorbitz (*18*) using x-ray crystallography and are known to form hydrophobic nanotubes. Their structures share a very similar, intricate, three dimensional network of hydrogen bonds. AI, with five carbon atoms in its side chain, packs to give a comparatively spacious channel, similar to the channel of AV which has dimensions $\sim 6.0 \times 4.5$ Å. Conversely, the five-carbon-atom side chains of IA pack to create a rather flat channel of dimensions, 5.0×2.5 Å. These two peptides provide an interesting contrast, since the large-channel structure of AI can contain water while its retroanalogue, IA, does not retain trapped water in its constricted channel. The standard 100 mg samples described were used here, prepared from the crystalline dipeptide as received, at 13% to 15% concentration. Spectra for AI and IA are shown in Figures 3 and 4,

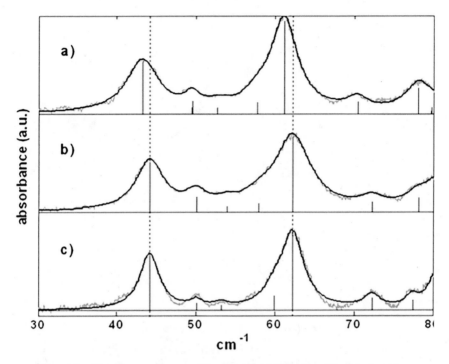

Figure 3. Absorbance spectra from the dipeptide alanyl isoleucine a) as received, b) after dehydration under vacuum c) after attempted rehydration. Water normally resides within the pores of the hydrophobic nanotube-forming crystalline dipeptide. Removal of water results in small shifts in the spectral lines, but attempted rehydration does not restore the original spectrum, as it does for trialanine.

respectively. For the dipeptides, the best non-linear least squares lineshape fits, shown superimposed on the data, are Lorentzian rather than Gaussian.

In Figure 3, the series of spectra of AI taken from 30 cm^{-1} to 80 cm^{-1} show the lowest frequency features. Figure 3a shows the spectrum from an untreated dipeptide sample. When the dipeptide crystals were treated by exposure to vacuum at 1.33×10^{-2} Pa (10^{-4} Torr) for 5 days, the spectrum of Figure 3b was obtained. The two strong peaks in this frequency interval are shifted from 43.31 cm^{-1} to 44.24 cm^{-1} and from 61.22 cm^{-1} to 62.33 cm^{-1}. In addition to shifts in the line centers, the apparent linewidths changed with dehydration from 4.95 to 4.24 cm^{-1} and from 4.20 to 5.45 cm^{-1}, respectively.

Finally, in Figure 3c, the dehydrated crystals were sealed in a desiccator with water for four days, in an attempt to rehydrate the dipeptide in the same way that *ap*-trialanine was rehydrated. After removing the peptide sample from

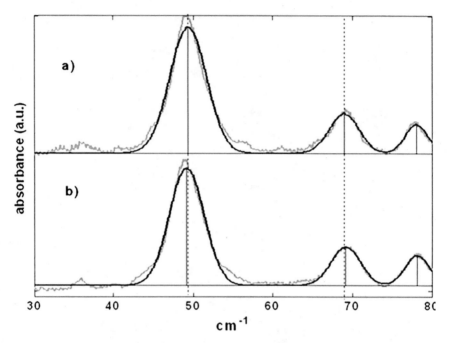

Figure 4. Absorbance spectra from the dipeptide isoleucyl alanine a) as received, b) after dehydration under vacuum. Water is not retained within the pores of this hydrophobic nanotube-forming crystalline dipeptide, and little if any shift is seen in the spectral lines after vacuum exposure.

the water-saturated atmosphere in the desiccator, the sample weight steadily decreased for over an hour, as excess water evaporated. Sample weight was allowed to reach equilibrium in atmosphere. The AI spectrum of Figure 3c was then obtained. Compared to the spectrum of dehydrated AI (Figure 3b), no measurable line shift is seen for the two strong peaks. However, the lines are narrower compared to Figure 3b, with widths of 3.02 cm^{-1} and 3.42 cm^{-1}, respectively.

The crystal structure for AI contains somewhat disordered co-crystallized water molecules in a 1:1 ratio with dipeptide molecules (*18*). This corresponds to a maximum 8% water content within the crystal pores, although it may be noted that solvent content can vary depending upon recrystallization conditions. An AI sample was found to lose 2% of its weight after 5 days of exposure to 1.33×10^{-2} Pa (10^{-4} Torr), corresponding to 25% or more of the water residing within the pores, depending on the initial occupancy. Solvents within the dipeptide nanotubes can be removed (*18,19*), necessarily changing the hydrophobic interactions between solvent and peptide molecules. Crystal dehydration under vacuum can therefore explain the line shifts observed in Figures 3a to 3b. Evidently, water trapped during crystal growth can be removed with relative ease, resulting in the shifted spectrum seen in Figure 3b. Figure 3c implies that water can not be driven back inside the hydrophobic-core nanotubes as easily as water was re-integrated into the trialanine crystal structure. Consequently, the AI spectrum, once shifted by water loss, remains shifted.

Figure 4 provides a comparison to spectra obtained from the retro-analogue, IA. This dipeptide loses its co-crystallized water within a few minutes when crystals are removed from solution (*18*) so that little if any water is expected to remain trapped in the pores of IA. Figure 4a shows a spectrum of the untreated sample while Figure 4b shows a spectrum from a sample treated by exposure to 1.33×10^{-2} Pa (10^{-4} Torr) vacuum for four days. The latter spectrum shows little change in the peak positions, with shifts of the three lines in this frequency interval measuring 0.2 cm^{-1} or less.

In summary, AI is known to retain water in its pores which interacts hydrophobically with the dipeptide molecules, while it is known that IA loses intra-crystalline water in air in a matter of minutes. Line shifts of 1 cm^{-1} accompany crystal dehydration in the AI spectrum compared to little noticeable shift in the IA spectrum on vacuum exposure. Therefore, the shifts in the AI spectrum are taken to indicate that water has been removed under vacuum from the AI structure whereas little if any water was present to remove from the IA structure. These results indicate that the low frequency THz modes are sensitive to changes in the weak hydrophobic interactions of water in these structures.

Crystal Defects in *p*-trialanine

The trialanine system is revisited to discuss inhomogeneous line-broadening effects from crystal defects. Such disorder in the crystalline sample may contribute to the difficulties in theoretical identification of modes by obscuring features in the experimentally obtained spectrum, as was previously suggested for anti- parallel β-sheet trialanine in particular (*17*). Spectra for trialanine in the parallel β-sheet form, also studied in previous work (*17*), are shown in Figure 5. For the spectrum of Figure 5a, 100 mg of trialanine were recrystallized by slow evaporation from 200 ml of a 20% DMF aqueous solution over a period of about a month, under mild vacuum of 0.51×10^{5} Pa (0.5 atmosphere). A standard 200 mg sample was prepared using 40 mg of the freshly recrystallized tripeptide and the spectrum in Figure 5a was acquired for this sample at 4.2 K in the usual manner. Although this spectrum is clearly identifiable as parallel sheet trialanine spectra, when compared to the spectrum discussed in previous work (*17*), the peaks are much broader and consequently many details of the spectrum are lost. In addition, line frequencies are shifted,

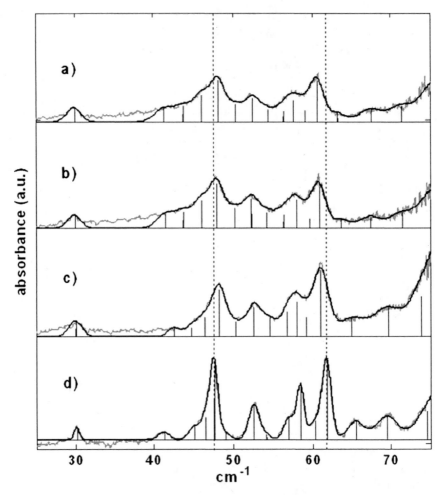

Figure 5. Absorbance spectra from the parallel β-sheet form of trialanine for a) recrystallized from 20% DMF solution b) the same sample, after vacuum exposure, c) a different sample, recrystallized from a 25% DMF solution, d) the same sample, after annealing at 40C for 72 hours. No appreciable change in the spectrum could be achieved by removing water under vacuum for the first sample, whereas annealing of the crystal defects greatly improved the spectrum of the second sample. Line broadening seen in the first three spectra are therefore attributed to the presence of crystal defects, rather than the presence of water traces.

with the two strongest, most easily identifiable peaks shifted by 0.5 cm^{-1} to 1.0 cm^{-1}, compared to peak positions found in prior work.

In Figure 5b, a second spectrum appears for a sample derived from the same recrystallization as Figure 5a. This sample was placed in the 1.33 × 10^{-2} Pa (10^{-4} Torr) vacuum chamber for 24 hours prior to freezing and data acquisition, to remove any traces of water. The spectrum is essentially the same as that in Figure 5a. The parallel sheet structure of trialanine contains no water and recrystallization was done under gentle vacuum to prevent trace water in the crystals. The lack of any significant change with vacuum exposure shows that the line shifts and broadening seen here, compared to the sharp spectrum seen previously (*17*), are not due to intracrystalline water.

The spectra of Figure 5c and 5d provide evidence that the line broadening and shifting are effects of crystal defects. For these spectra, trialanine was recrystallized from a 25% DMF aqueous solution under gentle vacuum. A standard 100 mg sample was pressed using 26 mg of recrystallized tripeptide and again, the spectrum was acquired for the freshly recrystallized sample at 4.2 K as shown in Figure 5c. After acquisition of the initial spectrum, the sample was removed from the vacuum chamber and heated at 40 C for 72 hours. After the sample was again cooled quickly to 4.2 K, the final spectrum appearing in Figure 5d was acquired.

Clearly, the spectrum of Figure 5c has much in common with those of 5a and 5b while the spectrum of Figure 5d is significantly narrower, and peak positions are within 0.1 cm^{-1} of those previously reported for parallel β-sheet trialanine (*17*). Gaussian widths obtained from fits for the lower frequency of the two strongest peaks are 2.12, 2.12, and 2.02 cm^{-1} for the spectra of Figure 5a,b and c, respectively. These values can be compared to 1.20 cm^{-1} for the spectrum in Figure 5d and 1.50 cm^{-1} reported previously (*17*). For the higher frequency of the two strongest peaks, the widths are 2.08, 2.24, and 2.32 cm^{-1} for spectra of Figure 5a,b and c, respectively, compared to 1.39 cm^{-1} for Figure 5d and 1.17 cm^{-1} reported previously (*17*).

Gentle heating of the sample evidently has caused changes in the crystal structure which are reflected in much narrower lines, and peak positions nearly identical to the spectrum previously reported, which was taken from a freshly recrystallized sample untreated by vacuum or heating. Since annealing of the crystal structure can "heal" crystal defects to some degree, these results provide evidence of line-broadening and shifting effects due to crystal structure defects, as opposed to the presence of trace solvent molecules which can cause similar effects.

Finally, it is worth noting that, although there are significant differences in the polyethylene powder used to obtain the final spectrum of Figure 5d and the spectrum obtained in prior work (*17*), the two spectra are nearly identical. For the prior work, polyethylene of ≈150 μm grain diameter was used to dilute the peptide samples and pressed at 6.89 × 10^{7} (10,000 psi). For the spectrum of Figure 5d, however, the polyethylene powder with a 4 μm mean grain diameter was used and the sample was pressed at 2.07×10^7 (3000 psi). The very different particle sizes appear to pack differently since the smaller grains require less pressure and give a pellet with a glassy surface compared to the rough surface for large grain polyethylene. We see no evidence here for pressure-dependent or particle-size dependent frequency shifts at the level of ± 0.1 cm^{-1}, as no significant difference can be seen in comparing the spectrum reported previously (17) to that of Figure 5d.

Conclusions

We have obtained THz vibrational spectra of a number of small peptides at 4.2 K. The THz spectra of *ap*-trialanine showed sensitivity to small changes in the hydration level as well as to the overall structural changes that occur when most of the water is removed. Sensitivity is observed both in the normally hydrated structure and in the nearly dehydrated structure of *ap*-trialanine. In addition, we have obtained spectra at 4.2 K for two hydrophobic nanotube-forming dipeptides with very similar crystal structures and hydrogen bonding networks. The dipeptide AI, like *ap*-trialanine, can contain water. Unlike *ap*-trialanine, its water is not bonded and interacts hydrophobically. The THz spectra of AI show sensitivity to the presence of co-crystallized water. In contrast, the THz spectra of the isostructural dipeptide IA, which does not retain water, showed little change after vacuum exposure. In contrast to the dramatic change seen for dehydrated *ap*-trialanine, the overall spectral pattern is preserved after vacuum exposure of the hydrated peptide AI, but lines are shifted up to 2 cm^{-1} and line widths narrow as much as 30 %. The small line shifts can be attributed to hydrogen bond deformation arising from the weak van der Waals interactions between hydrophilic solvent molecules inside the hydrophobic pore and the dipeptide molecules composing the pore. The THz spectra of this dipeptide implies that co-crystallized water molecules can be removed with relative ease under vacuum in a manner similar to *ap*-trialanine, but that rehydration of the hydrophobic nanotubes is not so easily accomplished. After attempted rehydration, the THz spectra of the AI dipeptide did not reflect the original spectra prior to dehydration. Instead, the spectral lines remained at the positions found immediately after dehydration although line widths broadened.

Finally, we have obtained spectra of *p*-trialanine of varying crystal quality. By comparing the spectrum from an annealed sample (originally of poor crystal quality) with the spectrum of another poor quality sample exposed to vacuum to remove any trace water, it was shown that line broadening and some spectral line shifts can be attributed to crystal defects rather than the presence of water. Overall, we have shown the extreme sensitivity of THz spectra to intermolecular interactions including variations in the level of hydration, hydrophobic interactions between solvent and peptide and the presence of crystal defects.

Acknowledgements

The authors wish to acknowledge the work of Professor Carl H. Gorbitz, University of Oslo, whose crystallographic studies of dipeptides have provided invaluable insight. We further wish to gratefully acknowledge the NSF SURF Program which supported two of us, C. Pfefferkorn and A. Schwarzkopf, and note that this material is based upon activities supported by the National Science Foundation under Agreement No PHY-0453430. Any opinions, findings, and conclusions or recommendations expressed are those of the authors and do not necessarily reflect the views of the National Science Foundation.

References

1. Kutteruf, M. R.; Brown, C. M.; Iwaki, L. K.; Campbell, M. B.; Korter, T. M.; Heilweil, E. J. *Chem. Phys. Lett.* **2003**, *375*, 337-343.
2. Yamamoto, K.; Tominaga, K.; Sasakawa, H.; Tamura, A.; Murakami, H.; Ohtake, H.; Sarukura, N. *Bull. Chem. Soc. Jpn.* **2002**, *75*, 1083-1092.
3. Walther, M.; Fischer, B.; Schall, M.; Helm, H.; Uhd Jepsen, P. *Chem. Phys. Lett.* **2000**, *332*, 389-395.
4. Zhang, C.; Tarhan, E.; Ramdas, A. K.; Weiner, A. M.; Durbin, S. M. *J. Phys. Chem. B* **2004**, *108*, 10077-10082.
5. Plusquellic, D. F.; Korter, T. M.; Fraser, G. T.; Lavrich, R. J.; Benck, E.C.; Bucher, C. R.; Domench, J.; Hight Walker, A. R. In *Terahertz Sensing Technology: Emerging Scientific Applications and Novel Device Concepts*; Wollard, D. L., Loerop, W. R., Shur, M. S., Eds.; World Scientific: Hackensack, NJ, 2003; Vol. 2, pp 385-404.
6. Chen, J.-Y.; Knab, J. R.; Cerne, J.; Markelz, A. G. *Phys. Rev. E* **2005**, *72*, 040901(R).
7. Knab, J. R.; Chen, J.-Y.; Markelz, A. G. *Biophys. J.* **2006**, *90*, 2576-2581.
8. Markelz, A. G.; Roitberg, A.; Heilweil, E. J. *Chem. Phys. Lett.* **2000**, *320*, 42-48.
9. Bykhovskaia, M.; Gelmont, B.; Globus, T.; Wollard, D. L.; Samuels, A. C.; Duong, T. H.; Zakrzewska, K. *Theor. Chem. Acc.* **2001**, *106*, 22-27.
10. Globus, T.; Bykhovskaia, M.; Wollard, D.; Gelmont, B. *J. Phys. D* **2003**, *36*, 1314-1322.
11. Nishizawa, J.; Suto, K.; Sasaki, T.; Tanabe, T.; Kimura, T. *J. Phys. D* **2003**, *36*, 2958-2961.
12. Korter, T. M.; Balu, R.; Campbell, M. B.; Beard, M. C.; Gregurick, S. K.; Heilweil, E. J. *Chem. Phys. Lett.* **2005**, *418*, 65-70.
13. Walther, M.; Fischer, B. M.; Jepsen, P. U. *Chem. Phys.* **2003**, *288*, 261-268.
14. Korter, T. M.; Plusquellic, D. F. *Chem. Phys. Lett.* **2004**, *385*, 45-51.
15. Fischer, B.; Hoffmann, M.; Helm, H.; Modjesch, G.; Jepsen, P. U. *Semicond. Sci. Technol.* **2005**, *20*, S246-S253.

16. Rutz, F.; Kleine-Ostmann, T.; Grunenberg, J.; Koch, M. *Proceedings of the Joint 29th International Conference on Infrared and Millimeter* Waves *and 12th International Conference on Terahertz Electronics*; **2004** p 737.

17. Siegrist, K.; Bucher, C. R.; Mandelbaum, I.; Hight-Walker, A. R.; Balu, R.; Gregurick, S. K.; Plusquellic, D. F. *J. Am. Chem. Soc.* **2006**, *in press.*

18. Görbitz, C. H. *New J. Chem.* **2003**, *27*, 1789-1793.

19. Görbitz, C. H. *Acta Crystallogr.* **2002**, *B58*, 849-854.

20. Görbitz, C. H.; Gundersen, E. *Acta Crystallogr. C* **1996**, *52*, 1764-1767.

21. Fawcett, J. K.; Camerman, N.; Camerman, A. *Acta Crystallogr. B* **1975**, *31*, 658-665.

22. Siegrist, K.; Plusquellic, D. F. Nat. Inst. Stand. Tech., Gaithersburg, MD *unpublished.*

23. McIntosh, K. A.; Brown, E. R.; Nichols, K. B.; McMahon, O. B.; DiNatale, W. F.; Lyszczarz, T. M. *Appl. Phys. Lett.* **1995**, *67*, 3844-3846.

24. Verghese, S.; McIntosh, K. A.; Brown, E. R. *Appl. Phys. Lett.* **1997**, *71*, 2743-2745.

25. Brown, E. R. *Appl. Phys. Lett.* **1999**, *75*, 769-771.

26. Duffy, S. M.; Verghese, S.; McIntosh, K. A.; Jackson, A.; Gossard, A. C.; Matsuura, S. *IEEE Trans. Microwave Theory Tech.* **2001**, *49*, 1032-1038.

27. German, K. R. *Appl. Opt.* **1981**, *20*, 3168-3171.

28. Vieira, N. D., Jr.; Mollenauer, L. F. *IEEE J. Quantum Electron.* **1985**, *QE-21*, 195-201.

29. Hempel, A.; Camerman, N.; Camerman, A. *Biopolymers* **1991**, *31*, 187-192.

30. Qian, W.; Bandekar, J.; Krimm, S. *Biopolymers* **1991**, *31*, 193-210.

Chapter 19

Changes of Near-UV Circular Dichroism Spectra of Human Hemoglobin upon the R→T Quaternary Structure Transition

Yayoi Aki-Jin[1], Yukifumi Nagai[1], Kiyohiro Imai[1,2], and Masako Nagai[1,2*]

[1]Research Center for Micro-Nano Technology, Hosei University, Tokyo 184–0003, Japan
[2]Department of Frontier Bioscience, Faculty of Engineering, Hosei University, Tokyo 184–8584, Japan

Human adult hemoglobin (Hb A) exhibits a distinct negative CD band at 287 nm in the deoxy-form (T, tense), but this band disappears in the oxy-form (R, relaxed). It was suggested that the environmental alteration of aromatic amino acids, Tyr-α42 and/or Trp-β37, at the $\alpha_1\beta_2$ subunit contact contributed to the negative CD band in the deoxy-form. However, precise assignment of aromatic amino acids responsible for the negative CD band still remains unsettled. To evaluate contribution of the aromatic amino acid residues to the negative CD band upon the R → T structure transition, we examined near-UV CD spectra of four mutant hemoglobins, two at the $\alpha_1\beta_2$ subunit contact, recombinant hemoglobin (rHb) (Tyr-α42→Ser) and rHb (Trp-β37→His), and the other two at the penultimate Tyr, rHb (Tyr-β145→Thr) and Hb Rouen (Tyr-α140→His). Environmental alteration in the penultimate tyrosine of both Tyr-α140 and Tyr-β145 primarily contributed to the negative CD band in the deoxy-form. Contributions of Tyr-α42 and Trp-β37 to the negative CD band were relatively small. Comparison of the negative CD bands of arithmetic mean of the isolated subunits with that of the recombined Hb A revealed contribution due to their tertiary structure change. The negative CD band of deoxyHb A was attributed to the sum of environmental alterations of aromatic amino acid residues induced by both tertiary structure change (longer wavelength region) and quaternary structure transition (shorter wavelength region).

The near-UV circular dichroism (CD) of human adult hemoglobin (Hb A) shows a characteristic change upon the quaternary structure transition: from a small positive band in the oxy (R)-form to a distinct negative band at 287 nm in the deoxy (T)-form (*1*). The near-UV CD spectrum is known to reflect environmental alteration of aromatic residues (Trp and Tyr) in the protein (*2*). An X-ray crystallographic study on Hb A (*3*) has demonstrated that upon the T→R transition a large movement takes place at the $\alpha_1\beta_2$ subunit contact and the anchored penultimate residues are ejected, freeing the C-terminal portion of each subunit. There are two aromatic residues (Tyr-α42 and Trp-β37) at the $\alpha_1\beta_2$ subunit contact and two residues (Tyr-α140 and Tyr-β145) at the penultimate position. It has been suggested that the negative CD band in the T-form arises from the hydrogen bond formation of Tyr-α42 and/or Trp-β37 upon deoxygenation (*1*). This negative CD band has served as a marker for the T-form (*1*). But the origin remains unsettled.

In order to evaluate the contributions of the aromatic residues at the $\alpha_1\beta_2$ sununit contact and the penultimate tyrosine residues to the negative CD band, three recombinant (r) hemoglobins, rHb Ser-α42 (Tyr-α42→Ser), rHb His-β37 (Trp-β37→His), and rHb Thr-β145 (Tyr-β145→Thr), were produced in *Escherichia coli* (*4, 5*). We compared the near-UV CD spectra of the three recombinant hemoglobins and Hb Rouen (Tyr-α140→His) (*6*) with those of Hb A under the conditions where the mutants had significant cooperativity and were able to undergo the T→R transition. We also examined the influences of the tertiary structure change on the negative CD band using the isolated α and β chains. We present here new finding for origin of the negative CD band in deoxyhemoglobin.

Materials and Methods

Preparation of Recombinant Hemoglobins

The Hb A expression plasmid pHE7 (*5*) containing human α- and β-globin genes and the *E. coli* methionine aminopeptidase gene was kindly provided by

Professor Chien Ho of Carnegie Mellon University. Plasmids for rHb Leu-α42, rHb Ser-α42, rHb His-β37, and rHb Thr-β145 were produced by site-directed mutagenesis using an amplification procedure of closed circular DNA *in vitro* (*7*). These plasmids were transformed into *E. coli* JM109. *E. coli* cells harboring the plasmid were grown at 30 °C in TB medium (*4, 5*). Expression of recombinant hemoglobin was induced by adding isopropyl β-thiogalactopyranoside to be 0.2 mM. The culture was then supplemented with hemin (30 μg/ml) and glucose (15 g/liter), and the growth was continued for another 5 h at 32 °C. The cells were harvested by centrifugation and stored frozen at –80 °C until needed for purification.

Recombinant hemoglobins were isolated and purified according to the method of Looker *et al.* (*8*) with some modifications. About 70 g frozen cell paste was routinely used as the starting material for a purification of recombinant hemoglobin. The thawed cell paste was suspended in the lysis buffer at 3 ml/g cells. The mixture was treated with lysozyme (1 mg/g of cells) and DNase I (30 μg/ml in 10 mM $MgCl_2$ and 1 mM $MnCl_2$) at 10 °C to disrupt the cells. The cell lysate was saturated with CO gas and stirred overnight in a cold room. The following procedures were carried out at 4 °C or on ice. After centrifugation, the supernatant was treated with polyethyleneimine (PEI) at a final concentration of 0.3 % to precipitate nucleic acids. The PEI-treated lysate was centrifuged at 23,000 g for 15 min. The supernatant was recovered, concentrated in an Amicon stirred cell concentrator and dialyzed against 20 mM Tris-HCl/0.1 mM triethylenetetraamine (TETA), pH 7.4, at 4 °C.

For the purification of recombinant hemoglobins, three chromatographic steps were employed (*8*). Before loading onto the column, the samples were always saturated with CO gas. The first column, Q-Sepharose fast flow column (2.5 x 20 cm), was equilibrated with 20 mM Tris-HCl/0.1mM TETA, pH 7.4, at 4 °C. This step captured a large amount of bacterial protein and the remaining nucleic acid, while recombinant hemoglobin passed through. The recombinant hemoglobin fraction was collected, concentrated and dialyzed against 20 mM Tris-HCl, pH 8.3, at 4 °C (Q_1 fraction). The second column, Q-Sepharose fast flow column (1.5 x 17 cm), was equilibrated with 20 mM Tris-HCl, pH 8.3, at 4 °C. After loading the Q_1 fraction and then washing the column with equilibration buffer, the bound recombinant hemoglobin was eluted with a linear gradient (total 300 ml) from 0 to 160 mM NaCl in equilibration buffer. Recombinant hemoglobin-containing fractions were pooled on the basis of absorbance at 576 nm and daialyzed against 10 mM sodium phosphate buffer, pH 6.8, at 4 °C (Q_2 fraction). The third column, SP-Sepharose fast flow column (1.5 x 45 cm), was equilibrated with 10 mM sodium phosphate buffer, pH 6.8, at 4 °C. The Q_2 fraction was loaded and then washed with one bed volume of equilibration buffer. A linear gradient of equilibration buffer *versus* 20 mM sodium phosphate buffer, pH 8.3, at 4 °C was used to elute the recombinant hemoglobin (total volume 400 ml). Fractions were pooled as purified recombinant hemoglobin on the basis of high ratio of ellipticity of CD at 260 nm and absorbance at 572 nm (ratio > 45).

Preparations of Hb A, Hb Rouen, and Hb Kansas

Hb A was purified by preparative isoelectric focusing electrophoresis (IEF) (*9*). Hb Rouen was isolated from patient's hemolyzate by preparative IEF on a Sephadex G-75 superfine gel flat bed containing 5 % Ampholine, pH 7-8 (*10*). Hb Rouen was focused on pI 7.55 relative to Hb A on pI 7.34. Hb Kansas was applied on a SP-Sepharose fast flow column (1.5 x 45 cm) equilibrated with 10 mM sodium phosphate, pH 6.8, at 4 °C. Hb Kansas was separated from Hb A by a linear gradient of the equilibration buffer *versus* 20 mM sodium phosphate buffer, pH 8.3. Purity of the isolated Hb Rouen and Hb Kansas was checked by analytical IEF on an Ampholine plate gel (pH range 3.5-9.5) and contamination of Hb A in Hb Rouen and Hb Kansas was confirmed to be less than 1 %. Isolated α and β chains of Hb A were prepared as reported previously (*11*).

Oxygen Equilibrium Measurement

Oxygen equilibrium curves of hemoglobin were determined by a spectrophotometric method according to Sugita and Yoneyama (*12*). Deoxygenation was carried out by repeating alternate evacuation and flushing with Q gas (helium-isobutane, 99.05:0.95) in a Thunberg type cell with l-cm light path. Absorption spectra were recorded with a Hitachi U-3210 spectrophotometer at 25 °C. Oxygen affinity of hemoglobin is expressed by the partial pressure of oxygen at half-saturation, P_{50}. Cooperativity is represented by the maximal slope of Hill plot, Hill's *n*.

CD Spectroscopy

CD spectra measurements were performed with a Jasco J-725 spectropolarimeter at 25 °C, using cell path length of 2 mm. The instrument was calibrated with (+)-10-camphorsulfonic acid. Spectra were acquired at a scan speed of 50 nm/min with a 1-nm slit and 1-sec response time, averaging 40-80 scans, and corrected by subtraction of the solvent spectrum obtained under identical condition. Unit of molar CD ($\Delta\varepsilon$) is $M^{-1}\cdot cm^{-1}$ (in heme basis). Spectra were measured several times using different preparations. The spectra of Hb A were measured in each series of experiments as a control. Deoxyhemoglobin was prepared by adding a small quantity of sodium dithionite powder to the oxyhemoglobin. The concentration of hemoglobin was determined after conversion to pyridine hemochrome using $E^{557}_{mM} = 34$.

Results

Near-UV CD Spectra of Hemoglobins with Low Cooperativity

In order to understand how the quaternary structure of hemoglobin correlates to the negative CD band at 287 nm, we investigated near-UV CD spectra of two mutant hemoglobins, rHb Leu-α42 (Tyr-α42→Leu) with a high oxygen affinity and low cooperativity and natural mutant hemoglobin, Hb Kansas (Asn-β102→Thr) with a low oxygen affinity and low cooperativity (*13*). Figure 1 shows CD spectra in the 240-320 nm region of these mutants in comparison with that of Hb A (dotted lines) in the oxy (R)- and deoxy (T)-forms. OxyHb A shows a large positive CD band at 260 nm and a small positive ellipticity near 285 nm. The positive CD band at 260 nm greatly decreases in ellipticity and a distinct negative CD band appears at 287 nm upon deoxygenation. The CD bands at 260 nm and at 280-300 nm have been attributed to the heme moiety and the aromatic amino acid residues, respectively (*1*). Perutz *et al.* (*1*) have suggested that the negative CD band at 287 nm is useful as a T-state marker.

In deoxyHb A, Tyr-α42 residue forms an intersubunit hydrogen bond with Asp-β99, which is free in the oxy-form and considered to stabilize the T-structure (*3*). Compared with Hb A (P_{50} = 11.3 mmHg, Hill's n = 2.8), rHb Leu-α42 exhibits a markedly increased oxygen affinity and low cooperativity (P_{50} = 1.1 mmHg, n = 1.4) (Table I). Hence, rHb Leu-α42 seems to remain in the R-structure even in the deoxy-form. As seen in Figure 1A, oxy-rHb Leu-α42 (solid lines) showed a CD spectrum similar to that for oxyHb A (dotted lines) although some minor differences were observed. On the other hand, besides a smaller positive CD band at 260 nm, deoxy-rHb Leu-α42 exhibited a smaller negative CD band at 287 nm than that of deoxyHb A. Hb Kansas (Asn-β102→Thr) is a natural mutant hemoglobin which loses the intersubunit hydrogen bond between Asn-β102 and Asp-α94 in the oxy-form (*3, 13*). Hb Kansas shows a low oxygen affinity and diminished coorperativity (P_{50} = 36.0 mmHg, n = 1.6) (Table I). Hb Kansas is thought to take the quaternary structure in the T-state even in the oxy-form. Figure 1B shows CD spectra of Hb Kansas (solid lines) and Hb A (dotted lines) in the 240-320 nm region. Hb Kansas exhibited similar but smaller positive CD bands at 260 nm than that of Hb A in both the oxy- and deoxy-forms. In the 280-300 nm region, deoxyHb Kansas gave an identical negative CD band with that of deoxyHb A, while oxyHb Kansas exhibited a small but distinct negative CD band. These results indicate definite correlation between the negative CD band at 287 nm and quaternary structure.

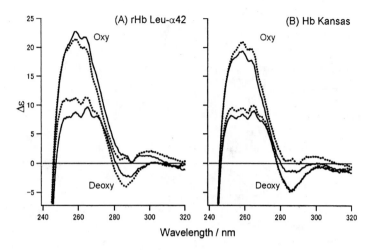

*Figure 1. Near-UV CD spectra of rHb Leu-α42 (Tyr-α42→Leu) (A)
and Hb Kansas (B) in the oxy- and deoxy-forms.*
*Dotted lines refers to Hb A and the solid lines to mutants. CD spectra were
acquired at a scan speed of 50 nm/min with a 1-nm slit and 1-sec response time,
averaging 40-80 scans, and corrected by subtraction of the solvent spectrum
obtained under identical condition. Unit of molar CD (Δε) is $M^{-1}·cm^{-1}$ (in heme
basis). Hemoglobin solution was 50 μM (in heme) in 0.05 M bis-Tris buffer,
pH 7.0 containing 0.1 M NaCl (rHb Leu-α42) or 0.1 M phosphate buffer,
pH 7.0 (Hb Kansas).*

*Near-UV CD Spectra of Mutant Hemoglobins Having a Significant
Cooperativity*

To address the origin of the negative CD band of deoxyHb A, we examined
near-UV CD spectra of Hb Rouen (Tyr-α140→His), rHb Ser-α42 (Tyr-
α42→Ser), rHb His-β37 (Trp-β37→His), and rHb Thr-β145 (Tyr-β145→Thr).
Oxygen equilibrium properties of these mutant hemoglobins are summarized in
Table I. Hb Rouen shows moderately increased oxygen affinity and significant
cooperatively at pH 7.3 (P_{50} = 7.0 mmHg, n = 2.1). In the absence of inositol
hexaphosphate (IHP), rHb Ser-α42 gives about 4 times higher oxygen affinity
than Hb A and significant coorperativity (n = 2.1). On the other hand, rHb His-
β37 and rHb Thr-β145 exhibit 7 or 26 times higher oxygen affinity than Hb A
and low cooperativity (n = 1.8 and 1.3, respectively). However, the addition of
IHP caused significant increase in cooperativity of these two recombinant
hemoglobins (n = 2.1 and 2.5, respectively) concomitantly with decrease in the
oxygen affinity. Based on these results, we measured near-UV CD spectra of

Table I. Oxygen equilibrium properties of mutant hemoglobins

Hemoglobin	IHP	pH	P_{50} (mmHg)[a]	P^A_{50}/P^X_{50}[b]	Hill's n
rHb Leu-α42	none	7.0	1.1	10	1.4
	2 mM	7.0	2.8	4.0	2.2
Hb Kansas	none	7.0	36.0	0.3	1.6
Hb Rouen	none	7.3	7.0	1.6[c]	2.1
rHb Ser-α42	none	7.0	2.7	4.2	2.1
	2 mM	7.0	11.0	1.0	2.1
rHb His-β37	none	7.0	1.7	6.8	1.8
	2 mM	7.0	6.9	1.6	2.1
rHb Thr-β145	none	7.0	0.4	26	1.3
	2 mM	7.0	1.6	7.2	2.5
Hb A	none	7.0	11.3		2.8

[a] Partial pressure of oxygen at half-saturation. [b] Ratio of P_{50} for Hb A to P_{50} for mutant Hbs. [c] Ratio of P_{50} for Hb A at pH 7.0 to P_{50} for Hb Rouen at pH 7.3. Hemoglobin, 50 μM (in heme) in 0.1 M phosphate buffer; 25 °C.

the mutant hemoglobins under the conditions where these mutants could undergo the T→R transition. The CD bands near 260 nm of all mutants were similar to that of Hb A (not shown), indicating that the aromatic residue replacement in the mutant hemoglobins scarcely had influence on the optical activity of the heme moiety.

Figure 2A shows the CD spectra of rHb His-β37 (a), rHb Ser-α42 (b), Hb Rouen (c) and rHb Thr-β145 (d) (solid lines) in the 275-305-nm region in comparison with those of Hb A (dotted lines). Three mutant hemoglobins, rHb His-β37, Hb Rouen, and rHb Thr-β145, displayed CD band with ellipticity values (Δε) different from those of Hb A in both the oxy- and deoxy-forms, but rHb Ser-α42 showed CD spectra similar to those of Hb A. Figure 2B shows the deoxy-minus-oxy difference CD spectra for rHb His-β37 (a'), rHb Ser-α42 (b'), Hb Rouen (c'), and rHb Thr-β145 (d') (solid lines) in comparison with that of Hb A (dotted lines) in the 275-305 nm region. Recombinant Hb Ser-α42 gave a difference spectrum nearly identical with that of Hb A. On the other hand, other mutant hemoglobins exhibited difference spectra clearly distinct from that of Hb A. The deoxy-minus-oxy difference spectrum represents the absolute ellipticity change of the CD band on going from the oxy-form to the deoxy-form. As ellipticity (Δε) at 287 nm of Hb A changes from 1.3 in the oxy-form to -4.3 in the deoxy-form, the absolute difference (ΔΔε) at 287 nm between the oxy-form and the deoxy-form amounts to 5.6. The ΔΔε values for rHb His-β37, rHb Ser-α42, Hb Rouen and rHb Thr-β145 are 4.6, 5.4, 3.8 and 4.1, respectively. From

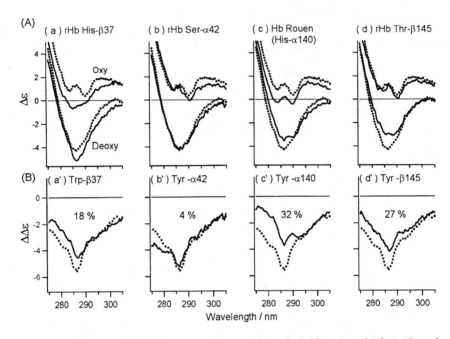

Figure 2. CD spectra of four mutant hemoglobins (solid lines) and Hb A (dotted lines) in the oxy- and deoxy-forms (A) and the deoxy-minus-oxy difference spectra of mutants (solid line) and Hb A (dotted line) (B).
Hemoglobin solution was 50 μM (in heme) in 0.1 M phosphate buffer, pH 7.3 (Hb Rouen) or pH 7.0 (the other hemoglobins). IHP was added to the solutions of rHb His-β37, rHb Ser-α42, and rHb Thr-β145 to be the final concentration of 2 mM. Percentage in (B) was calculated by the following: [(ΔΔε for Hb A) - (ΔΔε for mutant Hb)]/ (ΔΔε for Hb A) x 100, ΔΔε is the ellipticity of the deoxy-minus-oxy difference spectrum at 287 nm. The percentage of individual mutants represents the contribution of the replaced aromatic residue to the quaternary structure change. Other conditions are the same as those in Fig. 1.

the decrease in $\Delta\Delta\varepsilon$ value, the extent of contribution of Trp-β37 to the ellipticity at 287 nm in Hb A is estimated at 18 %. Similarly, contributions of Tyr-α42, Tyr-α140, and Tyr-β145 are calculated as 4, 32, and 27 %, respectively. These results show that the change in ellipticity at 287 nm upon deoxygenation is primarily due to environmental alteration of three aromatic residues, especially of the penultimate tyrosine residues. The double difference CD spectrum between the deoxy-minus-oxy difference of Hb A and that of mutant hemoglobin reveals contribution of replaced aromatic residue to the CD band in the 280-300 nm region. Therefore, the sum of double difference spectra in the four mutant hemoglobins is expected to be similar in shape to the deoxy-minus-oxy difference spectrum of Hb A. Figure 3 shows the sum of the double difference spectrum for the four mutant hemoglobins (A) together with the deoxy-minus-oxy difference spectrum of Hb A (B). The shape of negative CD band contributed by the four aromatic residues is approximately similar in the 280-290 nm region but markedly different in the 290-300 nm region. This result indicates that the environmental alteration of the four aromatic residues upon the quaternary structure transition chiefly contributes to appearance of the negative CD band at 280-290 nm region but not at the 290-300 nm region. It is very likely that oxygen dissociation causes the tertiary structure change of subunits resulting in the environmental alteration of aromatic residues. However, double difference spectrum can not reveal their substantial contribution if the tertiary structure in both Hb A and mutant hemoglobins does equally change upon deoxygenation. Thus, the negative CD band in the 290-300 nm region might be due to the tertiary structure change in subunits upon deoxygenation.

Effects of the Tertiary Structure Change on the Near-UV CD Spectra

To elucidate the origin of the negative CD band at 290-300 nm, effect of the tertiary structure change on the CD spectrum was examined using isolated subunits. Both α and β subunits in the oxy-form exhibit positive CD bands at 260 nm with different ellipticity which greately decrease upon deoxygenation (*11, 14*). In the 280-300 nm region, the ellipticity of the α subunit became more positive at 280 nm and more negative at 290 nm upon deoxygenation. The CD band of β subunit near 290 nm changed from positive to negative on going from oxy- to deoxy-form. These results prove that oxygen dissociation from subunits actually bring about the environmental alteration of aromatic residues through their tertiary structure change. Figure 4A shows the CD spectra of recombined Hb A (dotted lines) with those of the arithmetic mean of isolated α and β subunits (solid lines). The CD spectra of recombined Hb A are nearly identical with those of Hb A in the oxy- and the deoxy-forms. The CD spectra of the arithmetic mean of isolated subunits in the 250-270 nm region are similar to

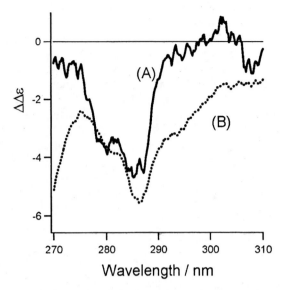

Figure 3. The sum of the double difference spectra between Hb A and the mutant hemoglobins (A) and the deoxy-minus-oxy difference spectrum of Hb A (B). Individual contribution of four aromatic amino residues was calculated by the double difference spectrum between the deoxy-minus-oxy difference spectrum of Hb A and that of mutant hemoglobin, as shown in Fig. 2 (B).

those of recombined Hb A in the oxy- and the deoxy-forms, indicating that $\alpha_2\beta_2$ tetramer formation from isolated subunits does scarcely affect the optical activity of the heme. In the 270-300 nm region, the CD band of the arithmetic mean in the oxy-form is the same as that of recombined Hb A, but the negative CD band of the arithmetic mean in the deoxy-form becomes much smaller and centered at longer wavelength compared with that of recombined Hb A. In isolated α and β subunits, their CD spectra of the arithmetic mean are due to the environmental alteration of Tyr and Trp residues not upon quaternary structure transition but through the tertiary structural change in response to oxygen binding or dissociation. Figure 4B shows the deoxy-minus-oxy difference spectrum of Hb A (a) and that for the arithmetic mean of the isolated subunits (b). The difference spectrum for Hb A exhibits a negative CD band with a peak at 287 nm, whereas that for the arithmetic mean gives a negative CD band centered around 290 nm. This result demonstrates that the longer wavelength region of the negative CD band of deoxyHb A is attributed to the environmental alteration of aromatic residues through the tertiary structure change.

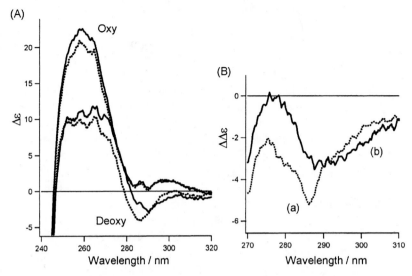

Figure 4. CD spectra of the recombined Hb A (dotted lines) and the arithmetic means of the spectra of isolated α and β subunits (solid lines) (A), and the deoxy-minus-oxy difference spectrum of Hb A (a) and that of arithmetic mean of isolated chains (b) (B). Recombined Hb A was prepared by mixing α and β subunits at equimolar concentration. The arithmetic mean of CD spectrum was calculated from 1/2 (CD spectrum of α + CD spectrum of β). Hemoglobin solution was 50 μM (in heme) in 0.1 M phosphate buffer, pH 7.0. Other conditions are the same as those in Fig. 1.

Discussion

Correlation between Near-UV CD Spectral Change and the Quaternary Structure Transition of Hemoglobin

Perutz et al. (1) have suggested that the negative CD band at 287 nm characteristic of deoxyHb A serves as a T-state marker from their observation that Hb Kempsey (Asp-β99→Asn) and a modified hemoglobin, Nes-des-Arg-Hb, lack the prominent negative peak at 287 nm. Crystallographic study has revealed that Hb Kempsey takes the R-structure in the deoxy-form (1). They interpreted the lack of the prominent negative CD band at 287 nm as remaining in the R-state even in the deoxy-form. We examined here the correlation between cooperativity and near-UV CD band due to the aromatic region of the globin using two mutant hemoglobins, rHb Leu-α42 (Tyr-α42→Leu) and Hb

Kansas (Asn-β102→Thr). As shown in Figure 1A, deoxy-rHb Leu-α42 showed a small CD band with negative peak around 290 nm different from deoxyHb A. But in the presence of IHP which causes the lowering of oxygen affinity and the increase in cooperativity (Table I), the negative CD band is closely similar to the negative CD band of deoxyHb A (data not shown). Similar spectral change in the negative CD band with and without IHP was observed in deoxyHb Kempsey (Asp-β99→Asn) with high oxygen affinity and diminished cooperativity (P_{50} = 0.23 mmHg, n = 1.1) (1, 15). DeoxyHb A is predominantly stabilized by hydrogen bond formations between α and β subunits. Hb Kempsey can not form the hydrogen bond between Tyr-α42 and Asp-β99 essential to stabilize the T-state in the deoxy-form (1, 3). Deoxy-rHb Leu-α42 can not form the hydrogen bond as well as deoxyHb Kempsey, suggesting that this recombinant hemoglobin remain also in the R-state even in the deoxy-form.

The low oxygen affinity and diminished coopeartivity of Hb Kansas were ascribed to the unstable R-structure owing to the lack of the hydrogen bond between Asn-β102 and Asp-α94 stabilizing the oxy (R)-structure (13, 16). DeoxyHb Kansas shows a negative CD band at 287 nm identical with deoxyHb A (Figure 1B). On the other hand, oxyHb Kansas exhibits a small and broad negative CD band centered at 287 nm unlike oxyHb A that give a small positive CD at 285 nm, indicating that Hb Kansas takes T-like structure even in the oxy-form. This result agrees with the finding based on NMR and near-UV CD measurements of NOHb Kansas (17).

Our present results provide another evidence supporting that the prominent negative CD band at 287 nm serves as a marker for the T-structure (1). As described above, deoxy-rHb Leu-α42 with and without IHP shows different negative CD band depending on the change in oxygen binding proterties. Therefore, this CD difference with and without IHP in aromatic region of globin seems to reflect actual environmental alteration of aromatic residues upon the R→T transition.

Changes of the Near-UV CD Spectrum by the Mutations of Tyr or Trp at the $\alpha_1\beta_2$ Contact

In order to address the origin of the T-state marker band, we measured CD spectra of rHb His-β37, rHb Ser-α42, Hb Rouen (Tyr-α140→His), and rHb Thr-β145 under the conditions where they showed significant cooperativity (n > 2.0), namely, they could undergo the R→T transition.

In deoxyHb A, Trp-β37 and Tyr-α42 at the $\alpha_1\beta_2$ subunit contact form intersubunit hydrogen bonds with Asp-α94 and Asp-β99, respectively (3). These hydrogen bonds are broken upon oxygenation with rearrangements of the $\alpha_1\beta_2$ contact. Therefore, it seems likely that these aromatic residues play a key role in the quaternary structure transition and are involved in the change of CD band at near UV region. However, in the present study using rHb Ser-α42 and rHb His-β37, contributions of Tyr-α42 and Trp-β37 to the ellipticity at 287 nm of negative CD band were estimated to be 4 % and 18 %, respectively (Figure 2B). These results indicate that contributions of these aromatic residues to the ellipticity at 287 nm are small in spite of their large movements in the quaternary structure transition.

A study of model compounds, *N*-acetyl-L-tyrosinamide and *N*-acetyl-L-tryptophanamide, dissolved in water or dioxane, has demonstrated that the CD spectra of Tyr and Trp do profoundly change depending on the solvent, that is, the hydrophobicity of their environment (*18*). However, Strickland *et al.* (*19*) have observed that the CD spectra of *N*-stearyl-L-tyrosine *n*-hexyl ester are hardly influenced by the formation of hydrogen bond in a nonpolar solvent. Tyr-α42 and Trp-β37 are located inside hemoglobin molecule, probably in a hydrophobic environment. Therefore, their observed contributions to the negative CD band at 287 nm suggest that movement at the $\alpha_1\beta_2$ contact and disruption of the two hydrogen bonds upon oxygenation do not undergo change of hydrophobicity in their environments. UV resonance Raman (UVRR) spectroscopy is known to give information about the hydrogen bond formation and the environment of tyrosine residue in the protein (*20-23*). UVRR study comparing rHb His-α42 (Tyr-α42→His) with Hb A has demonstrated that Tyr-α42 forms the hydrogen bond in the deoxy-form but hardly exhibits environmental alteration upon ligand binding (*24*).

Using Hb Rouen and rHb Thr-β145, contributions of Tyr-α140 and Tyr-β145 to the ellipticity at 287 nm of the negative CD band were also calculated to be 32 % and 27 %, respectively (Figure 2B). These results clearly demonstrate that development of the negative CD band upon the R→T transition is primarily due to the environmental alteration of Tyr-α140 and Tyr-β145. It has been shown that the penultimate Tyr residues of the α and β subunits are anchored in a pocket surrounded by the H and F helices and form intra-subunit hydrogen bonds with Val-α93 and Val-β98, respectively, in the deoxy (T) conformation (*3*). Upon oxygen binding, the salt bridges formed at the C-termini are broken and the penultimate tyrosine residues are expelled out of the pocket (*1*). A recent high-resolution X-ray study has indicated that these Tyr residues still stay within the F-H pocket in oxyHb A, but that the C-termini of both the α and β subunits are much more disordered in the R-state than in the T-state (*25*). This result also indicates the environmental alteration of the penultimate tyrosine residues. UVRR studies of Hb Rouen and des(His-β146, Tyr-β145)Hb have proved that the R→T transition induces actually an increase in hydrophobicity around the side chain of Tyr-α140 and Tyr-β145 (*10*). Hence, judging from these results, the negative CD band in the deoxy-form primarily reflects the environmental alteration of Tyr-α140 and Tyr-β145 caused by the salt-bridge formation at C-termini upon the R→T transition.

310

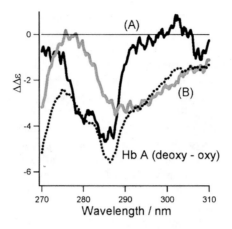

Figure 5. Superimposed CD spectra due to both the quaternary structure transition (A) and tertiary structure change (B) on the deoxy-minus-oxy difference spectrum of Hb A.
The CD spectrum due to the quaternary structure transition was the sum of the four double difference spectra between Hb A and the mutant hemoglobins, while that of the tertiary structure change was the deoxy-minus-oxy difference spectrum of the arithmetic mean of isolated chains.

The sum of the contribution for Tyr-α42, Trp-β37, Tyr-α140, and Tyr-β145 to the ellipticity at 287 nm was 81 %. In the present study, we compared the CD spectra of the arithmetic mean for the isolated α and β subunits with the CD spectra of the recombined Hb A and found that the tertiary structure change also contributed to the negative CD band in the deoxy-form as shown in Figure 4B. However, contributions of the environmental alteration of aromatic residues caused by the tertiary structure change to the negative CD band in the deoxy-form are different from those of the quaternary structure transition. Figure 5 shows separate contributions of environmental alteration of aromatic residues due to either the quaternary structure transition (A) or the tertiary structure change (B) to the deoxy-minus-oxy difference spectrum of Hb A. As can be seen, the former contributes to the band in the shorter wavelength region (275-290 nm) whereas the latter does to the band in the higher wavelength region (285-310 nm)

Acknowledgements

We thank Dr. C. Ho for the gift of plasmid, pHE7, Drs. H. Wajcman (Hb Rouen) and S. Ogawa (Hb Kansas) for their gifts of blood containing abnormal hemoglobins, Dr. H. Sakurai for his useful discussion, and Misses M. Satoh, S. Nanbu, and C. Nakasaka for their skillful assistance.

References

1. Perutz, M. F.; Ladner, J. E.; Simon, S. R.; Ho, C. *Biochemistry* **1974**, *13*, 2163-2173.
2. Strickland, E. H. *CRC Crit. Rev. Biochem.* **1974**, *2*, 113-175.
3. Baldwin, J.; Chothia, C. *J. Mol. Biol.* **1979**, *129*, 175-220.
4. Shen, T.-J.; Ho, N. T.; Simplaceanu, V.; Zou, M.; Green, B. N.; Tam, M. F.; Ho, C. *Proc. Natl. Acad. Sci. USA* **1993**, *90*, 8108-8112.
5. Shen, T.-J.; Ho, N. T.; Zou, M.; Sun, D. P.; Cottam, P. F.; Simplaceanu, V.; Tam, M. F.; Bell, D. A., Jr; Ho, C. *Protein Eng.* **1997**, *10*, 1085-1097.
6. Wajcman, H.; Kister, J.; Marden, M.; Lahary, A.; Monconduit, M.; Galacteros, F. *Biochim. Biophys. Acta* **1992**, *1180*, 53-57.
7. Chen, Z.; Ruffner, D. E. *Nucleic Acids Res.* **1998**, *26*, 1126-1127.
8. Looker, D.; Mathews, A. J.; Neway, J. O.; Stetler, G. L. *Methods Enzymol.* **1994**, *231*, 364-374.
9. Nagai, M.; Kaminaka, S.; Ohba, Y.; Nagai, Y.; Mizutani, Y.; Kitagawa, T. *J. Biol. Chem.* **1995**, *270*, 1636-1642.
10. Nagai, M.; Wajcman, H.; Lahary, A.; Nakatsukasa, T.; Nagatomo, S.; Kitagawa, T. *Biochemistry* **1999**, *38*, 1243-1251.
11. Li, R.; Nagai, Y.; Nagai, M. *J. Inorg. Biochem.* **2000**, *82*, 93-101.
12. Sugita, Y.; Yoneyama, Y. *J. Biol. Chem.* **1971**, *246*, 389-394.
13. Bonaventura, J.; Riggs, A. *J. Biol. Chem.* **1968**, *243*, 980-991.
14. Beychok, S.; Tyuma, I.; Benesch, R. E.; Benesch, R. *J. Biol. Chem.* **1967**, *242*, 2460-2462.
15. Bunn, H. F.; Wohl, R. C.; Bradley, T. B.; Cooley, M.; Gibson, Q. H. *J. Biol. Chem.* **1974**, *249*, 7402-7409.
16. Greer, J. *J. Mol. Biol.* **1971**, *59*, 99-105.
17. Salhany, J. M.; Ogawa, S.; Shulman, R. G. *Biochemistry* **1975**, *14*, 2180-2190.
18. Shiraki, M. *Sci. Pap. Coll. Gen. Educ. Univ. Tokyo* **1969**, *16*, 151-173.
19. Strickland, E. H.; Wilchek, M.; Horwitz, J.; Billups, C. *J. Biol. Chem.* **1972**, *247*, 572-580.
20. Takeuchi, H.; Watanabe, N.; Satoh, Y.; Harada, I. *J. Raman Spectrosc.* **1989**, *20*, 233-237.
21. Zhao, X.; Spiro, T. G. *J. Raman Spectrosc.* **1998**, *29*, 49-55.
22. Takeuchi, H.; Ohtsuka, Y.; Harada, I. *J. Am. Chem. Soc.* **1992**, *114*, 5321-5328.
23. Chi, Z.; Asher, S. A. *Biochemistry* **1998**, *37*, 2865-2872.
24. Nagai, M.; Imai, K.; Kaminaka, S.; Mizutani, Y.; Kitagawa, T. *J. Mol. Struc.* **1996**, *379*, 65-75.
25. Shaanan, B. *J. Mol. Biol.* **1983**, *171*, 31-59.

Chapter 20

Kelvin Physics of Protein Layers Printed in Microarray Format

Hong Huo, Larisa-Emilia Cheran, and Michael Thompson[*]

Department of Chemistry, University of Toronto, 80 St. George Street, Toronto, Ontario M5S 3H6, Canada

The electrical properties of protein molecules adsorbed onto a gold substrate are studied using a scanning Kelvin nanoprobe in a microarray format. The results demonstrate that this instrument can provide information on protein orientation, polarization, dimension and molecular interaction. Changes in the work function for absorbed neutravidin, bovine serum albumin and a complex of the two proteins on gold were measured. Also, the analogous changes with respect to the surface concentration of neutravidin were examined and the results indicate a saturation of work function values at a specific surface population. The results are discussed in terms of the Schottky model modified for protein semiconductive properties, such as depletion width, charge carrier density, band bending and permittivity.

Following rapidly on the heels of genomics, proteomics has become a very prominent research field in analytical biochemistry. In this area, the protein microarray represents a highly miniaturized parallel and multiplexed solid phase assay, which allows the production of high-throughput analyses, with low consumptions of reagent samples (nL level) and attractive manufacturing costs (*1*). This is why the technology has become a potentially powerful tool for diagnostic and therapeutic purposes, as well as for basic research in biology and medicine.

Well-established platforms, detection strategies and software tools are available for DNA microarray technology. However, this is not generally the case for protein chemistry, where there is tremendous molecular variability, coupled with complex chemical and structural properties. Moreover, significant challenges exist in detection strategies for the protein microarray, compared to the sister nucleic acid technology. Currently, labeled probe detection is the most common protocol for microarrays. Foremost amongst these is fluorescence analysis, which allows single molecule investigation. However, this method suffers from various drawbacks for protein studies, the most serious of which is the need to tag the biomolecule of interest with fluorophores or isotopes. This labeling process is both expensive and time-consuming. Additionally, the label may alter the native structure of a protein and potentially interfere with the bioaffinity interaction under study. Problems such as label stability and photo bleaching or unspecific binding of the fluorophore to the surface encountered with fluorescent reporters may also interfere with data collection and analysis. Detection using radiochemical labeling is another method of tagging, but is not widely used due to health and safety issues and long detection times (up to 10h).

Label-free strategies provide an alternative route for circumventing the above-mentioned difficulties with respect to protein microarray detection. Currently, relatively few label-free detection methods such as mass spectrometry (MS) (*2,3*), surface plasmon resonance (SPR) (*4*) and atomic force microscopy (AFM) (*5,6*) are applicable to the microarray format. These label-free technologies require the use of sophisticated tools that are not available in all bioanalytical laboratories.

In the present paper, we describe a scanning Kelvin nanoprobe (SKN) as an alternative label-free detection approach for work with protein microarrays (*7*). Based on the measurement of work function change, the SKN can measure inherent electrical properties of proteins on a solid surface, which depend on protein dimension, orientation, polarization and molecular interactions.

For many decades, the Kelvin probehas been extensively applied as a surface potential technique to study monolayers at air-water interfaces(*8-11*). Although there are three proposed models for estimating dipole moments in surface potential measurement(*9*), difficulties in interpreting experimental data represent the main reason why the technique is restricted to application to the air-solid interface(*12-16*). Such measurements and explanation of results based on dipole moments and Schottky models suggest that the SKN is a powerful method for the understanding of protein electrical properties offering, in turn, considerable potential for label-free microarray detection.

Experimental

Reagents and materials

Neutravidin biotin-binding protein was obtained from Pierce Biotechnology, Inc, Rockford, IL, USA and used without further purification. Bovine serum albumin and biotin labeled bovine serum albumin were purchased from Sigma Aldrich, Canada. Lyophilized proteins were dissolved in DPBS buffer solutions before use.

Dulbecco's phosphate buffered saline solution (DPBS) was purchased from Sigma Aldrich, Canada. Spectroscopy-grade acetone was obtained from Uniscience, Inc, Mississauga, Ont. Anhydrous ethyl alcohol and HPLC grade methanol were obtained from Commercial Alcohols Inc, Brampton, Ont., and VWR Canada (Mississauga, Ont.), respectively. All solvents were used as received.

Optically flat silicon wafers, 300-400 μm thick, were purchased from International Wafer Service, Canada.

Instrumentation

The scanning Kelvin nanoprobe

A prototype version of the scanning Kelvin nanoprobe (SKN) constructed in our laboratory was employed for all measurements(*17*). Figure 1 presents the main modules of the SKN block-diagram. The probe (tungsten, guarded, driven by a piezo actuator from Sensor Technology Ltd., Collingwood, Ont., Canada) vibrates over the sample, which is placed on an XY scanning table that can also be moved in the Z axis, using a second piezo actuator (Polytech PI, Auburn, MA, USA). The probe does not touch the surface, but follows its topography at an extremely small, constant distance. The signal is amplified by an ultra-low noise charge amplifier (Electro Optical Components, Inc. Santa Rosa, CA, USA) and converted to a voltage, which is fed into two lock-in amplifiers (Stanford Research Systems, Sunnyvale, CA, USA). One serves as a detector of contact

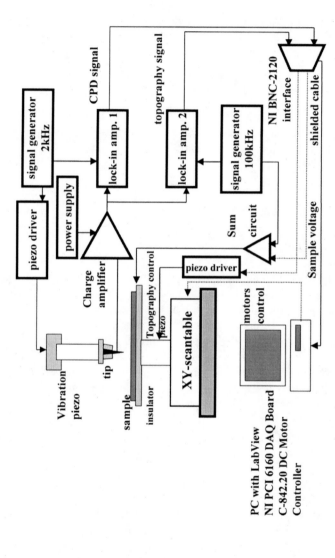

Figure 1. Block diagram of the scanning Kelvin nanoprobe. An ultra-low noise amplifier detects the Kelvin current. Two lock-in amplifiers are used for detecting the contact potential difference and the topography signals. The tip-sample distance is monitored by capacitative control, using a frequency well above the vibrating frequency and superposing a small AC signal to generate the control current. (Reprinted by kind permission of the Royal Society of Chemistry)

potential difference (the difference in work function between the tip material and the sample material) and the other for derivation of the topographic signal. The latter is extracted from the feedback loop that controls the constant distance between probe and surface, which is based on a capacitative measurement using a superimposed high frequency, low amplitude voltage. The whole instrument is controlled by a LabView program with a data acquisition board (National Instruments, USA). Data analysis is performed using a dedicated Origin 7 program (Origin Lab Corporation, MA, USA).

Microarrayer

Printing of protein spots was achieved using a Virtek Chipwriter Professional arrayer. (Waterloo, Ont., Canada).

Procedures

Gold substrate fabrication

Before coating, the silicon wafers were cut into 1mm x 2 cm pieces and cleaned in an ultrasonic bath for 15 min in acetone, ethanol and methanol, respectively. Following removal from solution, the substrates were dried in a stream of nitrogen. The clean silicon substrates were primed with a thin layer (10nm) of chromium, followed by the deposition of 200 nm of gold. The gold-coated silicon substrates were first rinsed with deionised water, and then sonicated in acetone, ethanol and methanol for 15 min, respectively. After sonication, the slides were extensively rinsed in deionised water again and dried under a stream of nitrogen.

Microarraying of proteins

All protein arrays were printed by contact printing, based on capillary action, using the robotic microarrayer. The micro-machined pins consistently delivered samples of approximately 1 nL onto the gold slides at designated locations. Typically, circular spots with diameters ranging from 150 to 250 μm with pitches of about 250-300 μm were produced. The printing was performed at a relative humidity of 65% and a temperature of 25°C. After spotting, the protein immobilized slides were kept in a humid environment (65% relative humidity,

25°C) for 120 min. After this time, the slides were washed with DPBS buffer and ultra pure water. The slides were then carefully inverted and gently immersed in DPBS buffer for 15 minutes to remove any unbound proteins. This was followed by rinsing with pure water, which removed any buffer salt. Finally, the arrayed slides were dried under a stream of nitrogen before being scanned with the SKN. All the samples were analyzed in air, at ambient temperature and pressure, using a lateral scanning step correlated with the dimensions of the investigated arrays.

Kelvin physics of gold-attached protein

Fermi levels and work function

In this work, The Kelvin nanoprobe has been used under ambient atmospheric conditions to measure the difference in work function between a tungsten probe and protein molecules attached to gold-coated silicon substrates. The work function, ϕ, is the minimum energy required to remove an electron from the Fermi level to a point just outside the metal, where the potential of the solid is negligible, or to an infinitely large distance from the surface, where the kinetic energy is zero, the so-called vacuum level. Electrons in a metal are distributed in available states following Fermi-Dirac statistics where the probability of occupancy $f_{(Ei)}$ of energy state E_i is given by(18):

$$f_{(E_i)} = \frac{1}{1 + \exp(E_i - E_F)/kT} \tag{1}$$

The Fermi level E_F is defined as the energy level at which the probability of being occupied by an electron is 0.5. If two metals (such as the tungsten Kelvin probe, and a gold surface) with different work functions, ϕ_{probe} and ϕ_{gold}, are electronically connected, electrons are free to flow from the metal with the lower work function (weaker electron binding) to the metal with the higher work function (stronger electron binding) until the two Fermi levels equalize. This distribution of electrons generates opposite charges on the metal surfaces resulting in a potential difference Φ, known as the contact potential difference (CPD), which is equal to the difference in the work functions:

$$\Phi_{bare-gold} = \phi_{probe} - \phi_{gold} \tag{2}$$

Thin layers of protein molecules adsorbed on a clean gold surface produce a change in work function ϕ_{ads}, which in turn causes a change in the measured CPD.

$$\Phi_{protein-gold} = \phi_{probe} - \phi'_{gold} = \phi_{probe} - \phi_{gold} - \phi_{ads} = \Phi_{bare-gold} - \phi_{ads} \qquad (3)$$

The change in CPD is often interpreted as the change in surface potential, ΔV. The work function is mainly comprised of two components: the chemical potential, which is a purely bulk property of metal, and the surface potential, which is a surface-sensitive parameter. Protein molecules absorbed on a gold surface have a profound effect on the surface potential and hence on the work function. Consequently, any change in work function caused by the protein molecules on the gold surface usually appear as changes in the surface potential:

$$\Delta V = \Phi_{protein-gold} - \Phi_{clean-gold} = -\phi_{ads} \qquad (4)$$

The change in surface potential ΔV is equal in magnitude and opposite in sign to the work function change ϕ_{ads}.

The protein monolayer is responsible for the electrostatic properties of the adsorbate-adsorbent (protein-gold) system. However, the potential change is not only a property of the protein layer, but is also affected by the modification of the electronic properties of the surface such as the covalent binding of any sulfur atom in the protein on the gold (Au^{+}-S^{-}). As a consequence, the change in surface potential involves two contributions: protein layers and protein-gold interface.

Dipole contribution

The adsorbed protein molecules have an intrinsic dipole moment and form a dipole sheet on the surface. According to the Helmholtz model(13), the electrostatic potential due to a protein monolayer as a dipole sheet on the gold surface causes a surface potential change, ΔV, given by

$$\Delta V = \mu / \varepsilon \varepsilon_0 A \qquad (5)$$

where A is the area per molecule, ε_0 the permittivity of the free space (8.85×10^{-12} Fm^{-1}), and ε the relative permittivity of the medium in which the dipoles are located, and μ is the perpendicular vector component of the dipole moment associated with each molecule in the monolayer. The direction of μ determines the sign of surface potential change. If the positive end of the dipole is further from the surface, ΔV will increase; if closer to the surface, it will decrease.. The molecular dipole moment μ normal to the substrate is given by

$$\mu = \frac{\mu_0}{1 + 9\alpha N^{1.5}} \tag{6}$$

where, μ_0 is the molecular dipole moment in the absence of depolarizing effects, α is the polarizability of a molecule and n is the number of dipoles per unit area.

In this model, it has been generally assumed that $\varepsilon=1$, although the definition of the dielectric constant for monolayers is still debatable. In this case, the mutual inductive depolarization effects that must occur when an ensemble of dipoles are closely packed, as well as the interfacial effects associated with the polarization of molecules at the sub-phase surface are neglected. An estimate of the surface potential change caused by protein dipoles on the gold surface, is given by the sum of two contributions(9) and can be written as

$$\Delta V = \Delta V_1 + \Delta V_2 = \mu_1 / \varepsilon_1 \varepsilon_0 A_1 + \mu_2 / \varepsilon_2 \varepsilon_0 A_2 \tag{7}$$

where ΔV_1 is the contribution of dipoles formed by some atoms in the protein and gold interface, such as Au^+-S^-, and ΔV_2 is the contribution of the protein molecules. μ/ε is defined as the apparent dipole moment such that μ_1/ε_1 represents the apparent dipole moment of the gold-protein interface region and μ_2/ε_2 is that of the protein layer (Fig. 2). In the case of several protein molecule interactions occurring on the surface, μ_2/ε_2 could be replaced by $(\mu_2/\varepsilon_2 + \mu_3/\varepsilon_3)$and the surface potential change would be given as,

$$\Delta V_2 = (\mu_2 / \varepsilon_2 + \mu_3 / \varepsilon_3) / \varepsilon_0 A \tag{8}$$

where subscripts 2 and 3 represent the different protein molecules.

The interpretation of the dipole layer contribution to the surface potential and work function changes are suitable for the protein monolayer or the two layers close to the gold surface, in which the charges associated with these layers are fixed within them.

Electronic characterization of the interface

The SKN images of the first protein monolayer or two layers can be explained using the Schottky effect, which is the image-force-induced lowering of the metal work function. When an electronic charge is at a distance x from the surface, an opposite charge is induced at the same distance under the surface

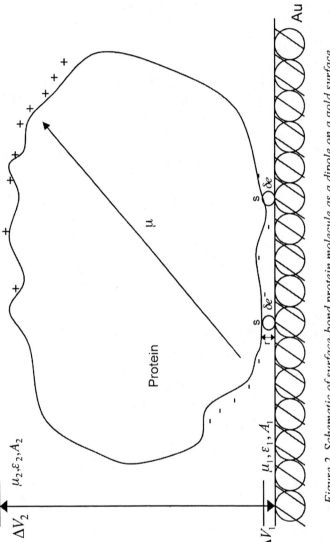

Figure 2. Schematic of surface-bond protein molecule as a dipole on a gold surface

of the metal, in a mirrored position. This opposite charge is named the image charge, and the two charges interact by a force of attraction named the image force. The image force depends on the potential energy of the electron and is extremely high at the surface of a metal, within a distance of two molecular diameters.

The primary effect of the monolayer on the surface potential is due to the dipole moment, as described in the previous section. For multiple protein layers, the Schottky model, which is connected to the semiconductive properties of proteins (19), gives a more appropriate description of the Kelvin physics of such macromolecules on metal. Since bulk protein layers have weakly conductive properties, the surface potential can be explained using a Schottky barrier model (20,21), which represents the rectifying characteristics of a metal-semiconductor junction. Proteins can be envisaged as a p-type semiconductor, exhibiting a depletion region (Fig. 3). At the metal-semiconductor (gold- protein film) junction, due to the electron transfer between the metal and the protein film, the Fermi level in the metal will tend to coincide with that in the semiconductive film. The result is a local change in energy of electrons at a junction giving rise to space charge effects and associated energy band bending at the interface. In a traditional semiconductor, the Schottky relation gives

$$V_b = \frac{enw^2}{2\varepsilon\varepsilon_0} \qquad (9)$$

where V_b is the built-in potential, e the charge of an electron, n is the dopant density in the semiconductor and w is the depletion width. Surface potential measurement for thicknesses less than the depletion width will represent band bending within the depletion region. Band bending will increase with the thickness of the protein multilayer, increasing the surface potential until the Fermi levels are coincident. For thin films (such as a monolayer), this coincidence will not be realized, and depletion will occur throughout the film. When the film thickness is greater than the depletion width, no further changes in the surface potential will occur as the thickness increases. The thickness required to produce the total change will be smallest for materials with the highest density of acceptors n.

Results and discussion

Kelvin physics and protein orientation

Protein molecular layers were formed on gold-coated silicon substrates by the adsorption of neutravidin molecules from DPBS buffer solution. The

322

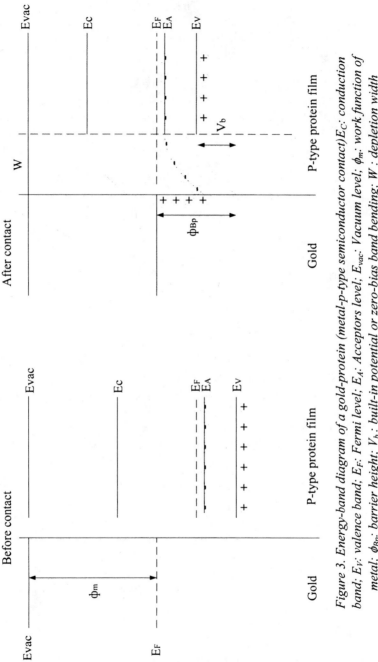

Figure 3. Energy-band diagram of a gold-protein (metal-p-type semiconductor contact)E_C: conduction band; E_V: valence band; E_F: Fermi level; E_A: Acceptors level; E_{vac}: Vacuum level; ϕ_m: work function of metal; ϕ_{Bp}: barrier height; V_b: built-in potential or zero-bias band bending; W: depletion width

adsorption of this protein on gold is believed to occur through a covalent binding of sulfur present in the protein to gold atoms. Neutravidin is a tetrameric protein and a deglycosylated derivative of avidin displaying similar binding affinity to biotin. However, it has a much reduced isoelectric point and exhibits less nonspecific binding than the parent molecule (22). The protein adsorbs strongly to gold surfaces without affecting its affinity for biotin. The neutravidin-biotin interaction is very strong and stable under most conditions ($K_d = 10^{-15}$) (23). A comparison of the CPD value for neutravidin attached to gold to that for a clean metal surface shows that there is an increase of 0.30V in surface potential caused by the protein layer (Fig. 4). This can potentially provide information regarding the orientation of protein molecules on the substrate. The neutravidin molecule on such a surface can be envisaged as a dipole orientated with the positive end of its dipole moment μ directed toward the protein-air interface, with the negative end being close to the metal surface (as shown in Fig. 2). The directionality of polarization of the neutravidin dipole is inferred from the positive sign of the change in surface potential of the protein treated surface with respect to that of the clean gold surface. The increase of surface potential of the neutravidin-attached surface implies a decrease of work function since the change in value, ΔV, is equal in magnitude and opposite in sign to the work function change $\Delta \phi$ according to equation 4.

However, the surface potential change not only originates from the protein layer, but also from Au^+-S^- dipoles. The latter generates a negative contribution to the potential, since the direction of the dipole moment in this case is negative (Fig. 2). In comparison with the dipoles of protein molecules, however, the contribution of Au^+-S^- dipoles is much smaller. As well, the Au^+ moiety is likely screened within a very short distance by the electrons within the metal, whereas this cannot occur for charges in the protein layer. As a result, the magnitude and sign of the surface potential change of neutravidin attached to the gold surface are largely due to the neutravidin dipole sheet. Here, it is important to emphasize that all SKN measurements were performed in ambient atmospheric conditions, under 25% humidity, which means that several layers of water molecules are present on the overall surface system. The work function change for a gold surface caused by water, and protein dipole moment variation with effects of water will be discussed in later work. In the case of the present work, the contribution from water can be omitted in view of the relatively high dipole moment values of the protein sheet.

Kelvin physics and protein interactions

The surface potential change caused by proteins can be employed to discern different protein molecules and interactions associated with their inherent electrical properties, such as dipole moment (μ) and dielectric constant (ε). In

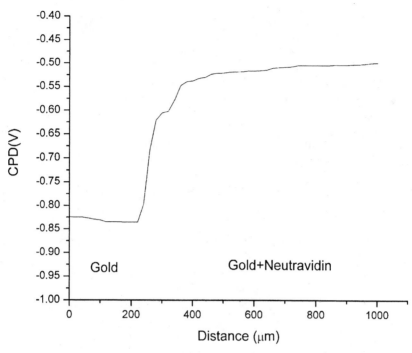

Figure 4. Variation in surface potential across a gold substrate onto a section of which a layer of protein molecules have been immobilized

our experiments, neutravidin and the binding of this molecule to biotinylated-BSA interaction was studied in microarray format. This was achieved by delivering nanoliter volumes of samples of the DPBS buffer, neutravidin, biotinylated-BSA and a mixture of neutravidin and biotinylated-BSA onto the substrate, in duplicates, using the high-precision contact-printing robot. Figure 5 shows the resulting surface potential image of the array as generated by the SKN instrument. For the buffer control area, there were no discernible spots detected on the slides by SKN, indicating the absence of nonspecific adsorption of the buffer on the gold surface. The surface potential images of the various protein domains exhibit reproducible signals. Neutravidin yields a smaller signal ($\Delta V \approx 0.13V$) than for BSA ($\Delta V \approx 0.20V$) and the neutravidin-biotinylated BSA complex ($\Delta V = 0.21$-$0.25V$). As equation (5) indicates, neutravidin should produce a larger signal since it has the smallest molecular size.

   A major factor that influences the surface potential is not only protein molecule size, but also the orientation and polarization of dipoles. Although neutravidin possesses more positive charges than BSA, the surface potential of BSA is larger than that of neutravidin, which implies that the dipoles of the two proteins are oriented in a different manner on the surface. Most of the neutravidin molecules are probably orientated with their positive charges close to the gold surface, whereas for BSA they are orientated with their negative charge close to the gold surface. Another cause for the difference in surface potential between the two proteins is probably the packing density of molecules on the surface. Compared to a neutravidin layer, the density for a BSA layer is likely higher since it possesses more sulfur atoms in its structure, which will contribute to the binding of the molecule to the gold surface.

The largest signal is obtained for the neutravidin-biotinylated BSA complex, a result which is expected since it has a larger dipole moment than any one of the individual protein molecules. However, the magnitude of surface potential for the complex is not composed of a simple combination of values for the two individual proteins. The explanation for this is that the interaction in the complex not only increases the dipole moment but also modifies the molecular properties, in which polar group reorientation will lead to a significant increase in the effective dielectric constant (24). Another issue that should be mentioned here is that the signals for two spots for the complex are not perfectly reproducible. The reason for this behavior likely originates from different orientations of the BSA-neutravidin complex on the surface. For instance, the complex will generate a different surface potential change if BSA were oriented towards the surface than if neutravidin were in this position.

In order to address the problem of the surface potential variation caused by different orientation of the protein complex, instead of spotting the protein mixture on the gold surface, the protein arrays were fabricated based on the procedure of Wingren et al. (25). First, BSA and biotinylated BSA were spotted on the slides, and following buffer and water washing, the protein layers were incubated with neutravidin. Here, the neutravidin was spotted on the top of the arrayed probe instead of incubating whole slides in large sample volumes. In this way, the amount of sample required was reduced by a factor of 4000. After washing, the scanned surface potential images were obtained, as shown in Fig. 6,. Before incubation (Fig. 6-A), the surface potential of the BSA ($\Delta V \approx 0.15$V) and biotinylated-BSA ($\Delta V \approx 0.16$V) are quite similar, which indicates that the small biotin moiety does contribute significantly to the Kelvin signal as expected. No significant change in the surface potential of BSA, $\Delta V \approx 0.17$V was found after interaction with neutravidin (Fig. 6-B), which means there is little or no interaction between the two proteins. However, for biotinylated-BSA, the signals for the neutravidin- biotinylated-BSA complex display an increased signal of $\Delta V \approx 0.22$V, and there is good inter-spot reproducibility. Compared to the BSA signal, the complex shows an increase of 50mV which originates from the neutravidin apparent dipole moment μ_3/ε_3, on equation 8. It should noted that the apparent dipole moment of BSA should be constant for spots with or without incubation. In this respect, the effective dielectric constant ε, which represents

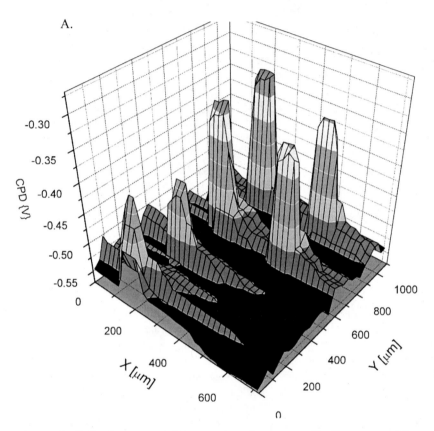

Figure 5. A) Surface potential image of a microarray of buffer, neutravidin, biotinylated-BSA and neutravidin- biotinylated-BSA complex. B) Array map showing the exact position of duplicates: B represents buffer without protein, N represents neutravidin, B-BSA represents biotinylated-BSA, C represents the neutravidin -biotinylated-BSA complex.

B.

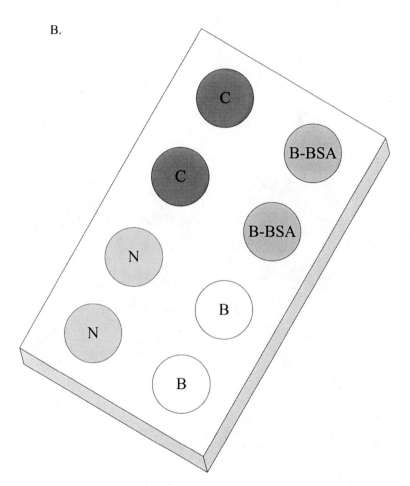

Figure 5. Continued.

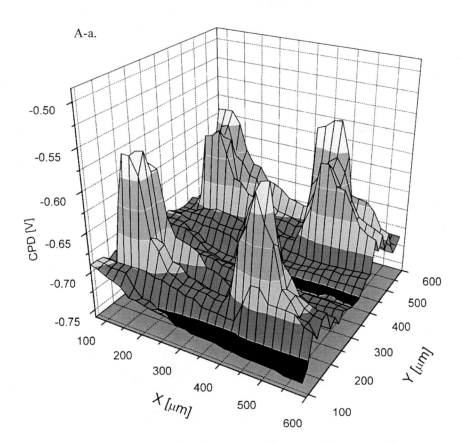

Figure 6. SKN detection of surface potential changes for neutravidin and biotinylated-BSA interaction. A: (a) Surface potential image of BSA and biotinylated-BSA microarray printed in duplicated spots. (b) Array map showing the exact position of duplicates: B-BSA represents biotinylated-BSA. B: (a) Surface potential image of a microarray of (A) incubated with neutravidin (b) Array map showing the exact position of duplicates: C represents the Neutravidin and biotinylated-BSA complex.
Figure 6A-a

A-b.

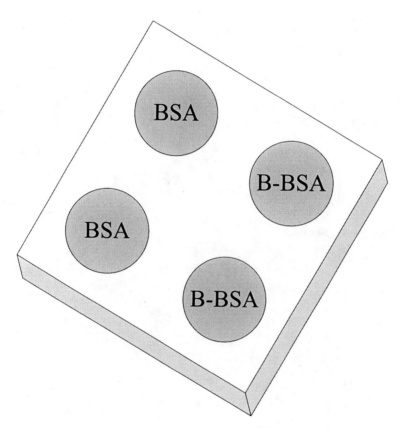

Figure 6. Continued. Figure 6A-b. Continnued on next page.

B-a.

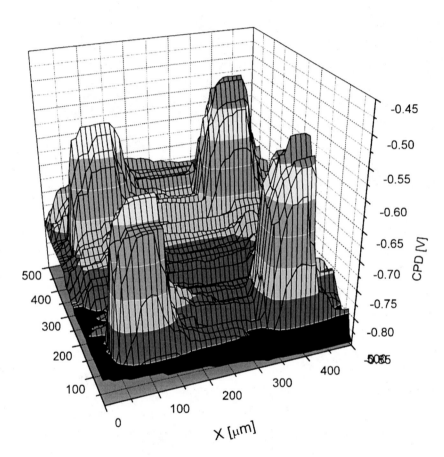

Figure 6. Continued. Figure 6B-a.

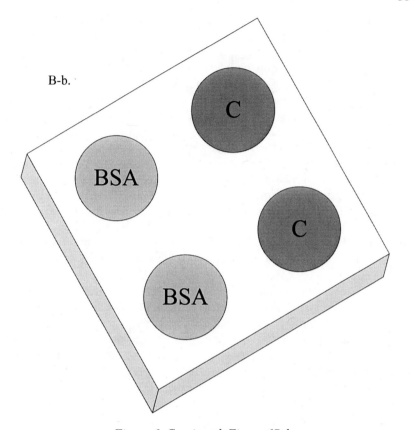

Figure 6. Continued. Figure 6B-b.

the reorientation of polar groups in the protein interior, increases considerably with protein relaxation, reorganization of polar group and changes in the degree of water solvent penetration to the protein during protein molecular interaction (*24*).

Kelvin physics and protein semiconductive properties

For any protein microarray, it is clear that spots of absorbed protein will not always form a monolayer. To investigate the Kelvin characteristic of the packed protein spots in multilayers, series of neutravidin solutions (0, 0.2, 0.4, 0.6, 0.8, 1.0 mg/ml) were spotted in duplicates on the gold surface. After carefully washing to remove buffer, the slides were scanned with the SKN. Figure 7 shows the relationship of the surface potential with the domain concentration (or thickness). As shown in Fig.7-A, the duplicated spots display excellent reproducibility with no discernible signal being detected for the buffer control spots. In addition, this result clearly shows that there is a trend of surface potential increase with the neutravidin concentration on the surface, until an apparent saturation plateau appears at a concentration of approximately 0.6 mg/ml.

We believe that this Kelvin behavior can be explained using the Schottky barrier model (Fig. 3). Neutravidin layers can be viewed as a semiconductor that contain p-type defects and with a depletion layer occurring in the film near the interface, as in conventional semiconductors. At the interface between neutravidin and gold, the Fermi level of the protein lies below that of the gold so that band bending occurs by virtue of the electrons transferring from the gold to the acceptor states of the neutravidin in order to equalize their respective Fermi levels. This causes a build-up of charges on both sides of the interface, resulting in an electric field and therefore a potential gradient according to Poisson's equation $\partial^2 V/\partial x^2 = \rho(x)$. According to equation 9, band bending increases with the thickness of the neutravidin layer, thus increasing the surface potential, until the thickness is greater than the depletion width. In Fig. 7-C, as the concentration increases to 0.6 mg/mL, the neutravidin layers represent different stages of a "frustrated" depletion width (20,21). Beyond that point, for film thicknesses greater than the depletion width (greater than 0.6mg/mL in Fig.7-C), no further changes in the surface potential occur as the thickness of the neutravidin layer is increased.

Conclusions

It has been shown that the scanning Kelvin nanoprobe can be used to discriminate between different types of proteins and protein interactions. This detection is associated with the electronic properties of the protein and protein complexes attached to a solid surface, demonstrating the potential of the technique for the label free detection of multiplexed protein microarrays. This study also demonstrates the feasibility of discerning biomolecules and their interactions using the SKN, and it is clear that the Kelvin technique is extremely useful in exploring the properties and behavior of protein films, especially in the microarray format. However this technique is not without its problems, the most prominent of which is the ambiguity involved in interpreting the origin of the measured changes. As the understanding of this type of Kelvin physics grows,

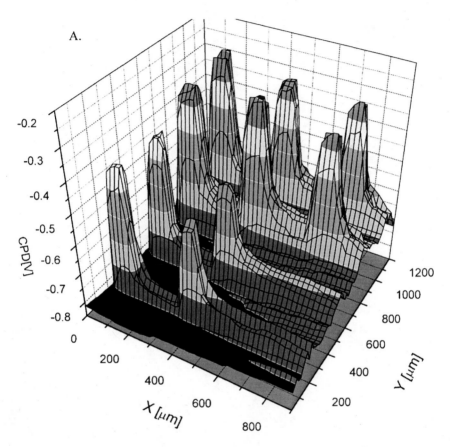

Figure 7. Kelvin characteristics of neutravidin concentration array.
A: Scanned CPD image of a series of neutravidin concentration domains on the
gold substrate. B: Array map showing the exact position of duplicate
neutravidin spots with concentrations of 0, 0.2, 0.4, 0.6, 0.8 and 1.0mg/ml.
C: Surface potential change as function of the neutravidin concentration
(thickness of spots) Continued on next page.

B.

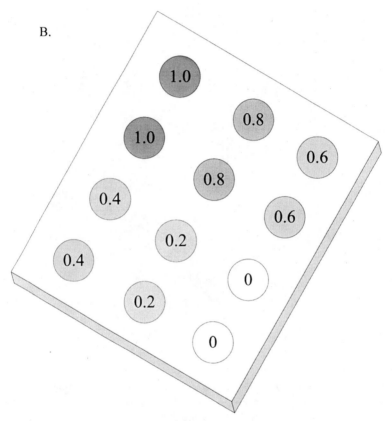

Figure 7. Continued.

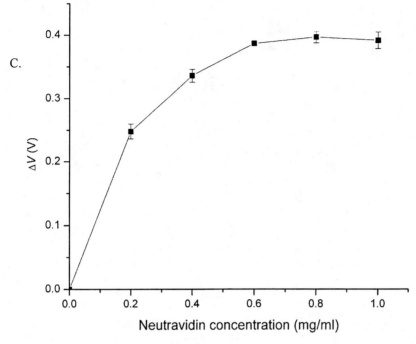

C.

Figure 7. Continued.

the SKN instrument could become a powerful tool in proteomics, providing a new type of information not readily obtainable from other techniques.

Acknowledgments

The authors are grateful to the Natural Sciences and Engineering Council of Canada and Maple Biosciences Ltd., Halifax, Nova Scotia for support of this work. Also, we wish to express our appreciation to Dr. Sherri Johnstone of the School of Engineering, Durham University, UK for much helpful discussion.

References

1. Lee, Y.; Lee, E. K.; Cho, Y. W.; Matsui, T.; Kang, I. C.; Kim, T. S.; Han, M. H., *Proteomics* **2003**, 3, (12), 2289-2304.

2. Batorfi, J.; Ye, B.; Mok, S. C.; Cseh, I.; Berkowitz, R. S.; Fulop, V., *Gynecologic Oncology* **2003**, 88, (3), 424-428.

3. Issaq, H. J.; Conrads, T. P.; Prieto, D. A.; Tirumalai, R.; Veenstra, T. D., *Analytical Chemistry* **2003**, 75, (7), 148A-155A.

4. Kim, M. G.; Shin, Y. B.; Jung, J. M.; Ro, H. S.; Chung, B. H., *Journal of Immunological Methods* **2005**, 297, (1-2), 125-132.

5. Jones, V. W.; Kenseth, J. R.; Porter, M. D.; Mosher, C. L.; Henderson, E., *Analytical Chemistry* **1998**, 70, (7), 1233-1241.

6. Dammer, U.; Hegner, M.; Anselmetti, D.; Wagner, P.; Dreier, M.; Huber, W.; Guntherodt, H. J., *Biophysical Journal* **1996**, 70, (5), 2437-2441.

7. Cheran, L. E.; Chacko, M.; Zhang, M. Q.; Thompson, M., *Analyst* **2004**, 129, (2), 161-168.

8. Demchak, R. J.; Fort, T., *Journal of Colloid and Interface Science* **1974**, 46, (2), 191-202.

9. Taylor, D. M.; Deoliveira, O. N.; Morgan, H., *Journal of Colloid and Interface Science* **1990**, 139, (2), 508-518.

10. Vogel, V.; Mobius, D., *Journal of Colloid and Interface Science* **1988**, 126, (2), 408-420.

11. Oliveira, O. N.; Taylor, D. M.; Morgan, H., *Thin Solid Films* **1992**, 210, (1-2), 76-78.

12. Evans, S. D.; Ulman, A., *Chemical Physics Letters* **1990**, 170, (5-6), 462-466.

13. Taylor, D. M.; Morgan, H.; Dsilva, C., *Journal of Physics D-Applied Physics* **1991**, 24, (8), 1443-1450.

14. Lu, J.; Delamarche, E.; Eng, L.; Bennewitz, R.; Meyer, E.; Guntherodt, H. J., *Langmuir* **1999**, 15, (23), 8184-8188.

15. Hayashi, K.; Saito, N.; Sugimura, H.; Takai, O.; Nakagiri, N., *Ultramicroscopy* **2002**, 91, (1-4), 151-156.

16. Hansen, D. C.; Hansen, K. M.; Ferrell, T. L.; Thundat, T., *Langmuir* **2003**, 19, (18), 7514-7520.

17. Thompson, M.; Cheran, L. E.; Zhang, M. Q.; Chacko, M.; Huo, H.; Sadeghi, S., *Biosensors & Bioelectronics* **2005**, 20, (8), 1471-1481.

18. Gundry, P. M.; Tompkins, F. C., Surface Potentials. In *Experimental methods in catalytic research*, eds.; Anderson, R. B.; Dawson, P. T., Academic Press: New York, 1968; pp 100-168.

19. Bone, S.; Eden, J.; Gascoyne, P. R. C.; Pethig, R., **1981**, 77, 1729-1732.

20. Tredgold, R. H.; Smith, G. W., *Thin Solid Films* **1983**, 99, (1-3), 215-220.

21. Jones, R.; Tredgold, R. H.; Hoorfar, A., *Thin Solid Films* **1985**, 123, (4), 307-314.

22. Hiller, Y.; Gershoni, J. M.; Bayer, E. A.; Wilchek, M., *Biochemical Journal* **1987**, 248, (1), 167-171.

23. Tassew, N.; Thompson, M., *Analytical Chemistry* **2002**, 74, (20), 5313-5320.
24. Sham, Y. Y.; Muegge, I.; Warshel, A., *Biophysical Journal* **1998**, 74, (4), 1744-1753.
25. Wingren, C.; Steinhauer, C.; Ingvarsson, J.; Persson, E.; Larsson, K.; Borrebaeck, C. A. K., *Proteomics* **2005**, 5, (5), 1281-1291.

Chapter 21

Vibrational Spectroscopy Studies on Biologically Relevant Molecules: From Anticancer Agents to Drugs of Abuse

M. P. M. Marques[1,*], F. Borges[1,2], A. M. Amorim da Costa[1], and L. A. E. Batista de Carvalho[1]

[1]Research Unit "Molecular Physical-Chemistry", University of Coimbra, 3000 Coimbra, Portugal
[2]Organic Chemistry Department, Faculty of Pharmacy, University of Porto, 4050-047 Porto, Portugal

Vibrational spectroscopy (both Raman and Inelastic Neutron Scattering), coupled to quantum mechanical calculations, is used in order to perform a thorough analysis of several biologically relevant molecules, such as chemotherapeutic agents, biogenic polyamines and their metal chelates, cardiovascular protectors, non-steroid anti-inflammatory drugs (NSAID's), drugs of abuse and phenolic compounds. The conjugation of these techniques yields valuable information regarding the structural preferences of the systems under investigation, which may help establish the structure-activity relationships (SAR's) ruling their biological function.

Introduction

Raman spectroscopy constitutes an invaluable method for the characterization and conformational analysis of biomolecules. Actually, one of the greatest, unsurpassed advantages of this technique is the fact that water is virtually transparent to the Raman effect, thus allowing the collection of good quality spectra for rather diluted aqueous samples, which is of the utmost importance in biochemical studies. Also, Raman spectroscopy has proved, in the last few years, to be a simple and reliable method for the determination of the composition profile of solid samples, either pure compounds or mixtures. Furthermore, due to its non-invasiveness, high sensitivity and good reproducibility, this technique, which does not need any special sample preparation, is becoming a valuable tool in the field of forensic science for the screening of illicit products, since it provides unique fingerprint spectra, specific for each compound. Therefore, from the detection of very small differences in chemical structure, Raman spectroscopy yields relevant information regarding the origin and/or synthetic route of a particular drug.

Inelastic Neutron Scattering (INS), in turn, is a vibrational spectroscopy technique well suited for the study of compounds containing hydrogen atoms (*1*), as opposed to optical methods (infrared and Raman) that are generally most sensitive to vibrations involving heavy atoms. Indeed, since neutrons have a mass similar to that of the hydrogen atom, an inelastic collision between them involves a significant transfer of both momentum and energy to the irradiated sample. The modes involving a significant hydrogen displacement will dominate the spectrum, since the scattering cross-section for hydrogen is 80 barns as opposed to *ca.* 5 barns for most other elements. Therefore, INS allows the detection of vibrational modes not available to the conventional Raman and infrared optical methods (namely in the low energy region), all vibrations being active (since INS is not subject to the photon selection rules). In fact, INS displays a clear and intense vibrational pattern in the frequency range below 600 cm^{-1}, while above *ca.* 1800 cm^{-1} its spectral quality begins to deteriorate. Raman spectroscopy comes into its own in this region, thus being particularly useful for the analysis of XH/XD (X=C, N, O) stretching modes. Moreover, theoretical Raman and INS band positions and intensities can be easily obtained through quantum mechanical calculations and the spectra can be accurately simulated (using dedicated programs such as Gaussian (*2*) and aCLIMAX (*3*)). Actually, INS is the only vibrational spectroscopic technique that allows a quantitative use of the experimental intensities to validate the calculated quantum mechanical results. It is then possible to link molecular geometry with the spectroscopic features and achieve a thorough conformational analysis of the systems under investigation.

We presently report the use of vibrational spectroscopy for the study of several molecules of recognized biological significance, namely chemotherapeutic agents, biogenic polyamines and polyamine metal chelates with potential anticancer activity, cardiovascular protectors, non-steroid anti-inflammatory drugs (NSAID's), drugs of abuse (*e.g.* amphetamines and opioids), and phenolic compounds of both natural and synthetic origin. The complementarity of the Raman and INS spectroscopies is exploited, in order to gain a better knowledge of the conformational behavior of these systems. This work has been developed in the Research Group "Molecular Physical-Chemistry" of the University of Coimbra (Portugal).

INS spectra were obtained in the Rutherford Appleton Laboratory (UK), at the world's leading pulsed neutron source (ISIS), on the TOSCA spectrometer (4). This is an indirect geometry time-of-flight, high resolution (($\Delta E/E$) *ca.* 2%), broad range, spectrometer, which links together both vibrational and structural capabilities.

Chemotherapeutic Agents

Since the biochemical role of transition metal complexes (*e.g.* anticancer agents) is strongly dependent on their conformational characteristics, the structural analysis of this kind of system is of the utmost importance for the understanding of the structure-activity relationships (SAR's) underlying their antineoplastic activity.

In view of obtaining this type of information, vibrational spectroscopy studies were undertaken for the well known platinum(II) chemotherapeutic drugs cisplatin (*cis*-$(NH_3)_2PtCl_2$) and carboplatin ($Pt(NH_3)_2C_6O_4H_6$), as well as for the analogous transplatin (*trans*-$(NH_3)_2PtCl_2$). Their overall vibrational pattern was found to be clearly dependent on their geometry (*e.g. cis* or *trans* relative to the metal ion), as well as on the nature of the ligand (*e.g.* degree of covalent character of the metal–ligand bonds). The experimental Raman and INS patterns were assigned in the light of the corresponding calculated vibrational frequencies and intensities (using the Density Functional Theory (DFT) and Effective Core Potential (ECP) approaches). INS allowed to expose the low lying librational modes of the NH_3 ligand, and the neutron scattering results for all the internal modes below *ca.* 1000 cm^{-1} were found to be in complete agreement with Raman studies presently being performed by the authors for similar systems, as well as with reported experimental (5-8) and calculated (9,10) data on several analogous dihalodiamine Pt(II) complexes.

For the three Pt(II) chelates studied, good quality INS spectra were obtained from *ca.* 250 mg of compound, which is the smallest sample of a hydrogenous compound for which a successful INS interpretation has been reported. The fact that these extremely small amounts of sample (250 mg as opposed to the usual 4-5 g) still allowed the acquisition of reliable data for this type of heavy metal complexes is rather hopeful for future INS studies of new compounds, for which larger quantities may be initially unavailable.

Biogenic Polyamines and Polyamine Metal Complexes

Biogenic polyamines – putrescine ($H_2N(CH_2)_4NH_2$), spermidine ($H_2N(CH_2)_3NH(CH_2)_4NH_2$) and spermine ($H_2N(CH_2)_3NH(CH_2)_4NH(CH_2)_3NH_2$)

– are ubiquitous in cells of higher organisms, and result from the decarboxylation of basic aminoacids. They are intrinsic polycations known to play an important role in many biological functions, being essential for most growth related processes in cells (*11-18*). On account of absolute polyamine requirement for cell differentiation, interference with polyamine biosynthesis can be a rather promising therapeutical approach against proliferative diseases. Moreover, it was found that linkage of some of these molecules to previously tested anticancer agents leads to a higher cytotoxic effect (*19-22*). In some cases, they can even enhance the efficacy of the long used first-generation drug cisplatin (*23*).

The highly sensitive structure-activity relationships that underlie and control polyamine function are very poorly known, and have been the target of intense research in the last few years. Over the last decade several studies on the cytotoxic properties of transition metal complexes with aliphatic polyamines have been carried out, aiming at obtaining new anticancer, third-generation drugs, displaying tissue specificity and enhanced efficacy relative to the clinically used compounds (*24-30*). In particular, a group of cisplatin-like complexes containing polyamine bridging ligands (*e.g.* putrescine, spermidine or spermine) have proven to display novel antitumor properties (*21,25,31-39*) (*e.g.* triplatinum BBR3464 (*29,35,36*)). Such agents were found to yield DNA adducts not available to conventional alkylating agents, through long-distance intra- and interstrand cross-links. Since DNA attack by this type of chelates (and consequent cytotoxicity) depends strongly on their chemical nature and structural preferences, a thorough conformational study of these potential antineoplastic compounds, at the molecular level, is essential for a rational design of new anticancer drugs.

Aliphatic Linear Polyamines

A conformational analysis of the homologous series of the linear α,ω-diamines ($H_2N(CH_2)_nNH_2$) (n=2 - 10, n=12) and the tri- and tetramines spermidine and spermine was undertaken (*40-48*). These molecules display a high conformational freedom, as well as an interdependence of the particular effects due to the nitrogens′ electronegativity and electron lone-pairs.

The complementary use of the Raman and INS techniques allowed the observation and assignment of the whole set of longitudinal acoustic vibrational modes (LAM′s) for these amines, both for their undeuterated and N-deuterated forms (*44*) (Figure 1A), as well as of the corresponding transverse modes (TAM′s). The INS experimental LAM′s are in good accordance with the LAM′s of the corresponding *n*-alkanes (*49*), which supports the idea of a significant conformational similarity (in the solid state) between these two sets of compounds.

The effect of both protonation and N-deuteration on the conformational behavior of these linear amines – and consequently on their vibrational pattern –

was investigated. In the totally protonated (physiological) state only the *all-trans* conformation was found to be present, and formation of intramolecular (N)H ⋯ :N or (C)H ⋯ :N hydrogen bonds is hindered *(41,42)*. In these conditions the linear polyamines behave as saturated alkanes, for which the *all-trans* geometry has long been recognized to be energetically favoured over the *gauche* ones *(50)*. Deuteration of the samples saw the observation of the anticipated shifts to lower frequencies in both INS and Raman (Figure 1B), as well as the loss of some INS intensity for the bands assigned to the vibrational modes of the ND_2 group, as a consequence of the variations in both mass (frequencies) and scattering cross-section (INS intensities). Also, a wavenumber shift to lower energies was found, in agreement with the theoretical results, for the LAM modes associated with a significant change in the CCN angle. The experimental data compared well with the calculated Raman wavenumbers and INS transition frequencies and intensities (Figure 2).

The non-observation of the NH_2 torsion modes in INS for most of the polyamines studied, however, was an unexpected result. The fact that 1,2-diaminoetane appeared as an exception, as both τ_{NH_2} A_u (409 cm^{-1}) and B_g (432 cm^{-1}) bands were distinctly detected as well as a third band at 522 cm^{-1} (Figure 3), was interpreted considering the presence of dimers in the solid for this small molecule. In fact, 1,2-diaminoetane having such a short alkylic chain between the two amine terminal groups will not be prone to form polymeric structures, due to an inefficient electronic charge delocalization through the carbon skeleton upon formation of intermolecular H_2N ⋯ H interactions simultaneously in both ends of the molecule, the dimeric species being the most favourable one in the condensed phase. The three distinct INS experimental features are thus assigned to the hydrogen bonded (central) NH_2 groups – 522 cm^{-1} – and to the (terminal) NH_2 moieties, not engaged in intermolecular hydrogen close contacts – 409 and 432 cm^{-1}. While the non-degeneracy of the lowest energy τ_{NH_2} modes was theoretically predicted for this particular molecule alone, the band at higher frequency was not found by the calculations. In the case of 1,3-diaminopropane and the larger diamines now investigated, in turn, oligomeric forms are probable to occur in the solid, both NH_2 groups being involved in intermolecular hydrogen close contacts, which would explain why only one band (H-bonded NH_2) was detected in the INS spectra (at *ca.* 530 cm^{-1}, for $H_2N(CH_2)_nNH_2$ n=3, 4 and 5). As the chain lengthtened, a clear and progressive intensity decrease of this feature was observed, until it completely disappeared for 1,6-diaminohexane (Figures 1A and 3), due to the smaller and smaller relative weight of the amine hydrogens in the molecule.

This study contributed to a better understanding of the conformational behavior of linear polyamines, which, in the solid state, is strongly determined by the occurrence of intermolecular R-HN-H⋯NH_2-R H-type interactions. This kind of close contacts was found to give rise to either infinite chain polymeric forms or to dimeric species, in accordance with the theoretical predictions *(48)*.

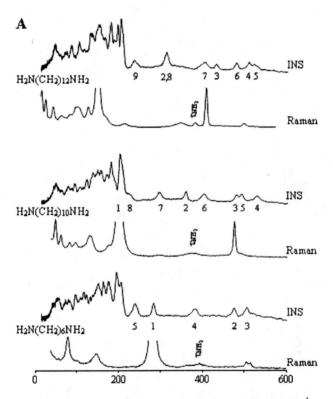

Figure 1. A – Experimental Raman and INS spectra (16 – 600 cm⁻¹, solid state) for the $H_2N(CH_2)_nNH_2$ (n=6, 10, 12) diamines. (The numbers refer to the LAM modes). B – Experimental INS spectra (16 – 2000 cm⁻¹, at 20 K) for $H_2N(CH_2)_6NH_2$ and spermidine, in their undeuterated and N-deuterated forms. (Reprinted with permission from reference 44. Copyright 2002.)

Continued on next page.

B

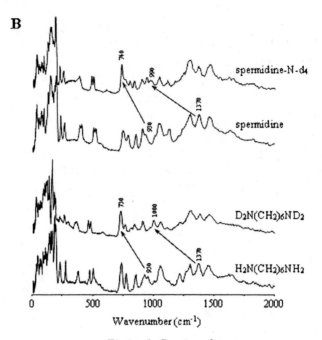

Figure 1. Continued.

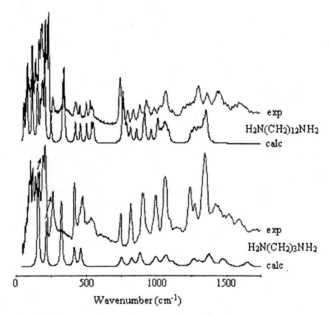

Figure 2. Experimental and calculated (B3LYP/6-31G*) INS spectra (16 – 1750 cm^{-1}, 20 K) for the H$_2$N(CH$_2$)$_n$NH$_2$ (n=3, 12) diamines. (Reprinted with permission from reference 44. Copyright 2002.)

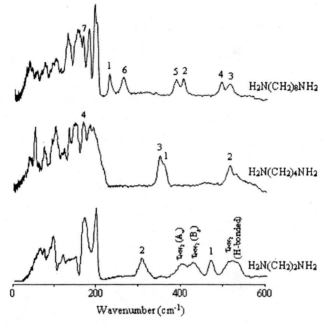

Figure 3. Experimental INS spectra (16 – 600 cm^{-1}, at 20 K) for the H$_2$N(CH$_2$)$_n$NH$_2$ (n=2, 4, 8) diamines. (The numbers refer to the LAM modes. (Reprinted with permission from reference 44. Copyright 2002.)

Agmatine

Agmatine (4-(aminobutyl)guanidine, AGM, Figure 4), a polyamine produced by decarboxylation of L-arginine, has a wide range of physiological functions (*51*). It was found to act as a neurotransmitter or neuromodulator (*52-54*), and to be able to stimulate insulin release (*55*). In addition, agmatine may behave as an antiproliferative and tumor suppressor agent (*56-61*). As accounted for other biogenic polyamines, the activity of agmatine as a transportable cation (*62*) and biological effector can be significantly influenced by its structural preferences. Thus, the study of its conformational characteristics is essential for understanding its diverse biochemical functions, namely its effect on the mitochondrial permeability transition (MPT) and its specific transport mechanism.

The study developed in our laboratory aimed at a better knowledge of the role of agmatine in biological systems, in the light of a thorough conformational analysis, both in the solid and in aqueous solution (for AGM′s distinct protonation states). Raman spectroscopy combined to DFT calculations were used (*63*), and the structural information thus gathered was analyzed in order to

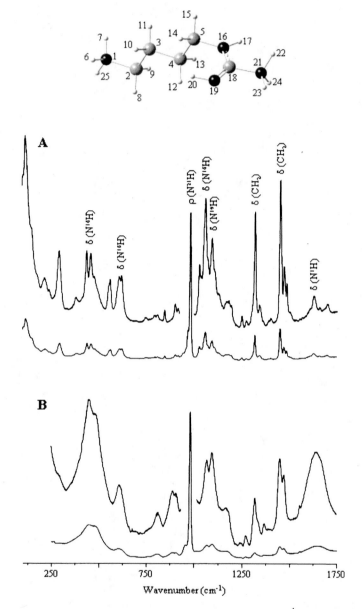

Figure 4. Experimental Raman spectra (100 – 1750 cm⁻¹, at 298 K) of diprotonated agmatine: solid state (A); aqueous solution (0.5 mol.dm⁻³, pH=7.3) (B). (δ and ρ represent deformation and rocking vibrational modes, respectively). The calculated (IEFPCM/B3LYP/6-31G**) most stable geometry for the agmatine molecule, at physiological conditions, is also represented. (Reprinted with permission from reference 63.)

explain the biological results simultaneously obtained in rat liver mitochondria (RLM) (*64*).

Raman spectra were collected for both solid agmatine and aqueous solutions, as a function of pH (Figure 4). A complete assignment of the experimental vibrational features was carried out, in the light of the theoretical results and the experimental spectra obtained for similar polyamines – dien ($H_2N(CH_2)_2NH(CH_2)_2NH_2$), propen ($H_2N(CH_2)_3NH(CH_2)_3NH_2$), putrescine, spermidine and spermine, (*42,44,46,47*). A very good agreement was found between the vibrational pattern of agmatine and its theoretically determined behavior. In fact, the expected changes due to *N*-protonation were clearly detected by Raman: as pH was increased (the amount of protonated species being lower and lower) a strong decrease in intensity was observed for the very intense band at 983 cm^{-1} (Figure 4), ascribed to a deformation (rocking) mode of the NH_3^+ guanidinium group (ρ_{NH_3}). Moreover, as the percentage of the unprotonated form raised the broad feature at *ca.* 1650 cm^{-1}, assigned to the NH_2 scissoring mode, increased in intensity. The occurrence of a strong ρ_{NH_3} mode is theoretically predicted, for both the mono- and dipositive forms of agmantine (displaying a protonated guanidinium group). The protonation state of agmatine can thus be easily and unequivocally determined through its Raman pattern in aqueous solution.

This study led to a thorough structural characterization of the most stable geometries of agmatine in aqueous solution, under different pH conditions: at physiological pH, the diprotonated (dipositive) species (Figure 4); at alkaline pH (*ca.* 7.3 to 9.0), the monoprotonated (monopositive) form; and in strong alkaline medium, the totally unprotonated (neutral) molecule. These results also made it possible to individualize the agmatine structures prone to interact with the mitochondrial site responsible for its transport and for protection against MPT induction. These studies are essential for the development of new agmatine-based therapeutic strategies (*e.g.* against drug addiction, pain-killing or tumor suppressing).

Polynuclear Polyamine Pt(II) and Pd(II) Chelates

Platinum and palladium-based antitumour drugs have been the target of intense research since Rosenberg′s discovery of an unexpected inhibition of cell division in the presence of cisplatin, in the late sixties (*23*). Several new polynuclear Pt(II) and Pd(II) polyamine chelates recently synthesized in our laboratory are being evaluated as to their conformational preferences by vibrational spectroscopy (Raman and INS) coupled to quantum mechanical calculations (*65*). These constitute a new class of compounds comprising cisplatin-like moities ([Pt(NH_3)_2Cl] or [Pt(NH_3)Cl_2]) linked by a variable length alkanediamine chain. In order to better understand the influence of the structural features on the antineoplastic properties of these systems, they were designed to differ in one of the following parameters: number of metal ions; coordination pattern and chemical environment of the metal(s); distance between the metal centers; structural properties of the ligand(s); total electric charge.

The INS spectra were simulated (3) from the normal mode eigenvectors yielded by DFT calculations (using the ECP approximation for the representation of the metal). Particular attention was paid to characteristic spectral regions, namely the ones comprising: (i) the NH_3 torsions; (ii) the N–M–Cl and N–M–N deformations (M=Pt, Pd); (iii) the vibrational features characteristic of the polyamine ligands. Apart from the CH and NH torsion and deformation modes, all the M-N and M-Cl oscillators, both stretching and bending, were observed and assigned. Davidov splittings of the NH_3 torsions were clearly detected by INS. Furthermore, the presently described study allowed to compare the low lying N–M–Cl and N–M–N deformations for both the Pd(II) and Pt(II) analogous chelates and thus determine the effect of the metal on these particular vibrational bands, which can be considered as a fingerprint of these kind of complexes.

The knowledge gathered by this type of study, together with concurrent biochemical assays for the quantification of the *in vitro* anti-proliferative and/or cytotoxic effects of the complexes towards distinct human cancer cell lines (66-68), may contribute to the determination of the structure-activity relationships ruling their biological properties. This will hopefully help to expose the molecular basis of toxicity, aiming at the design of new and more efficient anticancer agents for future clinical use.

Cardiovascular Protective Agents

Congestive heart failure (CHF) is presently one of the main causes of death worldwide, the overall one-year mortality rate remaining at 10%. Moreover, patients with CHF tend to experience a progressive decline in exercise tolerance and functional ability. Thus, there is a need for new therapies that will improve symptoms and survival rates regarding CHF. Carvedilol (1–[carbazolyl–(4)-oxy]-3-[2-methoxyphenoxyethyl)amino]-propanol-(2)) (Figure 5) is a compound displaying antioxidant properties (69,70) used in the clinical practice (COREG®) for the treatment of CHF, mild to moderate hypertension and myocardial infarction (71-75). Recent results propose that carvedilol exerts its effects by protecting cardiac mitochondria from oxidative stress events (76-81), since cardiac function is known to involve mitochondrial bioenergetics (82). Although the mechanisms underlying this protective role are still not completely understood, they are proposed to be closely related to the acid-base characteristics of the molecule, which underlies the importance of a correct determination of its pK_a value. In fact, carvedilol is known to behave as a weak protonophore, carrying protons through the mitochondrial membrane, thus causing a lowering of the electric membrane potential (83).

A conformational analysis of carvedilol as a function of pH was carried out in our group, by Raman spectroscopy combined to *ab initio* molecular orbital

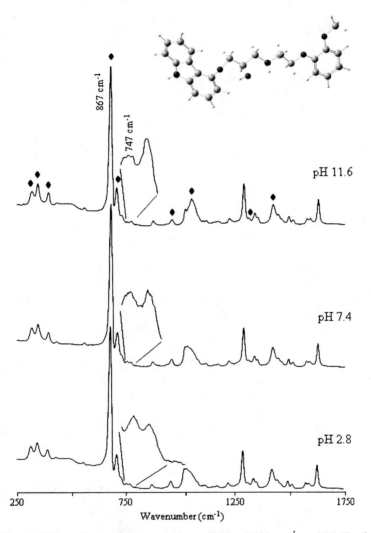

Figure 5. Experimental Raman spectra (250 – 1750 cm⁻¹, at 298 K) of 0.5 mol.dm⁻³ solutions of carvedilol in DMSO, for distinct pH values. (The bands due to the solvent are marked with the symbol ♦). The calculated (HF/3-21G(N)*) lowest energy conformer for carvedilol is also represented. (Reprinted with permission from reference 84. Copyright 2002.)

calculations (*83*), and the corresponding proton affinities were obtained. Moreover, the pK_a value of its secondary amine group was determined through Raman pH titration experiments (*84*), since a clear pH dependence was detected in the corresponding spectroscopic pattern (namely at 747 cm^{-1}, Figure 5) and an unequivocal correlation between the modes undergoing variation and a particular protonation degree of the molecule could be achieved. The pK_a value obtained (pK_a=8.25) is in quite good accordance with the one previously determined (pK_a=7.9) using analytical methods (*85*). Furthermore, the fact that the band at *ca.* 867 cm^{-1} (Figure 5), ascribed to a mixed mode due to the indole NH bending and to ring deformation (*86,87*), was not affected by the variation in pH is a good evidence that protonation occurs only in one of the nitrogen atoms of the carvedilol molecule (within its linear chain), the carbazole group remaining unchanged. In fact, the frequency of this Raman pattern is considered by some authors (*87*) as a probe of H-bonding at the indole nitrogen.

These results corroborate the relevance of the structural characteristics and acid-base properties of the amino side chain group of carvedilol on its proton shuttling activity across the inner mitochondrial membrane. BM910228, the carbazolyl-hydroxylated metabolite of carvedilol, was also studied. The diversity of biological effects displayed by carvedilol and its metabolite – carvedilol having a lower proton affinity than BM910228 – were shown to be associated to differences in their molecular conformation. This weaker affinity for protons enables carvedilol to release H$^+$ more easily within the mitochondrial matrix and thus to act as a slightly more efficient protonophore. The occurrence of intramolecular O $\cdots$ H(O) and O $\cdots$ H(N) hydrogen type interactions is determinant for the acid-base behavior of this kind of molecule. Actually, the lower proton affinity of carvedilol relative to BM910228 is most probably due to a higher stabilization of the protonated species (containing a NH$_2^+$ group) in the latter, through an O $\cdots$ H(N) interaction (O − H(N) distance = 1.71 Å *vs* 2.45 Å for the unprotonated molecule), that is favoured by the formation of a neighbouring strong O $\cdots$ H(O) close contact (O $\cdots$ H(O) distance = 1.60 Å, (OHO) angle = 170.5°).

Non-steroid Anti-inflammatory Drugs (NSAID's)

Ketoprofen (3-benzoyl-α-methylbenzeneacetic acid) is a well known non-steroid anti-inflammatory drug (NSAID) (Figure 6), which has been widely used in medicine for more than a decade as an analgesic and an antipyretic (*88,89*). One of the challenges facing drug delivery today is to obtain a steady concentration of the pharmacophore, instead of the dramatic variations that usually result from traditional administration. This goal can be achieved by a controlled release of the drug, which is presently a vital field of research in pharmacology and medicinal chemistry, and relies on the use of specially designed excipients and/or carrier systems. The presence of interactions between the therapeutic agents and the compounds used as excipients may be responsible

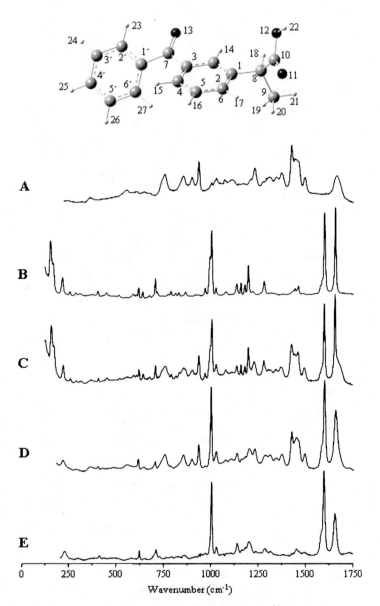

Figure 6. Experimental Raman spectra (75 – 1750 cm⁻¹, solid state, at 298 K) for PVP (A), ketoprofen (B), ketoprofen:PVP (1:1) fresh PM (C), ketoprofen:PVP (1:1) aged PM (D) and [ketoprofen:PVP (1:1) aged PM – PVP] (E). The calculated (B3LYP/6-31G*) lowest energy conformer for ketoprofen is also represented. (Reprinted with permission from reference 93.)

for changes in their stability, solubility and subsequent bioavailability, thus determining the *in vivo* drug release profile, as well as possible unwanted side-effects. Such interactions, however, do not necessarily imply an incompatibility, since the therapeutic activity of the drug may not be affected by the presence of the excipient. The occurrence of drug-excipient close-contacts in solid matrices is presently the object of some controversy (*90-92*). These couplings may occur through intermolecular hydrogen-like bonds, or by more subtle interactions such as Van der Waals contacts. The development of new drug-excipient binary mixtures implies a thorough characterization of the drugs, their solid state structure having a strong influence on both their stability and bioavailability, once their conformational preferences determine the mechanisms controlling their release from a particular delivery system, as well as their *in vivo* pharmacokinetics and pharmacodynamics.

Vibrational spectroscopy is especially appropriate to investigate this type of system, since it allows the detection and characterization of both intra- and intermolecular interactions. It does not require any special sample preparation, thus avoiding mechanical influences which may alter the physicochemical properties of the drug-excipient formulation. Inelastic Neutron Scattering (INS), in particular, is quite suitable for the study of these kinds of interactions involving hydrogen atoms.

For ketoprofen as the pharmacological agent, several binary drug-excipient physical mixtures (PM's) were studied (*93*) in our laboratory, using the following excipients: cellulose derivatives (swelling polymers), lactose (LAC), β-cyclodextrin (β-CD) and polyvinylpyrrolidone (PVP). Spectral evidence of drug-excipient close contacts was clearly detected upon ageing (24 h after sample preparation), both by Raman (Figure 6) and INS, for the (1:1, *w:w*) (ketoprofen:PVP) and (ketoprofen:LAC) PM's. Sample degradation was ruled out by stability studies on both components of the PM's (*94*). Therefore, these vibrational results should be taken as a clear proof of the occurrence of drug-excipient close-contacts. In the light of a complete conformational analysis performed for the isolated drug by DFT methods (*95*), it was possible to identify the groups involved in the drug-excipient interactions and to characterize these close contacts. In fact, the spectral changes observed for the ketoprofen:excipient aged PM's, as compared to isolated ketoprofen, involved two particular regions of the molecule – the methyl-carboxylic moiety (C^9H_3–C^8H–$C^{10}OOH$) and the inter-ring $C^7=O^{13}$ group (Figure 6). This is probably indicative of an interaction with the PVP and LAC excipients through hydrogen-type bonds with ketoprofen's terminal carboxylate, coupled to a decrease in the crystalline state of the drug. A slow dissolution-like effect of the drug into the excipient is thus proposed, leading to an amorphisation of ketoprofen and giving rise to new drug-excipient H-bonds.

The assessment of these drug-excipient interactions will hopefully lead to an understanding of their effect on the drug release process. This will provide a basis for the understanding of the microscopic mechanisms and interactions occurring within these mixtures, such that tomorrow's drug formulations might couple a higher efficacy and lower toxicity to a better patient compliance.

Amphetamine-like Drugs of Abuse

The abuse of psychoactive drugs such as amphetamines is known to produce serious health problems in users, which can ultimately result in death. While there has been much research on the effect of these drugs in humans, little has been investigated on the toxicity of the side products and synthetic reaction by-products. The illegal manufacture of amphetamine-like drugs of abuse relies upon the preparation of a diversity of chemical precursors, namely β-nitrostyrene precursors (96). Thus, ingestion of nitrostyrene-contaminated drugs of abuse (e.g. "ecstasy") is likely to have a considerable adverse effect on the user. Moreover, since different synthetic precursors and intermediates are usually found in illicit drugs of abuse (97), the determination of their presence in these products and their thorough characterization is of considerable forensic interest as a means of tracking the clandestine laboratories engaged in the synthesis of such drugs. Investigation of analytical tools aiming at such an identification is, however, very scarce. In our group efforts have been made in order to develop a rapid, efficient and sensitive method for the identification of some synthetic precursors of the most common amphetamine-like drugs, based on Raman spectroscopy. Indeed, this technique has lately been proposed as a reliable forensic method, quite useful for the determination of the composition profile of solid samples (e.g. seized "ecstasy" tablets) (98-102).

We report a vibrational spectroscopic study of several β-methyl-β-nitrostyrene derivatives that are important intermediates in the synthesis of illicit amphetamine-like drugs, such as 3,4-methylenedioxymethamphetamine ("ecstasy" or MDMA), 3,4-methylene-dioxyamphetamine (MDA), 4-methyl-thioamphetamine (MTA) and 4-methoxyamphetamine (PMA). A complete vibrational analysis of these systems was carried out by Raman spectroscopy and DFT calculations (103). The corresponding Raman spectra evidenced distinctive features for each of the β-nitrostyrene precursors investigated (Figure 7), allowing their ready identification in the final product (104).

These results show that even chemically similar intermediates are easily distinguished by Raman spectroscopy, which can also surpass other analytical methods currently used in criminal prosecutions once it concomitantly identifies the active compound and its by-products. The method has the additional advantage of allowing its extension to the main metabolites of amphetamine-like drugs. The final aim of these kinds of studies is to gather enough spectral data as to build a wide database, that will enable a rapid and unequivocal identification of synthetic precursors of illegally produced drugs of abuse. This will constitute an invaluable tool for both forensic control and toxicological studies.

Cocaine-opioid Interactions

Drug abuse is a serious health problem in our society and one of the greatest concerns of governments worldwide. Of particular interest is the increase in the

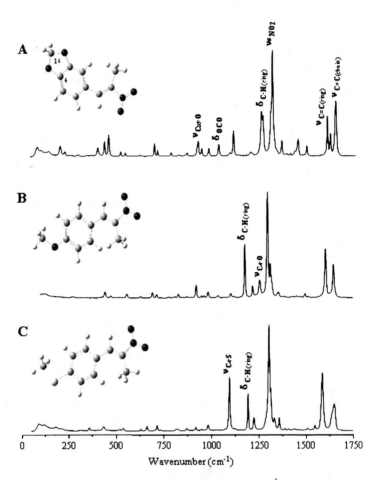

Figure 7. Experimental Raman spectra (75 – 1750 cm⁻¹, solid state, at 298 K)
for some precursors of amphetamine-like drugs: 3,4-methylenedioxy-β-methyl-
β-nitrostyrene (A); 4-methoxy-β-methyl-β-nitro-styrene (B); 4-methylthio-β-
methyl-β-nitrostyrene (C). (The main vibrational modes are assigned: δ_{OCO} and
δ_{CH} – CH (ring) and OH (ring) in-plane deformations; $v_{S\,NO_2}$ – NO_2 symmetric
stretching; $v_{C-O/C-S}$ and $v_{C=C}$ – C-O/C-S and C=C stretching). The most stable
calculated (B3LYP/6-31G**) geometries for each compound are also
represented. (Reprinted with permission of RSC from reference104.
Copyright 2004.)

number of drug addicts who report combined abuse of cocaine and opioid agonist heroin ("Speedball") (*105*). This drug combination is known to cause a more pleasurable drug experience than cocaine or heroin alone (*106,107*). This enhanced effect may contribute to the reduced motivation of "speedball" users to stop their drug habit, as well as their greater probability to relapse when compared to single drug users. Although the underlying biological basis for abuse of cocaine and opioid combinations is unclear, controlled clinical studies give insight into the desire for dual abuse of these substances. Moreover, electrochemical studies on the biological mechanisms of drugs of abuse evidenced a curious behavior for cocaine-opioid combinations (*108*). These findings, and the lack of chemical data on these kinds of drug mixtures, suggest the need for a more thorough investigation.

Consequently, we undertook a study of morphine (both in the basic (M) and N-protonated forms (MH)), heroin (H) and cocaine (C) (protonated species), with particular emphasis on the cocaine-opioid interactions. Raman spectra of both the free samples and the C:H and C:M (1:1, *w:w*) mixtures (with morphine either in the unprotonated or N-protonated forms) were obtained and analyzed in the light of the results obtained from quantum mechanical calculations, in view of understanding those interactions at a molecular level.

The Raman spectra of the cocaine:morphine (C:MH) and cocaine:heroin (C:H) samples reflected a clear interaction between the former, but not between cocaine and heroin. In fact, for the C:H mixture no variations in the Raman pattern were observed (either in the low or high frequency regions), when compared to the isolated molecules, even one week after preparation of the mixture. As to the C:MH sample, clear changes were detected (immediately after mixing) relative to the Raman features of the isolated components. The most affected bands were found to be the ones ascribed to the aromatic (oxygen-containing) moiety of the morphine molecule, namely the $v(C=C)_{ring}$ and $v(CH)_{ring}$ modes. This is easily understandable if the C:MH interplay is proposed to take place through the inner cavity of the morphine molecule, possibly *via* a (C=O)OH ⋯ O interaction, which is theoretically predicted to be favoured by the N-protonation process. In fact, no significant changes in the Raman spectrum of the basic form of morphine were detected in the presence of cocaine.

The conformational preferences of the drugs of abuse presently investigated (*e.g.* the sites where intermolecular interactions may occur), yielded by DFT calculations, were very helpful for interpreting the spectroscopic data. Actually, N-protonation of morphine was verified to lead to a slightly more open molecular conformation, which can favour the interaction of this opioid with cocaine, thus explaining the fact that no close contacts were detected for the C:M mixture. Furthermore, the presence of the two terminal $-O(C=O)CH_3$ groups in heroin were predicted to hamper the approximation of the cocaine molecule, while the approach to the morphine cavity defined by the two aromatic rings seems to be more favourable, since it does not involve any significant steric hindrance. Indeed, the calculated distances between the two components of the C:H and C:MH systems are considerably shorter for the latter. In the light of these results, a model for the C:MH interplay was proposed.

Phenolic Compounds

Interest in phenolic compounds has recently increased, owing to their role as antioxidants and their implication in the prevention of pathologies such as cancer (*109,110*), or cardiovascular and inflammatory diseases (*111-115*). Actually, this kind of compounds has been reported to display relevant biological activities such as antibacterial, antiviral, immune-stimulating and estrogenic properties (*116*), as well as antiproliferative and cytotoxic effects (*117-121*). Indeed, many phenols have lately been investigated for their potential use as cancer chemopreventive agents (*119,122*). Besides the natural phenolics (*e.g.* present in fruits and vegetables), some hydroxybenzoic derivatives are currently used as antioxidant additives in both food and pharmaceutical industry (*e.g.* E-310 (propyl gallate) and E-311 (octyl gallate)) (*123,124*). Although the specific role of dietary phenolic antioxidants in carcinogenesis has not been unequivocally elucidated (*122*), the main mechanism proposed for this protective action against deleterious oxidation is suggested to be associated to their free radical scavenging activity (*111,116,125-129*). This recognized cytotoxic activity is largely dependent on the structural characteristics of the compounds, which are intrinsically related to their antioxidant potency (*121,130,131*) and determine their lipophilicity and rate of incorporation into cells. Thus, the knowledge of the structural preferences and the hydrogen-bonding motif of these types of derivatives is essential for the understanding of the structure-property-activity relationships underlying their biological role.

Antioxidant phenolic systems with potential antibacterial and antineoplastic activity have been the object of research in our laboratory in the last few years (*103,104,132-138*). Several structurally related phenols, not available commercially, have been synthesized – from caffeic to gallic acid derivatives – and a thorough conformational analysis has been carried out by Raman spectroscopy and DFT calculations. Particular structural parameters were modified in order to determine their effect on the conformational behavior of the systems, namely: (i) the number of OH ring substituents; (ii) the length and degree of saturation of the carbon chain between the phenyl and the terminal functional group as well as the nature of this functional group (*e.g.* acid, ester, amide nitro). The structural preferences of these compounds were found to be mainly determined by electrostatic factors, as well as by the formation of (O)H ··· O and/or (C)H ··· O intra- and/or intermolecular interactions. A clear preference was found for a planar geometry, *i.e.* for the presence of a completely conjugated system, strongly stabilized through π-electron delocalization.

The Raman pattern of the phenolic derivatives investigated evidenced distinct features according to their particular geometrical differences (Figure 8), which allowed an unequivocal identification. A complete assignment of the spectra was performed in the light of the corresponding quantum mechanical calculations, thus yielding an accurate structural characterization.

This information greatly helped to interpret the biochemical results obtained for these systems, which were simultaneously screened for their antioxidant,

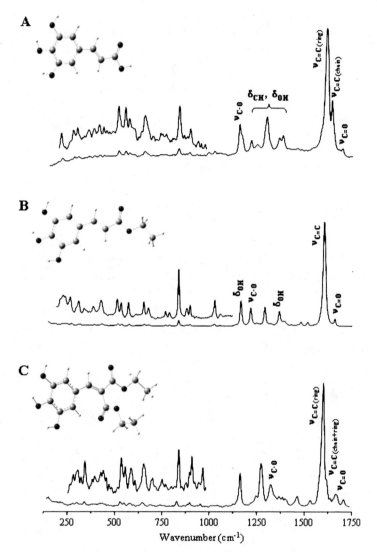

Figure 8. Experimental Raman spectra (50 – 1750 cm^{-1}, solid state, at 298 K) for the polyphenolic compounds: 3-(3,4,5-trihydroxyphenyl)-2-propenoic acid (A); ethyl 3-(3,4,5-trihydroxyphenyl)-2-propenoate (B); diethyl 2-(3,4,5-trihydroxyphenylmethylene)malonate (C). (The main vibrational modes are assigned: δ$_{CH}$ and δ$_{OH}$ – CH (ring) and OH (ring) in-plane deformations; ν$_{C-O}$ – C-O (ring) stretching; ν$_{C=C}$ and ν$_{C=O}$ – C=C (both chain and ring) and C=O stretching. The most stable calculated (B3LYP/6-31G**) geometries for each compound are also represented. (Reprinted with permission from reference 136. Copyright 2006.)

antibacterial, anti-inflammatory and antineoplastic activities. This work was designed as an interactive study, the synthesis of the phenolic derivatives (structurally characterized by vibrational spectroscopy and theoretical methods) being modified according to the data gathered on their biological activity.

Acknowledgements

The authors acknowledge financial support from FCT (Portugal) – Projects POCTI/33199/QUI/2000, POCTI/47256/QUI/2002, POCI/55631/QUI/2004 and POCI/SAU-FCF/58330/2004 (co-financed by the European Community fund FEDER), and from ESF – COST Action 922.

References

1. Mitchell, P.C.H.; Parker, S.F.; Ramirez-Cuesta, A.J.; Tomkinson, J. *Series on Neutron Techniques and Applications – Vol.3: Vibrational Spectroscopy with Neutrons*; World Scientific, 2005.
2. Frisch, M.J. *et al.* Gaussian 03, Revision B.04, Gaussian, Inc., Pittsburgh PA, 2003.
3. Ramirez-Cuesta, A.J. *Comp. Phys. Commun.* **2004**, *157*, 226.
4. www.isis.rl.ac.uk
5. Kazuo, N.; Paul, J.M.; Junnosuke, F.; Robert, A.C.; George, T.B. *Inorg. Chem.* **1965**, *4*, 36.
6. Perry, C.H.; Athans, D.P.; Young, E.F.; Durig, J.R.; Mitchell, B.R. *Spectrochim. Acta* **1967**, *23A*, 1137.
7. Degen, I.A.; Rowlands, A.J. *Spectrochim. Acta* **1991**, *47A*, 1263.
8. Baranska, H.; Kuduk-Jaworska, J.; Cacciari, S. *J. Raman Spec.* **1997**, *28*, 1.
9. Pavankumar, P.N.V.; Seetharamulu, P.; Yao, S.; Saxe, J.D.; Reddy, D.G.; Hausheer, F.H. *J. Comput. Chem.* **1999**, *20*, 365.
10. Zhang, L.; Wei, H.; Zhang, Y.; Guo, Z.; Zhu, L. *Spectrochim. Acta* **2002**, *58A*, 217.
11. *Polyamines in Cancer: Basic Mechanisms and Clinical Approaches*; Nishioka, K., Ed.; Springer Verlag: New York, 1996.
12. Pegg, A.E.; Feith, D.J.; Fong, L.Y.Y.; Coleman, C.S.; O'Brien, T.G.; Shantz, L.M. *Biochem. Soc. Trans.* **2003**, *31*, 356.
13. Bachrach, U. *Curr. Protein Pept. Sci.* **2005**, *6*, 559.
14. Seiler, N.; Raul, F. *J. Cell. Mol. Med.* **2005**, *9*, 623.
15. Igarashi, K.; Kashiwagi, K. *Clin. Exp. Metastasis.* **2005**, *22*, 255.
16. Gupta, R.; Krause-Ihle, T.; Bergmann, B.; Muller, I.B.; Khomutov, A.R.; Muller, S.; Walter, R.D.; Luersen, K. *Antimicrob. Agents Chemother.* **2005**, *49*, 2857.
17. Nasizadeh, S.; Myhre, L.; Thiman, L.; Alm, K.; Oredsson, S.; Persson, L. *Exp. Cell. Res.* **2005**, *15*, 254.
18. Manni, A.; Washington, S.; Hu, X.; Griffith, J.W.; Bruggeman, R.; Demers, L.M.; Mauger, D.; Verderame, M.F. *J. Biochem.* **2006**, *139*, 11 and refs therein.

19. Fogel-Petrovic, M.; Vujcic, S.; Miller, J.; Porter,C.W.; *FEBS Lett.* **1996**, *391*, 89.

20. Fogel-Petrovic, M.; Kramer, D.L.; Vujcic, S.; Miller, J.; Mcmanis, J.S.; Bergeron, R.J.; Porter,C.W. *Mol. Pharmacol.* **1997**, *52*, 69.

21. Rauter, H.; Di Domenico, R.; Menta, E.; Oliva, A.; Qu, Y.; Farrell, N. *Inorg. Chem.* **1997**, *36*, 3919 and refs. therein.

22. Ichimura, S.; Hamana, K.; Nenoi, M. Biochem. *Biophys. Res. Commun.* **1998**, *243*, 518.

23. Rosenberg, B.; Van Camp, L.; Trosko, J.E.; Mansour, V.H. *Nature* **1969**, *222*, 385.

24. Xiao, L.; Casero, R.A. *Biochem. J.* **1996**, *313*, 691.

25. Amo-Ochoa, P.; González, V.M.; Pérez, J.M.; Masaguer, J.R.; Alonso, C.; Navarro-Ranninger, C. *J. Inorg. Biochem.* **1996**, *64*, 287.

26. Wong, E.; Giandomenico, C.M. *Chem. Rev.* **1999**, *99*, 2451 and refs. therein.

27. Reedijk, J. *Chem. Rev.* **1999**, *99*, 2499 and refs. therein.

28. Van Boom, S.S.G.E.; Chen, B.W.; Teuben, J.M.; Reedijk, J. *Inorg. Chem.* **1999**, *38*, 1450.

29. Qu, Y.; Rauter, H.; Soares-Fontes, A.P.; Bandarage, R.; Kelland, L.R.; Farrell, N. *J. Med. Chem.* **2000**, *43*, 3189.

30. McGregor, T.D.; Hegmans, A.; Kaspárková, J.; Neplechová, K.; Nováková, O.; Penazová, H.; Vrána, O.; Brabec, V.; Farrell, N. *J. Biol. Inorg.Chem.* **2002**, *7*, 397.

31. Skov, K.A.; Adomat., H.; Farrell, N.; Matthews, J.B. *Anticancer Drug Design* **1998**, *13*, 207.

32. Bierbach, U.; Qu, Y.; Hambley, T.W.; Peroutka, J.; Nguyen, H.L.; Doedee, M.; Farrell, N. *Inorg. Chem.* **1999**, *38*, 3535 and refs. therein.

33. Brabec, V.; Kasparkova, J.; Vrana, O.; Novakova, O.; Cox, J.W.; Qu Y.; Farrell N. *Biochem.* **1999**, *38*, 6781 and refs. therein.

34. Jansen, B.A.J.; van der Zwan, I.; Reedijk, J.; den Dulk, H.; Brouwer, J. *Eur. J. Inorg. Chem.* **1999**, 1429.

35. Pratesi, G.; Perego, P.; Polizzi, D.; Righetti, S.C.; Supino, R.; Caserini, C.; Manzotti, C.; Giuliani, F.C.; Pezzoni, G.; Tognella, S.; Spinelli, S.; Farrell, N.; Zunino, F. *British J. Cancer* **1999**, *80*, 1912.

36. Perego, P.; Caserini, C.; Gatti, L.; Carenini. N.; Romanelli. S.; Supino, R.; Colangelo, D.; Viano I.; Leone, R.; Spinelli, S.; Pezzoni, G.; Manzotti, C.; Farrell, N.; Zunino, F. *Molec. Pharmacol.* **1999**, *55*, 528.

37. T. Servidei; Ferlini C.; Riccardi A.; Meco D.; Scambia G.; Segni G.; Manzotti C.; Riccardi R. *Eur. J. Cancer* **2001** *37*, 930 and refs therein.

38. Cox, J.W.; Berners-Price, S.J.; Davies, M.S.; Qu, Y.; Farrell, N. *J. Am. Chem. Soc.* **2001** *123*, 1316.

39. Qu, Y.; Scarsdale, N.J.; Tran, M.C.; Farrell, N. *J. Biol. Inorg. Chem.* **2003** *8*, 19 and refs therein.

40. Marques, M.P.M.; Batista de Carvalho, L.A.E.; Tomkinson, J. *The Rutherford Appleton Laboratory, ISIS Facility Annual Report*, 1999-2000.

41. Batista de Carvalho, L.A.E.; Lourenço, L.E.; Marques, M.P.M. *J.Molec.Struct.* **1999**, *482*, 639 and refs therein

360

42. Marques, M.P.M.; Batista de Carvalho, L.A.E. *COST 917: Biogenically Active Amines in Food*, Morgan, D.M.L.; White, A.; Sánchez-Jiménez, F.; Bardocz, S. Ed., European Commission, Luxembourg, 2000; Vol. IV, p. 122.

43. Marques, M.P.M.; Batista de Carvalho, L.A.E.; Tomkinson, J. *The Rutherford Appleton Laboratory, ISIS Facility Annual Report*, 2001-2002.

44. Marques, M.P.M.; Batista de Carvalho, L.A.E.; Tomkinson, J. *J. Phys. Chem. A* **2002**, *106*, 2473.

45. Amorim da Costa, A.M.; Marques, M.P.M.; Batista de Carvalho, L.A.E. *Vib. Spec.* **2002**, *29*, 61.

46. Amorim da Costa, A.M.; Marques, M.P.M.; Batista de Carvalho, L.A.E. *J. Raman Spec.* **2003**, *34*, 357.

47. Amorim da Costa, A.M.; Batista de Carvalho, L.A.E.; Marques, M.P.M. *Vib. Spec.* **2004**, *35*, 165.

48. Amado, A.M.; Otero, J.C.; Marques, M.P.M.; Batista de Carvalho, L.A.E. *Chem. Phys. Chem.* **2004**, *5*, 1837.

49. Braden, D.A.; Parker, S.F.; Tomkinson, J.; Hudson, B.S. *J. Chem. Phys.* **1999**, *111*, 429.

50. Mirkin, N.G.; Krimm, S. *J. Phys.Chem.* **1993**, *97*, 13887.

51. Grillo, M.A.; Colombatto, S. *Amino Acids* **2004**, *26*, 3.

52. Li, G.; Regunathan, S.; Barrow, C.J.; Eshraghi, J.; Cooper, R.; Reis, D.J. *Science* **1994**, *263*, 966.

53. Piletz, J.E.; Chikkala, D.N.; Ernsberger, P.J. *Pharmacol. Exp. Ther.* **1995**, *272*, 581.

54. Raasch, W.; Schafer, U.; Chun, J.; Dominiak, P. *British J. Pharm.* **2001**, *133*, 755.

55. Sener, A.; Lebrun, F.; Blaicher, F.; Malaisse, W.J. *Biochem. Pharmacol.* **1989**, *38*, 327.

56. Satriano, J.; Matsufuji, S.; Murakami, Y.; Lortie, M.J.; Schwartz, D.; Kelly, C.J.; Hayashi, S.I.; Blantz, R.C. *J. Biol. Chem.* **1998**, *273*, 15313.

57. Satriano, J.; Kelly, C.J.; Blantz, R.C. *Kidney Int.* **1999**, *56*, 1252.

58. Gardini, G.; Cabella, C.; Cravanzola, C.; Vargiu, C.; Belliardo, S.; Testore, G.; Solinas, S.P.; Toninello, A.; Grillo, M.A.; Colombatto, S. *J. Hepatol.* **2001**, *35*, 482.

59. Dudkowska, M.; Lai, J.; Gardini, G.; Stachurska, A.; Grzelakowska-Sztabert, B.; Colombatto, S.; Manteuffel-Cymborowska, M. *Biophys. Biochim. Acta* **2003**, *1619*, 159.

60. Gardini, G.; Cravanzola, C.; Autelli, R.; Testore, G.; Cesa, R.; Morando, L.; Solinas, S.P.; Muzio, G.; Grillo, M.A.; Colombatto, S. *J. Hepatol.* **2003**, *39*, 793.

61. Higashi, K.; Yoshida, K.; Nishimura, K.; Momiyama, E.; Kashiwagi, K.; Matsufuji, S.; Shirahata, A.; Igarashi, K. *J. Biochem.* **2004**, *136*, 533.

62. Grundemann, D.; Hahne, C.; Berkels, R.; Schomig, E. *J. Pharmac. Exp. Ther.* **2003**, *304*, 810 and refs. therein.

63. Toninello, A.; Battaglia, V.; Salvi, M.; Calheiros, R.; Marques, M.P.M. *Struct. Chem. in press.*

361

64. Salvi, M.; Battaglia, V.; Mancon, M.; Colombatto, S.; Cravanzola, C.; Calheiros, R.; Marques, M.P.M.; Grillo, M.A.; Toninello, A. *Biochem. J.* **2006**, *396*, *in press*.

65. Marques, M.P.M.; Batista de Carvalho, L.A.E.; Tomkinson, J. *The Rutherford Appleton Laboratory, ISIS Facility Annual Report*, 2003.

66. Marques, M.P.M.; Girão da Cruz, M.T.; Pedroso de Lima, M.C.; Gameiro, A.; Pereira, E.; Garcia, P. *Biochim. Biophys. Acta - Molecular Cell Research* **2002**, *61* 1589.

67. Teixeira, L.J.; Seabra, M.; Reis, E.; Girão da Cruz, M.T.; Pedroso de Lima, M.C.; Pereira, E.; Miranda, M.A.; Marques, M.P.M. *J. Med. Chem.*, **2004**, *47*, 2917.

68. Fiuza, S.M.; Amado, A.M.; Oliveira, P.J.; Sardão, V.A.; Batista de Carvalho, L.A.E.; Marques, M.P.M. *Letters in Drug Design and Development* **2006**, *3*, *in press*.

69. Ruffolo, R.R.Jr.; Gellai, M.; Heible, J.P.; Willette, R.N.; Nichols, A.J. *Eur. J. Clin. Pharmacol.* **1990**, *38*, 82.

70. Yue, T.L.; Cheng, H.Y.; Lysko, P.G.; McKenna, P.J.; Feuerstein, R.; Gu, J.L.; Lysko, K.A.; Davis, L.L.; Feuerstein, G. *J. Pharmacol. Exp. Ther.* **1992**, *263*, 92.

71. Ruffolo, R.R.Jr.; Boyle, D.A.; Brooks, D.P.; Feuerstein, G.Z.; Venuti, R.P.; Lukas, M.A.; Poste, G. *Cardiovascular Drug Reviews* **1992**, *10*, 127.

72. Pfeffer, M.A.; Stevenson, L.W. *N. Engl. J. Med.* **1996**, *334*, 1396.

73. Packer, M.; Bristow, M.R.; Cohn, J.N.; Colucci, W.S.; Fowler M.B.; Gilbert, E.M. *N. Engl. J. Med.* **1996**, *334*, 1349.

74. Packer, M.; Colucci, W.S.; Sackner-Bernstein, J.D.; Liang, C.; Goldscher, D.A.; Freeman, I.; Kukin, M.L.; Kinhal, V.; Udelson, J.E.; Klapholz, M.; Gottlieb, S.S.; Pearle, D.; Cody, R.J.; Gregory, J.J.; Kantrowitz, N.E.; LeJemtel, T.H.; Young, S.T.; Lukas, M.A.; Shusterman, N.H.; *Circulation* **1996**, *94*, 2793.

75. Dunn, C.J.; Lea, A.P.; Wagstaff, A.J. *Drugs* **1997**, *54*, 161.

76. Cleland, J.G.; Swedberg, K. *Lancet* **1996**, *347*, 1199.

77. Moreno, A.J.M.; Santos, D.J.S.L.; Palmeira, C.M. *Rev. Port. Cardiol.* **1998**, *17*, II-63.

78. Oliveira, P.J.; Santos, D.J.; Moreno, A.J.M. *Archiv. Biochem. Biophys.* **2000**, *374*, 279.

79. Abreu, R.M.V.; Santos, D.J.; Moreno, A.J.M. *J. Pharmacol. Exp. Ther.* **2000**, *295*, 1022.

80. Oliveira, P.J.; Coxito, P.M.; Rolo, A.P.; Santos, D.J.; Palmeira, C.M.; Moreno, A.J.M. *Eur. J. Pharmacol.* **2001**, *412*, 231.

81. Santos, D.J.; Moreno, A.J.M. *Biochem. Pharmacol.* **2001**, *61*, 155.

82. Ferrari, R. *J. Cardiovasc. Pharmacol.* **1996**, *28*, S1.

83. Oliveira, P.J.; Marques, M.P.M.; Batista de Carvalho, L.A.E.; Moreno, A.J.M. *Biochem. Biophys. Res. Commun.* **2000**, *276*, 82.

84. Marques, M.P.M.; Oliveira, P.J.; Moreno, A.J.M.; Batista de Carvalho, L.A.E. *J. Raman Spectroscopy*, **2002**, *33*, 778.

85. Cheng, H.Y.; Randall, C.S.; Holl, W.W.; Constantinides, P.P.; Yue, T.L.; Feuerstein, G.Z. *Biochim. Biophys. Acta* **1996**, *1284*, 20.

362

86. Takeuchi, H.; Harada, I. *Spectrochim. Acta* **1986**, *42A*, 1069.
87. Miura, T.; Takeuchi, H.; Harada, I. *Biochem.* **1988**, *27*, 88.
88. Julou, J; Guyonnet, J.C.; Ducrot, R.; Fournel, J.; Pasquet, J. *Scand. J. Rheumatol. Suppl.* **1976**, *14*, 33.
89. Avouac, B.; Teule, M. *J. Clin. Pharmacol.* **1988**, *28*, S2.
90. Gupta, M.K.; Tseng, Y.C.; Goldman, D.; Bogner, R.H. *Pharm. Res.* **2002**, *19*, 1663.
91. Granero, G.; Longhi, M. *Pharm. Dev. Technol.* **2002**, *7*, 381.
92. Akers, M.J. *J. Pharm. Sci.* **2002**, *91*, 2283.
93. Batista de Carvalho, L.A.E.; Marques, M.P.M.; Tomkinson, J. *Biopolymers in press*.
94. Vueba, M.L.; Batista de Carvalho, L.A.E.; Veiga, F.; Sousa, J.J.; Pina, M.E. *Eur. J. Pharm. Biopharm.* **2004**, *58*, 73.
95. Vueba, M.L.; Pina, M.E.; Veiga, F.; Sousa, J.J.; Batista de Carvalho, L.A.E. *Int. J. Pharm.* **2006**, *307*, 56.
96. Cason, T.A.D. *J. Forensic Sci.* **1990**, *35*, 675.
97. Poortman, A.J.; Lock, E. *J. Forensic Sci. Int.* **1999**, *100*, 221.
98. Bell, S.E.J.; Burns, D.T.; Dennis, A.C.; Speers, J.S. *Analyst* **2000**, *125*, 541.
99. S.E.J.; Burns, D.T.; Dennis, A.C.; Matchett, L.J.; Speers, J.S. *Analyst* **2000**, *125*, 1811.
100. Sägmüller, B.; Schwarze, B.; Brehm, G.; Schneider, S. *Analyst* **2001**, *126*, 2066.
101. Faulds, K.; Smith, W.E.; Graham, D.; Lacey, R.J. *Analyst* **2002**, *127*, 282.
102. Vankeirsbilck, T.; Vercauteren, A.; Baeyens, W.; van der Weken, G.; Verpoort, F.; Vergote, G.; Remon, J.P. *Trends Anal. Chem.* **2002**, *21*, 869.
103. Calheiros, R.; Milhazes, N.; Borges, F.; Marques, M.P.M. *J. Molec. Struct.* **2004**, *692*, 91.
104. Milhazes, N.; Borges, F.; Calheiros, R.; Marques, M.P.M. *Analyst* **2004**, *129*, 1106.
105. Frank, B.;Galea, J. *J. Addict. Dis.* **1996**, *15*, 1.
106. Ellinwood, E.H.Jr.; Eibergen, R.D.; Kilbey, M.M. *Ann. NY Acad. Sci.* **1976**, *281*, 393.
107. Ranaldi, R.; Munn, E. *Neuroreport*, **1998**, *9*, 2463.
108. Garrido, J.M.P.J.; Marques, M.P.M.; Silva, A.M.S.; Macedo, T.R.A.; Oliveira-Brett, A.M.; Borges, F. *J. Am. Chem. Soc.* submitted.
109. Gao, T.; Ci, Y.; Jian, H.; An, C. *Vib.Spec.* **2000**, *24*, 225 and refs. therein.
110. Silva, A.M.S.; Santos, C.M.M.; Cavaleiro, J.A.S.; Tavares, H.R.; Borges, F.; Silva, F.A.M. *Magnetic Resonance in Food Science – A View to the Next Century*; Royal Society of Chemistry: London, United Kingdom, 2001.
111. González, M.J.; López, D.; Argülies, M.; Riordan, N.H. *Age* **1996**, *19*, 17.
112. Kordel, W.; Dassenakis, M.; Lintelmann, J.; Padberg, S. *Pure Appl. Chem.* **1997**, *69*, 1571.
113. Lubal, P.; Siroky, D.; Fetsch, D.; Havel, J. *Talanta* **1998**, *47*, 401.
114 Bravo, L. *Nutrition Rev.* **1998**, *56*, 317.
115. Inoue, M.; Sakaguchi, N.; Isuzugawa, K.; Tani, H.; Ogihara, Y. *Biol. Pharm. Bull.* **2000**, *23*, 1153.

116. Rice-Evans, C.A.; Miller, N.J.; Paganga, G. *Free Rad. Biol Med.* **1996**, *20*, 933.

117. Rapta, P.; Misík, V.; Stasko, A.; Vrábel, I. *Free Rad. Biol Med.* **1995**, *18*, 901.

118 Lepley, D.M.; Li, B.; Birt, D.F.; Pelling, J.C. *Carcinogenesis* **1996**, *17*, 2367.

119. Agullo, G.; Gamet-Payrastre, L.; Manenti, S.; Viala, C.; Rémésy, C.; Chap, H.; Payrastre, B. *Biochem. Pharmacol.* **1997**, *53*, 649.

120. Skaper, S.D.; Fabris, M.; Ferrari, V.; Carbonare, M.D.; Leon, A. *Free Rad. Biol Med.* **1997**, *22*, 669.

121. Sergediene, E.; Jonsson, K.; Szymusiak, H.; Tyrakowska, B.; Rietjens, I.C.M.; Cenas, N. *FEBS Lett.* **1999**, *462*, 392.

122. Aruoma, O.I. *Chem.Toxic.* **1994**, *32*, 671.

123. Masaki, H.; Okamoto, N.; Sakaki, S.; Sakurai, H. *Biol. Pharm. Bull.* **1997**, *20*, 304.

124. Klein, E.; Weber, N. *J. Agric. Food Chem.* **2001**, *49*, 1224.

125. Nakamura, Y.; Ohto, Y.; Murakami, A.; Ohigashi, H. *J. Agric. Food Chem.* **1998**, *46*, 4545.

126. Khan, N.S.; Ahmad, A.; Hadi, S.M. *Chemico-Biol.Interac.* **2000**, *125*, 177.

127. Nakamura, Y.; Torikai, K.; Ohto, Y.; Murakami, A.; Tanaka, T.; Ohigashi, H. *Carcinogenesis* **2000**, *21*, 1899.

128. Rajan, P.; Vedernikova, I.; Cos, P.; B., Dirk, V.; Augustyns, K.; Haemers, A. *Bioorg. Med. Chem. Lett.* **2001**, *11*, 215.

129. Phan, T.T.; Wang, L.; Grayer, R.J.; Chan, S.Y.; Lee, S.T. *Biol. Pharm. Bull.* **2001**, *24*, 1373.

130. Passi, S.; Picardo, M.; Nazarro-Porro, M. *Biochem. J.* **1987**, *245*, 537.

131. Castellucio, C.; Paganga, G.; Melikian, N.; Bolwell, G.P.; Pridham, J.; Sampson, J.; Rice-Evans, C.A. *FEBS Lett.* **1995**, *368*, 188.

132. Gomes, C.A.; Girão da Cruz, M.T.; Andrade, J.L.; Milhazes, N.; Borges, F.; Marques, M.P.M. *J. Med. Chem.,* **2003**, *46*, 5395.

133. Fiuza, S.M.; Van Besien, E.; Milhazes, N.; Borges, F.; Marques, M.P.M. *J. Molec. Struct.,* **2004**, *693*, 103.

134. Fiuza, S.M.; Gomes, C.; Teixeira, L.J.; Girão da Cruz, M.T.; Cordeiro, M.N.D.S.; Milhazes, N.; Borges, F.; Marques, M.P.M. *Bioorg. Med. Chem.,* **2004**, *12*, 3581.

135. Teixeira, S.; Siquet, C.; Alves, C.; Boal, I.; Marques, M.P.M.; Borges, F.; Lima, J.L.F.C.; Reis, S. *Free Rad. Biol. Med.* **2005**, *39*, 1099.

136. Sousa, J.B.; Calheiros, R.; Rio, V.; Borges, F.; Marques, M.P.M. *J. Mol. Struct.* **2006**, *783*, 122.

137. Fresco, P.; Borges, F.; Diniz, C.; Marques, M.P.M. *Med. Res. Rev. in press.*

138. Milhazes, N.; Calheiros, R.; Marques, M.P.M.; Garrido, J.; Cordeiro, M.N.D.S.; Rodrigues, C.; Novais, C.; Peixe, L.; Borges, F. *Bioorg. Med. Chem. in press.*

Chapter 22

Correlative Tools for Measuring External Influences on Protein Folding–Denaturation Using Various Spectroscopic Methods

H. Bohr and S. Abdali

Quantum Protein Centre (QuP), Department of Physics, The Technical University of Denmark, DK-2800 Lyngby, Denmark

A set of novel instruments designed and constructed for experiments to study the folding and denaturation of proteins in conjunction with spectroscopy is presented along with data from selected experiments.

Introduction

One of the key features of life processes is the role of proteins whose structures guarantee the success of maintaining life in all cellular organisms. One of the unsolved challenges of protein research is to gain a greater understanding of protein folding and the related phenomenon of protein denaturation. At the heart of the problem lies in the fact that such studies require non-invasive techniques for exposing the protein folding process in the native environment and with sufficient time resolution (of the order less than milli - seconds) and atomic resolution. Some of these advantages can be provided by optical instrumentation, especially Raman scattering combined with surface enhancement.

In this report we present various novel instrument set-ups for studying both protein folding and the denaturation processes. These instruments can also be applied to study other bio-molecules such as DNA. The best known protein structure studies are those realized using X-ray crystallography, and they provide great details about protein structure, though in crystalline form. Similar details can be obtained using neutron scattering and NMR experiments for proteins in solutions but due to the long exposure times involved, crystallization and higher concentrations of the native forms, they lack the means of giving time-resolved expositions of dynamical events involving proteins such as the crucial protein folding and denaturation processes.

Optical measurements (*1*) such as Raman Scattering, Fluorescence techniques, Vibrational Circular Dichroism, (VCD), Optical Rotational Dispersion (ORD), Raman Optical Activity (ROA) and infrared absorption spectroscopy can overcome many of the obstacles mentioned above due to the fact that optical techniques are non-invasive and can monitor proteins in their native environment and with accurate time resolution. One disadvantage is the low sensitivity. However, the use of Surface Enhanced Raman Scattering (SERS), techniques (*2-4*) means that proteins can be observed down to the single molecule level. Thus, optical techniques hold great promise for the future investigation of protein dynamics processes provided that proteins can be maintained in a suitable and controllable sample cell.

In order to be able to design and construct a suitable cell for the elucidation of a particular protein process such as protein folding, albeit is necessary to understand the conditions of that process and the crucial parameters for maintaining it. We shall discuss these conditions before presenting the various cells.

Protein Folding and Denaturation Experiments

Ever since it was known that proteins do undergo a folding process resulting in a functional 3-dimensional structure after being synthesized in the cellular ribosome as linear chains of amino acids, attempts have been made to reproduce and study such folding processes *in vitro*. These often occur much slower than *in vivo* which is of the order of a few seconds. Since it is costly to synthesize large proteins in laboratories the usual approach is to isolate a protein in its native folded state and denature it and then refold it again. In this way the protein folding process can be controlled. Such procedures are commonly done in laboratories once a desired protein is produced by recombinant cloning techniques, isolated and purified or simply purchased, if it is available on the market. The next task is then to find denaturant agents that can denature or un-fold the protein without destroying the structure. For most proteins a common denaturant is urea or guanidinium chloride in 4-8 molar solutions. The next question is whether the denatured protein can refold again after removing the denaturant? That will often depend on the denaturation process. If concentration and temperature are too high one will often end up with aggregated forms of the proteins with no functional activity left. Hence, low temperature denaturation is preferable for refolding, where the native, active form of the protein is to be retrieved. Thus we are led to the next question of cold and hot denaturations of proteins, which will be discussed in the next section.

Let us briefly mentioned a theory (5) of how one can consider the denatured processes of proteins as a change in their intrinsic dynamics (6). Most biological chain molecules, as for example proteins, are not closed but linear. The linear molecule can be folded and packed into complex and dense geometrical structures, and they can also sustain collective excitations over a range that exceeds the persistence length of the molecular backbone, and possibly extend to the entire molecule. For a single molecule it takes a certain time before a representative set of conformations of the molecule has been scanned. Had the molecule been in vacuum and there have been no dissipative forces, the time between conformations could be very short, but this is normally not true because proteins are usually found in a viscous aqueous environment. Large motions of the backbone occur on the order of milliseconds. On shorter time scales the apparent persistence length is therefore significantly longer than the thermo dynamical average. In fact, the apparent persistence length diverges when the time scale goes to zero.

Hot and Cold Denaturation

Many proteins exhibit several ways of denaturing depending on the temperature, pressure or pH (7), and for example as a function of temperature many proteins exhibit three phases: Cold denatured protein, native (folded) phase, and hot denatured protein. Though it is well known that proteins denature when heated, it is generally less known that many proteins also denature upon cooling. In statistical science it is evident that order is decreased upon an increase in temperature, and thus order is enhanced upon cooling for the crystalline form. However, unlike crystals, proteins can undergo cold denaturation, and thus have an increased disorder when being cooled. Likewise, for many scientists it would be a surprise to learn that some proteins will denature when the pressure is raised to about 1000 atmosphere. A schematic phase diagram is shawn in Figure 1 (8). It illustrates the folded and denatured states of the protein in the temperature pressure phase space.

It is an empirical observation that heat is consumed during transition from the cold denaturation phase to the folded phase, as well as, from the folded phase to the hot denaturation phase (7). This is consistent with the situation encountered in the experiment with microwave radiation, where the protein in solution has three phases with exothermal and endothermal transitions connecting them. However, successive transitions of increased temperature will

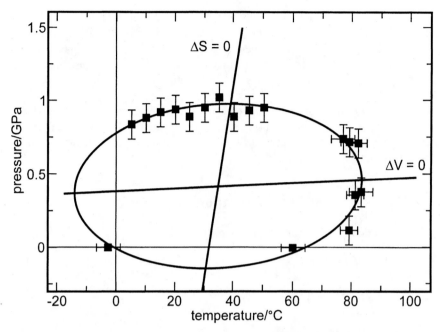

Figure 1. Shows a qualitative stability diagram of the protein, Zn-cytochrome c, in a buffer/glycerol mixture obtained from pressure experiments. The y-axis represents the pressure in MPa and the x-axis is temperature in Celsius. Inside the ellipse the protein is in its native state, outside in the denatured state. The $\Delta S=0$ (S being entropy) line cuts vertically at T=30, dividing the ellipse into two sections. The left part being $\Delta S<0$ and the right hand side $\Delta S>0$. The figure is depicted from ref. 8.

involve an increased intrinsic activity. It is important to realize that a phase analysis based on Figure 1 should not only be concerning folding and denaturation but also stability of protein conformation. In the cold denatured phase there is insufficient structural activity to maintain the folded structure, while in the hot denatured phase the structural activity will reverse the folding process leading to denaturation of the bended structure, and for some proteins it may make aggregation, among different molecules, favorable. The intrinsic structural activity of the proteins is, according to our theory (9), a result of topological excitations of linear chain molecules, so-called wringmodes, as shown in the Microwave section below.

Molecular resonators are thought to be molecules whose intrinsic modes can resonate, or transfer twist mode to the protein backbone. The latter phenomenon is an alternative way for an interaction to take place. These so-called molecular resonators can play an important role in the folding processes of certain proteins, and they have been seen in various refolding experiments, although the explanation, mentioned above for the role of twist modes is not fully proven. Usually such important "seeds" for the folding process are called pro-peptides; as they are synthesized in connection with the poly-peptide backbone.

A good example of such phenomenon is the refolding of the Alphalytic protenase, a Serine protease that only can attain its functional native state during refolding experiments, if samples of the corresponding pro-peptide are added to the solution in which the protein folding occurs (*10*).

Protein Denaturation Mechanisms

Protein denaturation can occur as a result of:

- Temperature
- Pressure
- Electromagnetic radiation
- pH changes
- Chemically induced

It is within these fields of denaturation phenomena that we have been making our experiments. There is, however, on the molecular level still not a clear understanding of protein denaturation. However, one might argue, thermodynamically speaking (as a complement to the topologically picture above), that the entropy plays an important role in particularly the chemical and pressure denaturation process. One expects that the water molecules are forming a tight caging (bound water) just around the protein, while further away the water molecules are structured in a kind of ordered network with low entropy. In the case of pressure denaturation the rate of water collision at constant volume increases as the temperature is raised. This is seen from the standard ideal gas equation: $PV = nRT$, which we use when treating protein and water solvent molecules as an ideal gas. With the breaking of H-bonds of the inner caging of bound water we obtain protein unfolding and exposure of hydrophobic side-chains. One might propose that at diminished pressure and low temperature one would have a similar effect of unfolding, but this time the lower collision rate of water causes the protein to fold out, but, apparently also maintaining order in the water structure.

Special Cells for Denaturation Experiments

In this section we shall present our design and construction of the sample cells for protein folding experiments. These sample cells can be used in combination with optical facilities such as lasers and detectors that can provide the desired spectra, e.g., Raman.

Thermo-bath

This sample cell is designed to be used in conjunction with a laser set-up for measuring the Optical Rotational Dispersion, ORD, by having polarizers and lenses on both sides of the cell, measuring and recording the rotation of the optical plane when the laser beam is passing through the cell. It has the dimensions of 5 centimeters. The cell is novel in the sense that it allows a careful adjustment and control of the temperature and thus being able to record a complete thermo-cycle of heating and cooling of an unstable protein in solution, thus going from cold denaturation through the folded native state to the hot denatured state and back again to the cold denatured state. Microwave radiation can also be applied to the protein sample in this cell, while maintaining a fixed temperature and monitoring the heat transfer.

Figure 4 (9) shows a heating-cooling cycle measured from the thermo-bath cell, explained above. The y-axis is showing the polarization while the x- axis is representing the temperature ranging from the cold denatured state (10 degrees Celsius) through the folded native state (25 degrees) to the hot denatured state (45 degrees) and back again to the cold denatured state. The Betalactoglobulin protein sample is brought in a solution of 0.5 M KCl and 4 M Urea, and the pH of the sample is 3. In such solution the protein exhibits hot, native and cold denaturation states depending on the temperature. The full line represents ORD data obtained from the effect of applying 2.24 GHz microwave irradiation for 15 seconds at 8 and 20 degrees Celsius, while the dashed line represents ORD data without irradiation. . The heating starts at 4° C and ends at 48° C and back again. Microwave radiation is applied at 8° C in the heating and at 20° C in the cooling cycle The curves clearly show that the microwave irradiation has a profound non-thermal effect on the structure of the protein, enhancing respectively the folding and denaturing processes in the heating cooling cycle. The little change in activity corresponds to an increase of a factor 100 in the reaction rate.

Pressure Cell

This cell is designed for pressures of several thousand atmospheres so that the effects of pressure on protein denaturation can be studied (11). The pressure can be brought to decrease rapidly from 1000 to 1 atmosphere pressure by breaking a window in the cell as illustrated in Figure 2. The effect of such high pressure on protein folding is illustrated in Figure 1. The high pressure is generated outside the cell, and led through a tube like cavity to the central chamber, which contains the sample. The chamber is surrounded by steel blocks

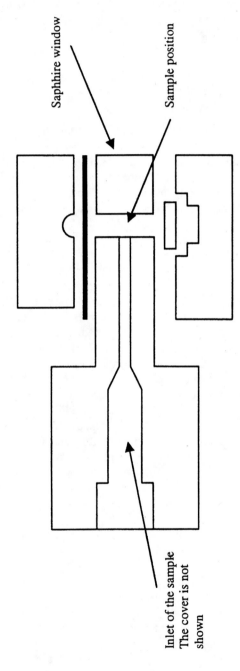

Saphhire window

Sample position

Inlet of the sample
The cover is not
shown

Figure 2. Pressure jump cell. The dimensions are 4 cm long and 1.5 cm in diameter.

that press the cell together with a sapphire window, through which a laser beam can be transmitted for detection. The dark plate on the upper part is broken at critical presuure above the denaturation point and then leading to refolding of the protein to be observed by Raman scattering.

Microwave Cell

Since we have previously (9) demonstrated that proteins can undergo structural conformational changes when microwave radiation is applied at frequencies around 2 GHz, a cell for protein refolding experiments was also constructed. This cell is designed to operate with electromagnetic radiation in the region of a few Gigahertz and with temperature changes.

The microwave cell, shown in Figure 3, is designed to be employed in variable applied electromagnetic fields, i.e., for different frequencies (1-4 GHz). This is controlled by varying the central cavity length (A+C and B+D) and the frequency range, which could be employed here, without firstly losing the efficiency of the microwave and secondly in such a way that the heat effect is controlled. This is achieved by having two quartz tubes, mounted coaxially, and where the protein is inserted in the inner one, while water is running constantly in the outer tube. A laser beam is focused on the centre of the inner cell and a Raman spectrometer and detector are placed 90° to the tube window. Raman spectra were recorded using a BioTools ChiralRaman instrument, which utilizes 532 nm laser source and a CCD detector. A laser power of 250 mW was used and the spectrum was recorded for 20 min.

Figures 5a and 5b show Raman spectra recorded during experiments with the microwave cell. The spectrum of the protein before applying microwave is shown in Figure 5a. The same conditions were applied after having the protein exposed to MW for different powers: 100, 440 and 800 W, all for 40 seconds, as can be seen in Figure 5b.

The different curves of Raman spectra represent microwave exposure of the Betalactoglobulin protein sample at a frequency of around 2 GHz. Q-mill water was used since a solution containing urea would cause fluorescence. The curves show minor changes as well as an intensity variation, especially around the band 420-440 cm^{-1} (signifying secondary structure changes), depending on the power.

Conclusion

Specially constructed sample cells for optics measurements of protein folding can in principle give insight into the structural changes of proteins occurring in the folding/denaturation processes. These cells can be constructed specifically to combine the variation of thermodynamical parameters, e.g.,

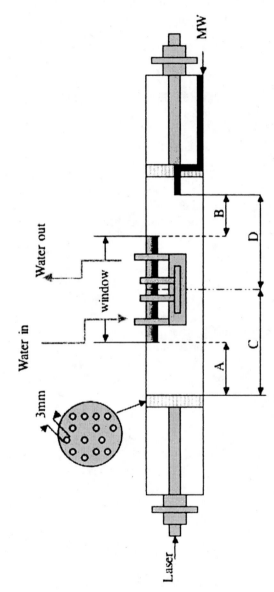

Figure 3. A microwave cell constructed to be employed in variable applied microwave field, in a frequency range of 1-4 GHz. This can be achieved by varying the cavity (A+C and B+D) of the tube.

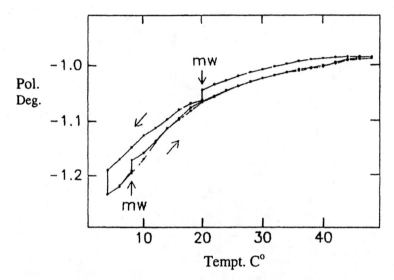

Figure 4. Shows a heating cooling cycle of Betalactoglobulin in a urea solution (see text). The y-axis represents the polarization of the light going through a thermal cell, and the x-axis the temperature in Celsius. The dashed line corresponds to a heating-cooling cycle with no microwave irradiation, while the solid line is with microwave (12).

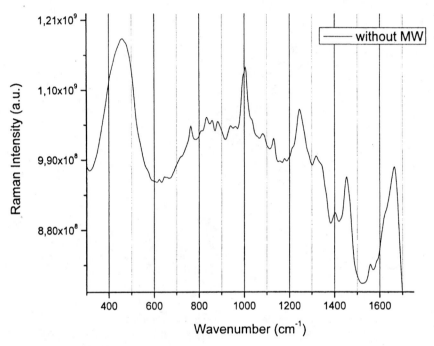

Figure 5a. Raman spectrum of Betalactoglobulin, measured by a. BioTool ChiralRaman instrument, utilizing 532 nm laser source and a CCD camera.

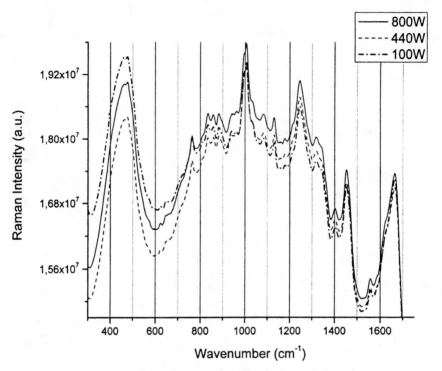

Figure 5b. Raman spectra of Betalactoglobulin recorded after exposure to microwave radiation of: 100, 400 and 800 W for 40 seconds.

temperature and pressure. The valuable information obtained from these cells can be improved when combined to *in situ* optical spectroscopy of a non-invasive nature, such as ORD and Raman scattering.

Acknowledgment

We would like to acknowledge the Danish National Research Foundation (DG) for supporting this work.

References

1. L.D. Barron, L. Hecht, E.V. Blanch and A.F. Bell, Progress in Biophysics & Molecular Biology **2000**, *73*, 1-49

2. K. Kneipp, H. Kneipp, S. Abdali, R.W. Berg and H. Bohr, Spectroscopy **2004**, *18*, 433-440

3. S. Abdali, Proc. ICORS XIXth., Brisbane, Australia **2004**.

4. S. Abdali, Jour. Raman Spec., online May **2006**.

5. J. Bohr, H. Bohr and S. Brunak, Biophys. Chem. **1997**, *63*, 97-105

6. H. Bohr, Mathematical and computer modeling, **2000**, *31*, 1-9

7. T.E. Creighton, *Proteins*, 2nd edition, Freeman, New York, NY, 1993

8. C. Scharnagl, M. Reif and J. Friedrich, Biochemica et biophysica Acta, **2005**, *1749*, 187-213

9. H. Bohr and J. Bohr, Phys. Rev. E, **2000**, *61*, 410-414

10. D. Baker, K.S. Shiau and D.A. Aggard, Current Biol. **1993**, *5*, 966.

11. T. Emilsson, M. Grubele and H. Bohr, Proposal for a pressure jump experiment, Carlsberg Funding, **1996**

12. H. Bohr and J. Bohr, *Chemical topology*, Editors. D. Bonchev and D. H. Rouvray, Gordon and Breach **2000**, 239-280.

Chapter 23

Prediction of Bovine Cartilage Proteoglycan Content Using Energy Dispersive X-ray Analysis or Optical Absorbance and a Multivariate Techniques–Fourier Transform Infrared Microspectroscopy Model

J. C. Bowden[1], L. Rintoul[1], T. Bostrom[1], J. M. Pope[1], and E. Wentrup-Byrne[1,2,*]

Schools of [1]Chemical and Physical Sciences and [2]Tissue Repair and Regeneration, Institute of Health and Biomedical Innovation, Queensland University of Technology, GPO Box 2434 Brisbane, Queensland 4001, Australia

Two analytical techniques: optical absorbance of Safranin-O stained cartilage sections and energy dispersive X-ray analysis have been used to construct partial least squares models from Fourier transform infra red spectral data. These models can be used to semi-quantitatively predict proteoglycan content in cartilage sections. Valuable information about the spatial distribution of the constituents in native, degraded or even engineered cartilage, can be obtained when these are used in conjunction with infra red imaging techniques.

As part of our larger study on the changes in articular cartilage associated with the development of osteoarthritis (OA), a suite of analytical and spectroscopic techniques are proving useful. Our long-term goal is to develop a spectroscopic-based diagnostic technique for the monitoring and analysis of the structural and biochemical changes occurring during the course of OA disease. A better understanding of the inter-relationships between different factors affecting cartilage function is essential for this goal to be realised. For example, diffusion tensor magnetic resonance imaging (MRI) has been used to observe differences in the magnitude and anisotropy of water diffusion in cartilage (1). Data from a combination of multivariate and spectroscopic techniques, polarized Fourier transform infra-red microspectroscopy (FTIRM) and polarized light microscopy (PLM), are being systematically compared and correlated with MRI data from the same cartilage samples to ascertain changes in the collagen fibers (2). In addition, this multi-techniques approach has made it possible to develop an easily accessible method combining multivariate statistical analysis and FTIRM data to accurately predict the proteoglycan (PG) content using data obtained from energy dispersive X-ray microanalysis (EDX) and optical absorbance data from Safranin-O stained specimens (2).

Articular cartilage consists of an extracellular matrix comprised mainly of a three-dimensional hydrated network of collagen fibres in which are embedded chondrocytes and PGs. Proteoglycan is the name given to a broad range of macromolecular, heterogeneous compounds formed by combining polymeric sugars called glycosaminoglycans (GAGs) (Figure 1) with a protein (3). Damage to the collagen network has been postulated as a key initiating event in the development of OA and many studies concentrate on examining changes to this collagen matrix (4) since they ultimately lead to changes in the mechanical behaviour and ability of the cartilage tissue to function. Proteoglycans, (Figure 1) as the other major component of the intercellular cartilage matrix, are responsible for the resilience and elasticity of the tissue (5-6). Changes in the PG content in diseased cartilage have been associated with concomitant structural changes in the collagen fibrils (7-8). Hence, quantification of the PG content in cartilage samples should prove useful in monitoring not only molecular changes but also changes in the collagen architecture during the cartilage degradation process.

Because of the lack of availability of cartilage samples possessing a range of degradation states for *in vitro* studies, it is common practice to use enzymes to create models of degraded articular cartilage. The intention is that the model samples produced resemble cartilage with early-stage OA including loss of PGs and changes in the collagen fibres (9-10). In theory, the degree of artificial degradation can be controlled by judicious choice of enzymes and conditions. In practice, however, there are some issues with respect to the consistency of the samples produced (*10*). The suite of samples analysed in this study consisted of "normal" bovine articular cartilage as well as a series of trypsin-degraded samples.

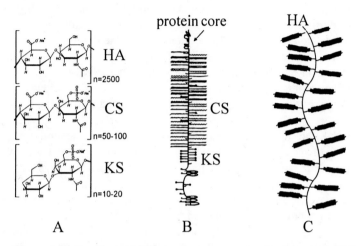

Figure 1.A. Cartilage GAG sugar monomer: hyaluronic acid (HA), chondroitin sulphate (CS), keratin sulphate (KS). B. Aggrecan proteoglycan monomer showing protein core and areas of GAG attachment. C. Aggrecan monomers bound to HA.

The primary aggregating proteoglycan found in human cartilage is aggrecan. Aggrecan contains about 85% w/w sulphated GAG molecules (Figure 1) mainly chondroitin sulphate (CS) and keratin sulphate (KS). The negatively charged sulphate groups allow specific binding of positively charged dyes such as Safranin-O in a quantitative manner. Safranin-O is commonly used as a quantitative stain since it was found to have consistent spectral characteristics at various concentrations. However, one of the long-recognised dangers of using this technique is the presence of metachromasia (*11*) which can affect the quantification of the stain. Metachromasia is the term given to the shifting of the absorbance maximum (usually) to shorter wavelengths through dye-dye interactions and generally occurs when a dye is at higher concentrations. The resulting separate populations of dye molecules will affect quantification due to loss of the dye molecule whose extinction coefficient was of interest. Thus, it is essential that a consistent staining and mounting methodology of the cartilage sections to minimise metachromasia is used if Safranin-O is to be used qualitatively in conjunction with digital photomicroscopy (*11*). This allows for relative amounts of the stain to be observed, along with its distribution within a cartilage section. Quantification between cartilage sections requires a more rigorous procedure such as the use of microspectrophotometry or absorbance measurements. Optical absorbance measurements using absorbance at a single wavelength allow for accurate determination of stain concentration, providing the stain does not show metachromasia. Although microspectrophotometry allows for the complete spectral characteristics of the dye to be measured and has been shown to accurately quantify amounts of Safranin-O within different cartilage sections (32), it does require specialised equipment such as a monochromator or a high-resolution filter set. One of the advantages of using FTIR microspectroscopy is that the problems associated with metachromasia can be avoided.

Although the individual constituents of cartilage such as collagen (14-15) and PGs (16) have been extensively examined by infra red spectroscopy it was not until more recently that FTIR microscopic (FTIRM) imaging was used to study intact cartilage (17). Because the spectra of the individual constituents displayed considerable overlap the authors used mixtures of collagen and PGs to determine quantitative values. They emphasise the potential of FTIRM imaging techniques in detecting changes in cartilage - more specifically in collagen and PGs – in OA. In addition, the fact that various IR techniques can be coupled with fibre optic probes is of great clinical interest (18).

In recent years FTIRM in conjunction with multivariate analysis has been shown to be a powerful tool in the study of various animal and human tissues including cartilage (19). Although IR techniques lack the specificity of immunohistochemical techniques, they do allow for the estimation of component ratios and the spatial distribution of the components of complex tissues such as cartilage, at different stages of the disease process. This in turn may prove extremely valuable in establishing the aetiology of OA.

In order that the full clinical potential of these IR techniques be realised, the strengths and limitations of the combined techniques approach must be examined in order to establish the reliability of the quantitative data generated. In this paper we outline our results using the combined tool of IR and multivariate techniques to develop a model to predict PG content in bovine cartilage sections by EDX and absorbance photo-microscopy.

Experimental

Methods: Sample Preparation

Nine "normal" intact bovine knee patellae from animals aged 18-30 months were harvested from a local abattoir within 24h of slaughter. Samples were maintained prior to analysis in Minimum Essential Medium with Earle's salts

(Sigma, M-0275) supplemented with 100 U/mL penicillin and 100 μg/mL streptomycin (Sigma, P4458), 70 μg/mL L-ascorbic acid (Sigma, G3126), and 200mM L-glutamine (Sigma, A4034). Three cartilage sections on bone were taken from each knee and prepared according to the specific protocol required for each of the experimental techniques. In six of the nine knees two sections were trypsin-treated and the third was examined as a "normal" sample. Sections from the final three knees were examined as "normal" cartilage.

Trypsin degradation: Degraded samples were prepared by enzymatic treatment with trypsin (Sigma, T4665, 1 g/L) for 14 hours on cartilage harvested as above, at 37 °C in an orbital incubator. Samples were MRI scanned (*1*) then frozen until required for FTIRM and histology.

Sample sectioning:

A rectangular piece of cartilage was isolated and freed from the soft tissue but left attached at one end to the bone. This was mounted onto a microtome stub and the bone end embedded in "optimal cutting temperature" (OCT) medium and frozen in liquid nitrogen. Care was taken that the cartilage itself did not come into contact with the OCT. Serial 15μm thick slices were microtomed at -14° C. Slices were placed on tin oxide-coated IR reflective transparent slides (MirrIR, Kevley Technologies, OH USA) and stored in a constant humidity environment. Edges of the cut sections were fixed with wax to reduce lifting and moving of samples during analysis.

Safranin-O samples for absorbance measurements: After cryomicrotoming the FTIR samples were fixed in 95% ethanol. Samples were then stained using a standard procedure with 0.1% w/v Safranin-O. (*10*) Post-staining images of the same areas were processed to obtain absorbance images.

EDX analysis: The tissue sections were coated with a thin layer of carbon in a high vacuum carbon evaporator and mounted on the stage of a scanning electron microscope (SEM) for elemental analysis.

Instrumentation

FTIRM: A Nicolet Continuμm microscope, with an MCT/A detector, attached to a Nexus 870 spectrometer was used in line scan mode. The analysis aperture was set at 75x150 μm. Spectra were taken in 75 μm steps from the bone to the articular surface: 16scans at a resolution of 4cm^{-1} were ratioed with a 128 scan background using Happ-Genzel apodization. The 650-4000cm^{-1} range was used.

EDX: EDX spectra were acquired using an EDAX microanalysis system fitted to an FEI Quanta 200 SEM. Line scans perpendicular to the articular surface were made using a defocussed beam to minimise electron beam-induced sample damage and to average each area of analysis. The sample working distance was 10.0 mm and the focal distance 7.0mm. The scans approximately corresponded to those made using FTIRM. Net element peak counts at each point on the scan were collected at an accelerating voltage of 20kV, using beam spot size 5.0, 60 s dwell time per analysis point, and an amplifier time constant of 17 ms.

Absorbance measurements: A Nikon Labophot-pol microscope using the 5x objective and fitted with a blue 1W LED light source (λ = 480 nm,) λ_{fwhm} = 25 nm, LumiLED, USA) and 10 bit digital camera (SV Micro, Sound Vision, Framingham, MA) was used. Images were taken both before and after staining. These images were then processed (ImageJ v1.34s, NIH USA) according to our published protocol (*10*).

Data Analysis

Multivariate techniques: Partial Least Squares (PLS) regression was carried out on collected data using 'The Unscrambler' multivariate analysis package (v7.5, 1999, CAMO ASA). Spectral data was mean centred and the 800-1550 cm^{-1} region analysed. Sample sets: Optical absorbance, 15 sections, 265 spectra; EDX S/C model, 15 sections, 264 spectra.

Results and Discussion

Some of the advantages of using modern IR techniques such as mapping and imaging - as opposed to classic histological and bulk chemical analysis methods – are: their minimally destructive capabilities, the possibility of visualising the spatial distribution of the principal constituents, the clinically important option of coupling with fibre optic probes plus the fact that when used in combination with other techniques it becomes possible to obtain semi-quantitative data. Spencer et al (*19*) determined the relative concentrations of chondroitin sulphate and collagen in tissue sections using multivariate least-squares analysis on their spectral sets on the assumption that each spectrum was a linear combination of these two constituents. They found that the CS to collagen ratios were lower than their biochemical results obtained previously *(20)*. They attributed this to the non-specificity of the FTIR technique to different types of proteins and suggest that by including other minor matrix constituents it may be possible to better differentiate and quantify them.

In our approach we have applied PLS techniques to IR data in conjunction with EDX and optical absorbance data to develop a model which makes it possible to predict sulphur content, ie PG content, in both native and trypsin-digested cartilage. We are currently extending the use of this model to IR imaging data. Figure 2 shows the infra red spectra of representative (A) untreated (UT) and (B) trypsin-treated (TT) cartilage specimens. As can be seen from the spectra, spectral differences are subtle even when PG content is depleted as a result of trypsin treatment. In tissues such as cartilage, there is

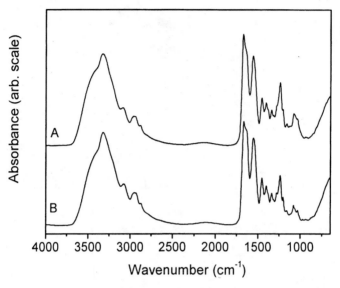

Figure 2. Typical infrared spectra of A: untreated cartilage;
B: trypsin-treated cartilage.

significant spectral overlap of the constituents of interest, even in the native state, which is why more complex approaches are needed to extract semi-quantitative data from the spectra.

Figure 3 shows results for the PLS-predicted vs measured optical absorbance data for the series of untreated cartilage samples. As a visual aid the perfect 1 to 1 calibration is shown as a solid line. From the data it can be seen that the model reasonably predicts absorbance values up to about 2.5 absorbance units but at higher values the model underpredicts. This, we believe, could be attributed to the fact that at higher concentrations the Safranin-O stain may be subject to metachromasia-type artifacts. Calibration gave an R^2 of 0.87 and a Standard Error of Calibration (SEC) of 0.35 using eleven PCs.

Figure 4 shows the wavenumber dependence of the coefficient weightings, sometimes referred to as the correlation spectrum (B), from the PLS calibration. There is good agreement between the peaks in the reference CS spectrum (A) and the regions of high weighting in the correlation spectrum. The CS spectrum

384

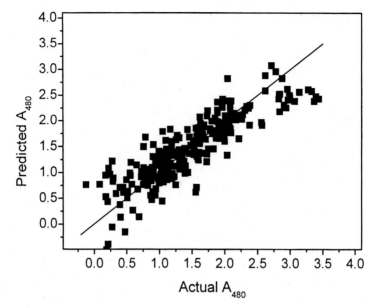

Figure 3. Predicted vs Actual Optical Absorbance from the PLS model of infrared spectra and optical absorbance data for untreated cartilage

shows bands due to the sulphate stretch area at 1245 cm^{-1} and strong absorption due to sugar C-O-C, C-OH and C-C ring vibrations at 1125-920 cm^{-1} (*17,19*). Thus, we conclude that Safranin-O optical absorbance measurements are indeed a good measure of GAG and hence PG content and that infra red spectra from any cartilage specimen can be used in conjunction with the PLS model to predict the PG content.

Figure 5 shows optical absorbance data from a cartilage section, not included in the calibration set, as a function of distance from the articular surface to the bone. Also shown is the PLS-predicted PG content calculated from the infra red data. There is good agreement except for the under prediction in the higher absorbance region.

Although EDX has previously proven useful in the analyses of cartilage (*21,22*) to the best of our knowledge it has not been used in conjunction with either infra red or multivariate techniques. Figure 6 shows results for the PLS-predicted vs measured EDX sulphur to carbon intensity ratios (S/C) for the same untreated cartilage samples. Ratios to carbon were used to minimise variability in peak intensities due to potential variations in beam current and specimen thickness. Using twelve PCs this calibration gave an R^2 of 0.78 and a SEC of 0.013. Figure 7 shows the EDX S/C profile from the same cartilage section as in Fig. 5. Also shown is the PLS-predicted S/C ratio derived from the infra red

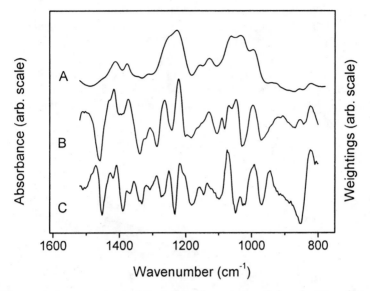

Figure 4. A comparison of the infrared spectrum of chondroitin sulfate (A, lhs y axis); the coefficient weightings spectrum of the PLS calibration of optical absorbance (A_{480}) (B, rhs y axis) the coefficient weightings spectrum of the PLS calibration of the C/S ratio (C, rhs y axis).

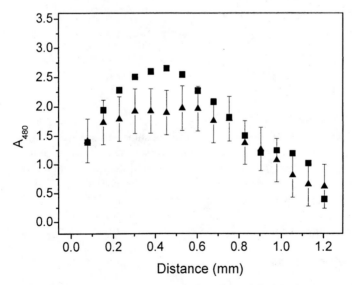

Figure 5. Independent validation comparison of optical absorbance transect of Safranin-O stained cartilage section from articular surface (origin) to bone ■ measured; ▲ PLS predicted.

data. The profiles show the same general trend with the model data somewhat high. During EDX sample preparation a small section of tissue folded and this is reflected in three values (↑) and hence the EDX data from these three samples must be disregarded.

The coefficient weightings for the S/C model are shown in Figure 4B. Again, the areas of high weighting correspond to the peaks in the CS reference spectrum although not as well as the optical absorbance weightings. Both the

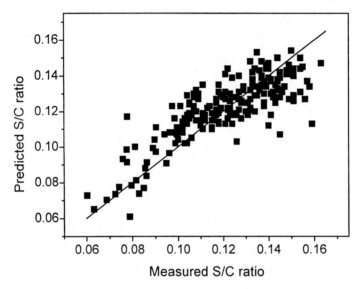

Figure 6. Predicted vs Actual EDX derived S/C ratio from the PLS model of infrared spectra and optical absorbance data for untreated cartilage.

optical absorbance and the S/C model regression coefficients show similar features indicating that they are related to the same or similar constituents. This finding supports our choice of techniques selected for their relevance and sensitivity, however, inspection of the R^2 and SECs reveals that the S/C calibration is not as accurate as that for the optical absorbance.

Some of the potential difficulties encountered in EDX are possible electron beam-induced mass loss of some constituents in the tissue, lack of consistency in sample coatings (though the carbon coating is usually too thin to significantly affect carbon measurements), and a mismatch in sampling areas between techniques. Further the sulphur and carbon measurements are of total element content, and do not take into account the different compounds in which these elements are present. Although the EDX model correctly predicts S content trends, it is difficult to relate this quantitatively to the actual CS content. Optical absorbance, on the other hand, opens up the possibility for quantitative CS determination using the molar extinction coefficient of Safranin-O (*13*). Another advantage is its high resolution which allows it to be matched more readily with FTIRM imaging techniques.

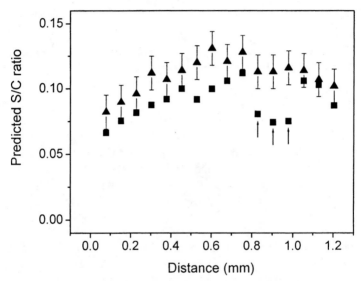

Figure 7. Independent validation comparison of EDX S/C ratio from a transect of 'normal' cartilage section from bone to articular surface ■ measured; ▲ PLS predicted.

These analyses were extended to include the trypsin treated samples. Figure 8 shows sample variability between UT and TT sections for the two techniques: (A) optical absorbance and (B) EDX. The TT sections show a reduction to almost zero in optical absorbance but with EDX, although reduced, there is still a variable sulphur content due the residual sulfur in other protein constituents. PLS model predictions for the TT samples showed similar trends but with significant zero offset errors (data not presented). Inclusion of TT treated samples in the calibration data set improved the prediction of these low-S samples but at the cost of increased SEC overall.

388

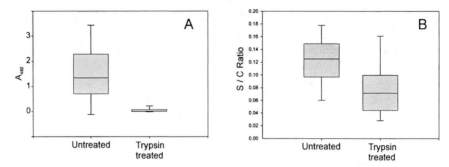

Figure 8. Comparison of UT to TT cartilage. A. Optical absorbance data.
B. EDX S/C data. Maximum value upper quartile; median; lower quartile
and lowest values for all points analyzed.

Conclusions

Infra red spectroscopy and PLS have been used in conjunction with optical absorbance and EDX data to develop models which can be ultimately used to predict PG content in cartilage sections. This opens up the possibility of using such a model for the monitoring and analysis of the structural and biochemical changes occurring during the course of OA disease. These methods can be easily extended to IR imaging to provide high spatial resolution distribution data without the need for constituent reference spectra. Both our PLS models could be further improved, by not only increasing the number of samples in the calibration set, but also by investigating the effect of including sections of different thicknesses. The wealth of information in the IR spectra of complex biological materials such as cartilage is slowly being unraveled by studies such as these.

Acknowledgements

We thank QUT for a Strategic Collaborative Research Grant to support this research. Thanks are also due to other project collaborators: Professor R Crawford, Dr R Meder, H. Moody, Professor K Oloyede and S.K. de Visser. In addition, particular thanks are due to S.K. de Visser for initial sample preparation. The Vibrational Spectroscopy Group and the Faculty of Science at QUT were also generous in their financial support of JB. EWB thanks the TRR program (IHBI, QUT) for conference travel support.

References

1. Meder, R.; de Visser, S.K., Bowden, J.C., Bostrom, T.; Pope, J. M. *Osteoarthritis and Cartilage* **2006**, *14*, 875-881.
2. Bowden, J. C.; de Visser, S. K.; Wentrup-Byrne, E.; Rintoul, L.; Bostrom, T.; Meder, R.; Pope, J. M. *Multivariate Exploration of Spectroscopic Data from Bovine Cartilage.* Proceedings of ACOVS-6, Sydney, **2005**.
3. Goldring, M. B.; In *Primer on the Rheumatic Diseases;* Klippel, J. H.; Weyand, C. M.; Wortmann, R. L.; Eds.; The Arthritis Foundation: Atlanta, GA, **1997**; p14-18.
4. Brandt, K. D.; Doherty, M.; Lohmander, L. S. *Osteoarthritis;* NY, Oxford University Press, **1998**.
5. Kempson, G. E.; Spivey, C. J.; Swanson, S. A. V.; Freeman, M. A. R. *J.Biomech.* **1971**, *4*, 597-609.
6. Scott, J. E. *Philos. Trans .R. Soc. B.* **1975**, *271*, 235-242.
7. Carney, S. L.; Billingham, M. E. J.; Muir, H.; Sandy, J. D. *J.Orthop.Res.* **1984**, *2*, 201-206.
8. Buckwalter, J. A.; Mankin, H. J. *J. Bone Joint Surgery.* **1997**, *79A (4)*, 612-633.
9. Rieppo, J.; Toyras. J.; Nieminen, M. T. *Cells Tissues Organs,* **2003**, *175*, 121-132.
10. Moody, H. R.; Brown, C. P.; Bowden, J. C.; Crawford. R. W.; McElwain, D. L. S.; Oloyede, A. O.; *J. Anat.* **2006**, *209*, 259-267.
11. Rosenberg, L. *J. Bone & Joint Surg.* **1971**, *53A, (1)* 69-82.
12. Martin, I.; Obradovic, B.; Freed, L. E.; Vunjak-Novakovic, G. *Ann. Biomed. Eng.,* **1999**, *27*, 5, 656-662.
13. Kiviranta, I.; Jurvelin, J.; Tammi, M.; Saamanen, A.-M.; Helminen, H. J. **1985**, *Histochemistry, 82*, 249-255.
14. Lazarev, Y. A.; Grishkovsky, T. C.; Khromova, T. B. *Biopolymers,* **1985**, *24*, 1449-1478.
15. Fraser, R. D. B.; MacRae, T. P. In Conformation in Fibrous Proteins and Related Synthetic Polypeptides.; Horecker, B.; Kaplan, N. O.; Marmur, J.; Scheraga, H. A.; Eds. Academic Press, NY, **1973**, p344-402.
16. Bychkov, S. M.; Bogatov, V. N.; Kuz'mina, S. A. *Biull. Eksp. Biol. Med.* **1981**, *92*, 302-305.
17. Camacho, N. P.; West, P.; Torzilli, P. A.; Mendelsohn, R. *Biopolymers,* **2000**, *62*, 1-8.
18. West, P.; Bostrom. M. P. G.; Torzilli, P. A.; Camacho, N. P. *Appl. Spectrosc.* **2004**, *58, (4)* 376-381.

19. Potter, K.; Kidder, L.H.; Levin, I. W.; Lewis, E. N.; Spencer, R. G. *Arthritis & Rheumatism,* **2001**, *44, (4),* 846-855.
20. Potter, K.; Spencer, R. G.; McFarland, E.; W. *Biochem.Biophys. Acta.* **1997**, *1334,* 129-139.
21. Wroblewski, J.; Makower, A-M, *Scanning Microscopy,* **1988**, *2,(2),* 1103-1111.
22. Middleton, J. F. S.; Hunt, S.; Oates, K. *Cell Tissue Res.,* **1988**, *253,* 469-475.

Chapter 24

The Effect of Synthetic Conditions on the Free Volume of Poly(2-hydroxyethyl methacrylate) as Studied by ^{1}H NMR, ^{129}Xe NMR, and Position Annihilation Spectroscopy

Kylie M. Varcoe[1,2,3], Idriss Blakey[1,2], Traian V. Chirila[1,4,5], Anita J. Hill[6,7], and Andrew K. Whittaker[1,2,*]

[1]Australian Institute for Bioengineering and Nanotechnology,
The University of Queensland, St. Lucia, Queensland 4072, Australia
[2]Centre for Magnetic Resonance, The University of Queensland, St. Lucia,
Queensland 4072, Australia
[3]School of Molecular and Microbial Sciences, The University
of Queensland, St. Lucia, Queensland 4072, Australia
[4]Queensland Eye Institute, South Brisbane, Queensland 4101, Australia
[5]School of Physical and Chemical Sciences, Queensland University
of Technology, Brisbane, Queensland 4001, Australia
[6]CSIRO Manufacturing and Infrastructure Technology, Clayton,
Victoria 3168, Australia
[7]School of Chemistry, Monash University, Clayton, Victoria 3800, Australia

The porous structure of PHEMA hydrogels prepared under a range of conditions has been examined using three distinct probes. Positron annihilation lifetime spectroscopy and ^{129}Xe NMR are sensitive to pores in the range 0.1-10 nm, while ^{1}H NMR of water within the hydrogel provides information on a range of sizes up to several microns. PHEMA samples were prepared in solution with from 5-30 wt. % of water in the polymerization mixture. Below 30 wt. % the water exists in nanometer-sized pores, and above this a substantial proportion reside in micron sized pores. The PALS and ^{129}Xe NMR results support the existence of relatively hydrophobic domains of sub-nanometer size.

Free volume in materials was first introduced as an intuitive theoretical notion by Batschinski (*1*). As atoms are spherical, it is understandable that at any temperature, 100 % of the volume will not be occupied by the atoms. Free volume (V_f) was therefore defined as the regions within a material that are unoccupied by the molecules. These regions are of a dynamic nature, with segmental motions of the polymer chains causing local fluctuations in the free volume (*2-5*). Therefore, the size of the free volume pores vary in both time and space. Free volume is a characteristic of amorphous or semi-crystalline polymers that plays an important role in explaining many macroscopically-observable properties such as ion conductivity, viscous behavior, the glass transition temperature, gas permeation and barrier properties, and many mechanical properties. Thus, the ability to measure free volume size and distribution, is of interest to aid in the design and tailoring of materials for specific applications.

Hydrogels have been extensively used in medicine as biomaterials due to their bio- and blood-compatibility (*6*). Poly(2-hydroxyethyl methacrylate) (PHEMA) is one such biomaterial and has been used in many applications including drug delivery systems, contact lenses and intraocular implants. PHEMA hydrogel biomaterials are open to many different types of penetrants that may diffuse into the material through the free volume voids. For many applications, diffusion of particular molecules into the material is often desirable. For example, the diffusion of oxygen through contact lenses is a necessary property for clinical success of these devices. However, in other cases, the diffusion of certain molecules is undesirable, such as the diffusion of calcium solutes into intraocular lenses, which can cause precipitation of calcium phases rendering the lenses opaque. The transport of these penetrants occurs through the accessible space, identifiable with the free volume, so a measure of this space is expected to provide information useful for the tailoring of these materials to better suit an application.

A variety of methods, both direct and indirect, have been developed for the measurement of free volume. Most simply the change in free volume can be measured from the macroscopic change in volume of a polymer on a change in temperature. Small angle X-ray and neutron diffraction are methods used to directly measure free volume by the determination of static density fluctuations in the materials (*7*). The rates of photo-isomerization of photochromic dyes imbedded within a polymeric material has been shown to be highly dependent upon the available free volume (*3, 8*). The techniques used in this study,[129]Xe NMR and positron annihilation lifetime spectroscopy (PALS) have recently been successfully applied to the measurement of pore sizes in nanoporous solids and more recently in polymers.

PALS measures the rate of annihilation of positrons injected into a material from a radioactive [22]Na source. The positron lifetime is inversely proportional to the overlap of the positron density and the electron density of the surrounding medium. The positron within the polymer matrix will pick up an electron from its surroundings to form a positronium atom in one of two spin states, namely *ortho* (*o*-Ps) (triplet state) or *para* (*p*-Ps) (singlet state) (*9*). The lifetime of the *p*-Ps is very short (order of 125 ps) with annihilation occurring before interaction with the polymer free volume (*10*). The lifetime of the *o*-Ps is much longer (140 ns in vacuum, shortened to a few ns in condensed matter) and is directly related to the probability of collision with an electron and therefore, depends on the size of the pores in which they are confined. The annihilation rate (= $1/\tau_3$) depends on the probability of the overlap of the *o*-Ps wave function with the wave functions of the surrounding electrons, which is dependent on the cavity size. Positroniums have a diameter of 1.06 Å and therefore PALS is particularly sensitive to free volume pores in the angstrom size range that persist for times on the scale of 10^{-10} s and longer (*11*). This time scale is sufficient to allow time for the annihilation of the ortho-positronium (a few nanoseconds).

^{129}Xe NMR is a powerful technique that allows the measurement of the size of pores within material in the angstrom to nanometre size range (*9, 12, 13*). ^{129}Xe atoms have a highly-symmetrical electron cloud which is readily distorted, and the ^{129}Xe chemical shift is therefore extremely sensitive to the surrounding environment. Thus xenon NMR can be used as an inert probe of the angstrom-sized free volume within a material. Any change in the free volume will result in changes to the chemical shift. The ^{129}Xe NMR chemical shift is usually represented by the sum of the various interactions experienced by the xenon atoms, as indicated in equation (1) (*14*).

$$\delta = \delta_0 + \delta_{wall} + \delta_{Xe} + \delta_E + \delta_M \qquad (1)$$

where δ_0 is the shift due to gaseous xenon and is generally used as the reference shift (0 ppm), δ_{wall} is the shift due to the interactions between the xenon atoms and the pore wall, δ_{Xe} arises from collisions between xenon atoms and δ_E and δ_M correspond to the shift due to the local electric and magnetic fields, respectively. These last two terms, δ_E and δ_M, can be ignored for polymers that do not contain charged groups (*15*). In some materials the value of δ_{Xe} is pressure dependent, and hence can be accounted for by extrapolation of the observed chemical shift to zero pressure, although in polymeric systems at low xenon pressures the term has been shown to be negligible (*15*). This leaves δ_{wall}, a parameter that is characteristic of the region of the material in which the xenon atoms reside. As the pore size decreases, and the xenon environment become less like that of a free gas, the ^{129}Xe chemical shift will move further from the reference peak at 0 ppm.

The ^{1}H NMR relaxation times of water within the porous structure of materials can be used in many cases to determine properties of the material including the average pore size. A detailed examination of the application of ^{1}H NMR relaxometry to the study of hydrogels has been provided by McBrierty *et al.* (*16*) who discuss the complementary approaches to analyzing the decay of the magnetization of water protons within hydrogels including PHEMA. One method of analysis is to describe the relaxation decays in terms of a sum of discrete populations of water molecules. This approach was adopted by Ghi *et al.* (*17*) in their study of bulk-polymerized PHEMA and copolymers of HEMA with hydrophilic monomers. The total water content within PHEMA could be divided into three types, namely bound, intermediate and free, in order of decreasing probability of interactions between the water molecules and the polymer chains. It is possible to estimate from these measurements the average pore size experienced by the populations of water molecules relaxing with the longest relaxation times.

As with most hydrogels, PHEMA possesses a network of pores, the size and connectivity of which determine the transport properties of molecules so important for many applications of this polymer. It is well known that the structure of PHEMA hydrogels is sensitive to the polymerization conditions and in particular the presence of additives or co-solvents during reaction of the monomer. While numerous studies of the pore structure of PHEMA polymerized from HEMA in the bulk, or from HEMA in aqueous solution have been conducted, a number of questions remain. As will be apparent from the brief review below, there remains uncertainty of the details of the pore stucture of PHEMA. Also it is unclear whether within the so-called homogeneous PHEMA hydrogels there exist relatively hydrophobic domains separated from the hydrated porous network. The experiments reported here address these issues.

PHEMA polymerized from bulk monomer is a dense glassy polymer which when swollen in water to equilibrium contains approximately 40 % of its final weight as water. It was demonstrated over 40 years ago (*18*) that this water content is remarkably insensitive to the presence of additives such as water and ethylene glycol in the polymerization mixture. Hodge and coworkers (*19*) have examined the free volume in dry bulk-polymerized PHEMA and copolymers with 2-ethoxyethyl methacrylate (EEMA) using PALS. The free volume in the absence of water varied in a regular manner from 0.25 nm for PHEMA to 0.31 nm for PEEMA, consistent with a change in the proximity of the glass transition temperature to the measurement temperature. Recent ^{1}H NMR measurements on bulk-polymerized PHEMA swollen to equilibrium in water demonstrated a complex pore structure with water residing within two main populations of pores having approximate radii of 10 and 500 nm (*17*).

Polymers prepared in the presence of water, or other additives, are likely to have more complex structures. The literature refers to either "homogeneous" or "heterogeneous" hydrogels depending on whether the materials are optically transparent or not. The structure of heterogeneous hydrogels, formed in the presence of around 40 wt. % of water in the polymerization mixture, results from phase separation or precipitation during the polymerization reaction as the quality of the solvent decreases on depletion of the monomer. These materials have been studied in detail principally by microscopic techniques, and have been revealed to consist of polymer particles up to approximately five µm in diameter fused together to form a porous network with channels up to approximately 80 µm in diameter. The effects of polymerization conditions on the porous structure have been examined in detail (20-27).

The structure of homogeneous PHEMA, that is materials polymerized in solutions with less than approximately 45 wt. % water in the reaction mixture (28), has been examined in greater detail. The earliest estimate of pore size in these PHEMA materials was 0.4 nm for a polymer prepared in the presence of water and ethylene glycol was provided by Refojo (29) who used the relationship between water permeability and average pore diameter developed by Ferry (30). The materials investigated by Refojo contained 39 wt. % water. This method was later applied by Haldon and Lee (31) to similar PHEMA samples of 41-42 wt. % hydration to obtain a pore radius between 0.4 and 0.8 nm. It was acknowledged that the assumptions implicit in the use of the Ferry equation resulted in underestimation of the pore size. This was highlighted by the observation that sodium fluorescein, with a radius of 0.55 nm, could readily diffuse into PHEMA. Later Kou et al. (32) demonstrated that solutes of radius 0.6 nm were able to penetrate PHEMA and copolymers of HEMA with methacrylic acid, and that the rate of diffusion was consistent with free volume theory.

Refojo and Leong (33) used optical observations and measurement of the diffusion of macromolecules of known size and shape into PHEMA to gain an estimate of the pore size. It was shown that dextrans and lysozyme penetrated PHEMA leading to the conclusion that the pore sizes must be at least equal to 3 nm.

In a series of papers Kimmich et al. (34-36) prepared interpenetrating networks of PHEMA and poly(ethylene glycol) (PEG) by UV curing of microphase-separated mixtures of the monomer with PEG. The PEG content was 20 wt. %. The authors were able to visualize the isolated channels of PEG using electron microscopy and estimate the diameter of the PEG channels using field-gradient diffusometry, and obtained an estimate of the pore radius in PHEMA of 8-10 nm.

The results described above are somewhat conflicting and demonstrate that PHEMA has a range of pore sizes, and that a porous network probably exists on a number of size scales. Of particular relevance to this study is the knowledge that the pore structure of PHEMA is a sensitive function of the method of preparation, for example the amount of water, and crosslinking monomer in the polymerization mixture, as well as the rate of polymerization determined by the initiator fraction (20-23, 25, 26). It is clear that no one single technique will provide a complete picture of such a complex morphology, and therefore in this study, we use the three complementary techniques, [129]Xe NMR, PALS and ^{1}H NMR, to examine the free volume and pore structure in PHEMA prepared under a range of different conditions. These techniques allow the study of porous structures over four orders of magnitude in size, *i.e.* from the angstrom to the micron level. In addition two of these probes are more sensitive to porous structures within relatively hydrophobic domains, and one measures directly the environment of water in pores.

Experimental Section

Synthesis of Polymers

Optical-pure HEMA monomer was supplied by Bimax, USA and used without distillation. All other chemicals were purchased from Sigma-Aldrich, USA. PHEMA specimens were cut from sheets prepared by the polymerization of solutions of respectively 70, 80, 90 and 95 wt. % HEMA in water, containing 0.5 wt. % (to HEMA) ethyleneglycol dimethacrylate as a crosslinking agent, and 0.1 wt. % (to HEMA) of a 10 % aqueous ammonium persulfate as an initiating component. Following the mixing of the above agents in a beaker, 0.1 wt. % (to HEMA) of N,N,N',N'-tetramethylethylenediamine was added as the second initiating component and the mixtures were rapidly stirred in a vortex. Each solution was then poured between glass plates lined with 3M transparency sheets and spaced at about 1.2 mm by a silicone gasket, which were placed in an oven, where they were maintained for 24 hours at 50 °C, followed by another 24 hours at 80 °C.

The samples have been given the designations 5, 10, 20 and 30 to show the water content within the polymerization mixture, and examined without further addition of water. The designations 5H, 10H, 20H and 30H have been given to samples subsequently hydrated to equilibrium.

Sample Preparation

High purity, sterile and non-pyrogenic water, with osmolality zero (Viaflex, Baxter) was used in all experiments. The hydrated samples were equilibrated in distilled water for several weeks, removed from the water just prior to measurement, the surface was dried with a paper tissue and the sample placed in a sealed tube and then inserted into the NMR resonator. The samples that had not been equilibrated in water were doubly heat-sealed in poly(ethylene) bags immediately after synthesis until just prior to measurement, when they were quickly removed and placed into sealed tubes and then into the NMR resonator.

¹H NMR Measurements

The proton spectra were acquired on a Bruker MSL300 spectrometer using a 90° pulse time of 5 μs and a recycle delay of 10 sec. A spectral width of 125 khz was used and a total of 4096 data points were collected with 16 free induction decays (FIDs) being co-added to improve the signal-to-noise ratio. The ¹H T₂ relaxation times were measured using the Carr-Purcell-Meiboom-Gill (CPMG) pulse sequence with a 90° pulse time of 5 μs, a 180° pulse of 10 μs and a value of τ equal to 100 μs. A total of 4096 points in the T₂ decay were collected. A repetition time of 6 sec was used and 16 scans were co-added to improve the signal-to-noise ratio. The experimental relaxation decays were fitted to a sum of exponential decays using the Marquardt-Levenberg minimization scheme, as described in our previous work (17).

¹²⁹Xe NMR Measurements

For the measurement of each sample, approximately 1 g of sample in pieces of approximately 1 mm in diameter was placed into a specially-constructed glass cell. The cell was connected to the gas line and evacuated to approximately 0.01 MPa then filled with 99.98 % xenon gas at natural abundance (^{129}Xe ~ 26.44 %) to a pressure of 1 MPa. The system was then allowed to come to equilibrium over 24 hours. The ^{129}Xe NMR measurements were performed at room temperature on a Bruker MSL300 spectrometer with a ^{129}Xe resonance of 82.974 MHz. The signals were acquired with a 41 ms acquisition time, a recycle delay of 10 s and a π/2 pulse time of 13 μs. Typically, on the order of 15 000 scans were required to achieve a good signal-to-noise ratio. During processing of the data a line broadening factor of 100 Hz was used. The chemical shift of the xenon dissolved within the polymer was determined by reference to the signal of free xenon gas in the cell (0 ppm).

PALS Measurements

PALS measurements were performed using an automated EG&G Ortec fast–fast coincidence system with a resolution of 240 ps equipped with a Peltier temperature control system. Two disc shaped samples (diameter 10 mm and ~1.1 mm thick) were placed on either side of a ^{22}NaCl source. The ^{22}NaCl source was a 2 mm diameter 25 μCi spot sandwiched between two sandwiched between two waterproof Mylar films. Data were analysed using the PFPOSFIT program. At least five spectra of 30,000 peak counts were collected, with each spectrum taking approximately 1.5 h to collect. No source correction was used. Three lifetimes were fitted with the shortest fixed to 125 ps, characteristic of p-Ps self-annihilation.

Results and Discussion

^{1}H NMR T_2 measurements

NMR relaxometry is widely used to determine the size of pores in which water resides in porous media. Fundamentally the rate of relaxation of the NMR magnetization depends on the probability of the water molecules encountering a pore wall, or a macromolecule in polymeric hydrogels, and thus provides a measure of pore size. A typical ^{1}H T_2 relaxation decay curve obtained for sample 30H is displayed in Figure 1. The decay of the spin-spin relaxation for all of the hydrogel samples was found to be best fitted to the sum of three exponential functions. The validity of this method of analysis was confirmed by analysis of the data with a non-negative least squares data fitting program based on a Laplace transform algorithm (37). This unconstrained fit of the data showed that three distinct relaxation populations were present in all data sets. Furthermore, the observation of three distinct populations is consistent with results of analysis of bulk-polymerized PHEMA hydrogels reported previously (17). Note that the relaxation times of the polymeric protons are short compared with the relaxation times of the water protons, and therefore the decay curves are due to relaxation of the signal of the water only. The observation of three distinct T_2 relaxation times has previously been interpreted to be due to the presence of different types of water within the structure of the PHEMA samples (17). These three types of water are often referred to as "bound", "intermediate" and "bulk" water, in order of decreasing probability of the water molecules interacting with the polymer (16). Molecules that are free to rapidly move and tumble, i.e. the bulk water, will tend to have long relaxation times, whereas less mobile species, i.e. the bound or intermediate water, will experience stronger dipole-dipole interactions with the hydrogel and hence undergo more rapid spin-spin relaxation.

The relaxation of water molecules within a hydrogel matrix occurs through dipole-dipole coupling with the solid or semi-solid polymer chains. The rate of relaxation therefore depends on the probability of exchange of magnetization through dipole-dipole coupling and hence on the proximity of the water molecules to the polymer chains. Zimmerman and Brittin (38) were the first to examine the effects of exchange and exchange rate on the relaxation times of water molecules in contact with a gel. These authors suggested that within the population of mobile spins, i.e. protons on water molecules, the observed rate of relaxation depends on the rate of exchange between spins strongly interacting

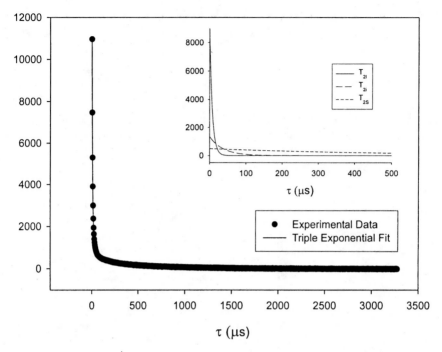

Figure 1. Typical $^1H\ T_2$ decay curve of PHEMA sample 30H with the insert showing the decay of the three components of T_2 calculated from the three-exponential fit. The three components are labelled T_{2s}, T_{2i} and T_{2l} in increasing order.

with the gel protons and spins resident on water molecules unaffected by the presence of the gel. In the "two phase fast-exchange" model the rate of exchange between these two populations is assumed to be rapid compared with the intrinsic relaxation time of the bulk water. Of course there may be a continuum of "states" within any population of water, however, in the limit of fast exchange the two-phase model has great utility in describing the average relaxation properties of the spins (*39*). The model allows the calculation of the pore size from the T_2 relaxation times (*40, 41*). In this analysis the pore can be considered to consist of a region bounded by a layer of thickness λ at the surface of the pore. This layer has been reported to be in the order of one or two monolayers. If the pore is large compared to λ, then the proportion of the volume influenced by the surface, p_s, is given by:

$$p_s = \lambda \frac{s}{v} \tag{2}$$

where s is the surface area and v is the volume of the region. The observed relaxation time within the pore is therefore described by:

$$\frac{1}{T_2} = \frac{p_s}{T_{2\,surface}} + \frac{p_b}{T_{2b}} \tag{3}$$

given that in the other phase, for which $p_b = 1 - p_s$, bulk relaxation occurs. $T_{2surface}$ is the relaxation time of proton spins with the "surface-affected" layer and T_{2b} denotes the relaxation time of the water protons within the pore and is characteristic of "free" or "bulk" water.

For pores having cylindrical geometry, the observed relaxation time is thus given by:

$$\frac{1}{T_2(r)} = \frac{1}{T_{2b}} + \left(\frac{2\lambda}{T_{2\,surface}} \right)\frac{1}{r} \tag{4}$$

where r is the radius of the cylindrical pore.

The T_2 values obtained for each sample are listed in Table 1, along with the relative fractions of each component of T_2, the calculated pore sizes, and the water content (WC), determined from integration of the broad-line proton NMR spectra.

The hydrated samples all had an equilibrium water content within the range of 38-40 wt. %. This was found to be in agreement with the values obtained by both Hodge et al. (19) and McConville et al. (42), who obtained values of 35 wt. % and 38-39 wt. % respectively. These results indicate that the total amount of water absorbed by the hydrogels at equilibrium does not depend on the initial water content in the polymerization mixture. However, the initial water content does have a significant effect on the types or water and degree of interaction that occurs between the water molecules and the polymer chains, as is evidenced by the variation in the T_2 data in Table 1.

The short component of the T_2 relaxation decay has been assigned to "bound" water and is due to water having relatively restricted mobility interacting strongly with the polymer chains. The value of T_{2s} remained approximately constant for the samples hydrated to EWC and the samples obtained straight from synthesis, although a small increase in T_{2s} on hydration to equilibrium (5H-30H) is probably due to an expansion of the network on addition swelling. For samples prepared with less than 30 wt. % water present in the polymerization mixture during synthesis (samples 5, 20), the majority (>75 %) of the water existed as bound water. The proportion of this type of water for these samples, expressed as a weight fraction of the polymer samples, ranged from 11-17 wt. %. This was found to be in good agreement with Sung et al. (43) and Lee et al. (44), who found using NMR and DSC, respectively, that up to

Table 1. The value and fraction of each component of ^1H NMR T_2 based on a three-exponential fit of the experimental data (the three components are labeled s = short, i = intermediate, l = long).

Sample	WC (%)	T_{2s}	T_{2i} (ms)	T_{2l}	% of T_{2s}	% of T_{2i}	% of T_{2l}	Pore Size (T_{2s}) (μm)	Pore Size (T_{2i})	Pore Size (T_{2l})
5	12.0	1	20	174	90	4	6	0.08	1.63	16.9
20	18.0	2	17	148	95	3	2	0.16	1.36	13.9
30	24.3	2	9	161	9	90	1	0.16	0.72	15.4
5H	38.6	4	54	142	5	5	89	0.32	4.32	13.2
10H	39.0	2	18	116	6	2	91	0.16	1.44	10.5
20H	38.2	2	14	84	6	3	90	0.16	1.12	7.3
30H	40.2	6	20	155	62	28	6	0.48	1.60	14.7

approximately 20 wt. % of the total water exists as bound water, for samples prepared in a similar manner.

In a previous study of the properties of water in bulk-PHEMA the intermediate T_2 values have been interpreted as being due to water molecules residing within micro-pores within the polymer matrix (*17*). For samples with less than 20 wt. % water present during synthesis, it was observed that relatively little intermediate water was observed within the polymer matrix. This observation is consistent with a three-step absorption process suggested by Sung *et al.* (*43*), in which the water initially binds to the hydrophilic sites within the polymer matrix and that once these sites are filled intermediate water begins to fill the micro- pores. The results suggest that the hydrophilic sites within PHEMA are able to accommodate between 20-30 wt. % of water before the porous structure starts to fill. The values of the relaxation times of the water molecules within these pores increase on hydration to equilibrium, again indicative of a further expansion of the pore diameters. The values of the pore diameters were calculated using the observed values of T_2 and the "two-phase fast exchange" model assuming a surface relaxation time, $T_{2surface}$, of 100 μs (*45*), and a spin-spin relaxation time of water, T_{2b}, of 1.0 s at 307 K. The thickness of the surface affected layer, λ, was assumed to be 4 nm (*41*).

The results presented here indicate that there is an important change in the pore structure on going from 20 to 30 wt. % of water in the polymerization mixture. Previously it has been suggested that the critical water content leading to heterogeneous PHEMA is closer to 45 % (Chirila et al. and references therein) (*28*) on consideration that the equilibrium water content of PHEMA hydrogels is related to the polymer-solvent interaction parameter and hence solvent quality.

While this may indeed be the case, the results indicating that a lower water content may be more appropriate as the critical content. The ^{1}H NMR results presented here and reported previously (*17*) show clear evidence of additional fractions of water molecules residing in larger micron-sized pores, which are interacting less strongly with the PHEMA chains.

The long component of T_2 is due to highly mobile water molecules, referred to as free or bulk water. These molecules have been proposed to exist within the larger pores of the polymer as well as within cracks and defects in the structure (*17*). The proportion of water protons decaying with the long relaxation time, T_{2l} was small for the samples measured straight from synthesis (series 5-30), *i.e.* less than 1 wt. % of the polymer sample. For the hydrated samples however, this population became dominant with greater than 89 % of the water present within the structure existing as free or bulk water. The large amount of water absorbed in these hydrated samples causes the hydrogel to swell, resulting in swelling stresses and hence the formation of macro-pores and perhaps swelling-induced cracks (*17*). The exception to this is sample 30H, that is the PHEMA prepared with 30 wt. % water and subsequently equilibrated to equilibrium in an excess of water. In this material the majority of water exists as bound water (62 %), a relatively large amount of intermediate water (28 %) in comparison to the other hydrated samples (2-5 %) and only a small amount as bulk water (6 %). This could be due to the larger amount of water in the initial synthesis resulting in a larger proportion of micro-pores. Thus any further absorbed water obtained after synthesis would then occupy these pores rather than creating larger pores and cracks, as occurred with the other hydrated samples. This suggests that the water may be acting as a porogen during the PHEMA synthesis.

The radii of the pores containing the water giving rise to the long component of T_2 could be calculated using equations (2)-(4), and were found to range from $7 - 15$ μm. For the samples studied straight from synthesis there was no observable relationship between pore size and water content in the synthetic mixture, indicating that the water content does not affect the size of these macro-pores. It is also observed that the pore radius for the samples hydrated to equilibration does not differ significantly from the samples prior to the second hydration step. This implies that only the number of these larger pores or cracks increases with hydration and not the size of the features.

Recently Ghi and coworkers have examined in detail the ^{1}H T_2 relaxation times of water protons in fully-hydrated PHEMA and HEMA copolymers (*17*). The polymers were polymerized in the absence of water and crosslinking agent, and subsequently hydrated to equilibrium in pure water. The authors also found three populations of water molecules within the PHEMA hydrogel with values of $T_{2s} = 0.11$ ms (35 %), $T_{2i} = 6$ ms (63 %) and $T_{2l} = 44$ ms (1 %). These values correspond to respective pore radii of approximately 9 nm, 480 nm and 3.5 μm. The first of these pore radii is in good agreement with previous values reported for bulk polymerized PHEMA discussed above. However, the ^{1}H NMR measurements appear to be identifying additional larger pore structures not accessible by simple measurement of transport properties previously used. In addition the results show that for bulk-polymerized PHEMA the average pore sizes are significantly smaller than in the PHEMA samples examined in this study, and that the presence of water in the polymerization mixture results in a larger proportion of larger pores at full hydration.

[129]Xe NMR Measurements

As mentioned above the [129]Xe NMR chemical shift can be used to calculate the average diameter of the pores in a material using the equations developed by Demarquay and Fraissard (46). Equation (5) describes the relationship between the [129]Xe chemical shift (δ_{wall}) and the mean free path length (*l*) of the xenon atoms within the confined space of the pores:

$$\delta_{wall} = \frac{\delta_a . a}{a + l} \tag{5}$$

Here *a* and δ_a are constants dependent on the material type. Values applicable to polymeric materials have been determined by comparison of the xenon chemical shift with the pore sizes determined using PALS results (9), and are a = -0.00217 nm and δ_a = -8640 ppm for spherical pores and a = -0.205 nm and δ_a = -151 ppm for cylindrical pores. The pore diameter can then be found using $l = D_C - D_{Xe}$ for cylindrical pores and $l = (D_S - D_{Xe})/2$ for spherical pores, where D_{Xe} is the van der Waals diameter of xenon (0.44 nm) and D_C and D_S are the diameters of cylindrical and spherical pores, respectively.

Due to the low solubility of xenon in water (Ostwald solubility of 0.11 (47)) it is likely that the overall solubility of xenon in the hydrogel will be low, and that the xenon will tend to partition into regions of the hydrogel of low water content, for example hydrophobic polymer domains comprised of aggregated portions of the polymer backbones (48). Therefore, long acquisition times were required to obtain reasonable NMR spectra, with 15 000 scans generally being needed to clearly distinguish the peaks. A typical [129]Xe NMR spectrum obtained for one of the hydrated PHEMA samples is shown in Figure 2. The chemical shifts of xenon within the hydrogel and the pore sizes calculated as described above are given in Table 2. Spectra could not be obtained for fully-dehydrated samples, due to the very low solubility of xenon gas in a polymer well below the glass transition temperature (Tg ~ 383 K). Note that the width of the peak assigned to xenon gas within the polymer is of the order of 3 ppm at 82.3 MHz. This contrasts with typical linewidths in spectra of xenon within glassy polymers,

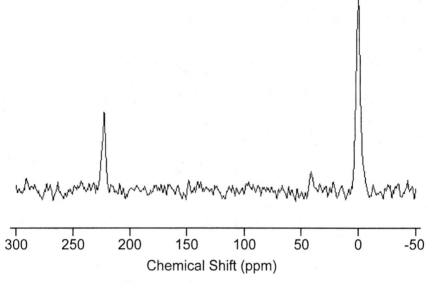

300   250   200   150   100   50   0   -50

Chemical Shift (ppm)

Figure 2. Typical ^{129}Xe NMR spectrum obtained for PHEMA sample 30.

for example poly(methyl methacrylate) of the order of 60 ppm, due to a distribution of pore sizes accessed by the xenon atoms (*49*). The relatively narrow linewidth in the hydrated polymer is consistent with plasticization of the polymer chains by the presence of water.

Table 2. ^{129}Xe chemical shifts, oPs lifetimes and calculated pore diameters for selected PHEMA samples.

Sample	Chemical Shift	D_c (Å)	D_s (Å)	PALS, τ_3 (ns)	D_{PALS} (nm)
5	243	0.772	0.599	1.694	0.511
30	221	0.785	0.614	2.065	0.583
5H	217	0.788	0.617	2.039	0.578
30H	215	0.789	0.619	2.037	0.578

Xenon dissolved in water is known to have a ^{129}Xe chemical shift of 196 ppm (*50*). Due to the lack of a ^{129}Xe signal at or close to this position, we can conclude that in this system the solubility in water is too low to result in an observable signal and the NMR signal that is observed is due to xenon in weakly- or non-hydrated domains within the hydrogel. Kariyo and coworkers

(48) have reported a similar result in their ^{129}Xe and ^{1}H NMR study of poly(N-isopropylacrylamide) across the lower critical solution temperature. Above the LCST the authors report a single peak in the ^{129}Xe NMR spectrum at 215 ppm, and only a weak peak due to xenon in water. On the other hand, below the LCST, when the polymer is solvated by the water molecules, two separate peaks are assigned to interaction of the xenon with the polymer chains in either relatively hydrophilic or hydrophobic environments.

As can be seen from Table 2 the observed chemical shift showed a shift towards the reference peak at 0 ppm as the water content was increased, indicating an increase in the pore size. For the samples analyzed prior to swelling to equilibrium (samples 5 and 30), a decrease in chemical shift of approximately 20 ppm was observed as the proportion of water was increased from 5 to 30 wt. %. This corresponds to an increase in pore size of 0.13 or 0.15 Å, depending upon the assumed pore geometry. Such an increase (51) is consistent with this material passing from a glassy to a rubbery state on the increase in water content fro 5 to 30 wt. %.

The chemical shift of the xenon in the two samples hydrated to equilibrium (5H and 30H) are different by just 2 ppm. This corresponds to a pore size difference of only 0.01 – 0.02 Å. It is seen therefore, that the initial water content does not significantly affect the pore size of the swollen samples and that the samples appear to swell to approximately the same water content.

The samples which contained 5 wt. % water in the polymerization mixture show an increase in pore size of 0.16 – 0.18 Å on hydration to equilibrium (samples 5 and 5H), while the sample prepared with 30 wt. % water shows an increase of only 0.04 – 0.05 Å on hydration to equilibrium.

PALS measurements

As mentioned above the *ortho*-positronium becomes localized in free volume cavities and measures the free volume indirectly through sensitivity to the local electron density. The lifetime of the o-Ps is related to the average free volume of the material, with longer lifetimes corresponding to larger free volume elements. The PALS spectra obtained for the hydrogel samples in this studied were analyzed to provide a single o-Ps lifetime, τ_3, the values of which are listed in Table 2. If an o-Ps is located at the core of a spherical void, an average free volume element radius, r, can be estimated from the oPs lifetime, τ_3, as shown in equation 6:

$$\frac{1}{\tau} = 2\left[1 - \left(\frac{r}{r_0}\right) + \frac{1}{2\pi}\sin\left(\frac{2\pi r}{r_0}\right)\right] \qquad (6)$$

Here r_0 is equal to $r+\Delta r$, with the best value of $\Delta r = 1.66$ Å determined empirically (*52*). The uncertainties in the absolute values of radius calculated using this equation are relatively large, however it is generally considered that useful relative values of pore size can be obtained on analysis of a series of materials. The values of diameter, D_{PALS}, calculated for the samples studied here are listed in Table 2. It is evident from Table 2 that the PALS technique is sensitive to similar pore sizes as [129]Xe NMR, and that the values obtained are broadly consistent with those obtained by NMR, considering the assumptions used in calculation of pore diameters for both techniques. The results indicate that the sample 5 has substantially smaller free volume cavities than both sample 30 and the same sample hydrated to equilibrium. The value of the lifetime for sample 5 is very close to the value of 1.7 ns reported previously by this group for bulk PHEMA (*19*).

Measurements were also made on PHEMA prepared in the bulk in the absence of water in the polymerization mixture. The values of τ_3 and D_{PALS} obtained for this material were 1.606 ns and 4.924 Å. This indicates that the addition of 5 wt. % water in the polymerization mixture leads to an approximately 4 % increase in diameter of the free volume cavity, and that substantially higher increases are seen on addition of further water.

Conclusions

The results presented here show that the PHEMA hydrogels prepared in the presence of water contain a porous network on a number of length scales and with varying structure. On the angstrom level, substantial free volume has been identified by both [129]Xe NMR and positron lifetime annihilation spectroscopy. It is likely that the free volume cavities detected by these two techniques exists within relatively hydrophobic, i.e. non-hydrated domains. On the micron level and larger a network of water-filled pores was identified by ^{1}H NMR relaxation time measurements. It is this porous network that is responsible for the transport properties of PHEMA confirmed in numerous previous studies. The results are consistent with previous NMR studies of bulk-polymerized PHEMA.

The ^{1}H NMR T_2 relaxation decay of water in PHEMA hydrogels was best described by a three exponential function for all of the samples studied. The three components have been assigned bound water, an intermediate component, and water residing within larger pores and cracks, in order of increasing T_2 relaxation time. The relative and absolute intensities of the three components indicate that the amount of water in the polymerization mixture determines the types of water existing within the structure but not the amount of water absorbed during hydration. Up to ~20 wt. % water content in the polymerization mixture, the water exists mainly as bound water. As the water content in the polymerization mixture is increased, intermediate and bulk water begin to be formed.

The ^{129}Xe NMR and PALS results show that these methods measure pores on the angstrom scale. With increasing water content in the polymerization mixture there is an increase in pore size, and hydration to equilibrium further increases the size of these pores. Both of these methods use probes that partition to the relatively-hydrophobic domains of the hydrogels.

In summary the complex structure of PHEMA can only be described by a combination of techniques which probe a variety of length scales. Previous estimations of pore sizes based on measurement of the penetration of various sized solutes are limited to determining the minimum pore size; as has been demonstrated here a substantial fraction of the population of water molecules does reside in larger pores, and the proportion of the respective populations depends on synthetic conditions.

Acknowledgments

The authors would like to thank the Australian Research Council for funding under the Discovery grants scheme (grant DP0208223). We thank the Queensland State Government for providing IB with a Smart State Fellowship, and to Sylvia and Charles Viertel Charitable Foundation (Queensland) for an unrestricted grant to TVC.

References

1. Batschinski, A. J. Zeitschrift fur Physikalische Chemie **1913**, *84*, 643-706.
2. Hill, A. J.; Freeman, B. D.; Jaffe, M.; Merkel, T. C.; Pinnau, I. Journal of Molecular Structure **2005**, *739*, 173-178.
3. Vallee, R. A. L.; Cotlet, M.; Van der Auweraer, M.; Hofkens, J.; Muellen, K.; De Schryver, F. C. Journal of the American Chemical Society **2004**, *126*, 2296-2297.
4. Cohen, M. H.; Turnbull, D. Journal of Chemical Physics **1959**, *31*, 1164-1169.
5. Cohen, M. H.; Grest, G. S. Physical Review B: Condensed Matter and Materials Physics **1979**, *20*, 1077-1098.
6. Hoffman, A. S. Advanced Drug Delivery Reviews **2002**, *54*, 3-12.
7. Song, H. H.; Roe, R. J. Macromolecules **1987**, *20*, 2723-2732.
8. Victor, J. G.; Torkelson, J. M. Macromolecules **1988**, *21*, 3490-3497.
9. Nagasaka, B.; Eguchi, T.; Nakayama, H.; Nakamura, N.; Ito, Y. Radiation Physics and Chemistry **2000**, *58*, 581-585.
10. Shantarovich, V. P.; Azamatova, Z. K.; Novikov, Y. A.; Yampolskii, Y. P. Macromolecules **1998**, *31*, 3963-3966.
11. Liu, J.; Deng, Q.; Jean, Y. C. Macromolecules **1993**, *26*, 7149-55.
12. Suzuki, T.; Yoshimizu, H.; Tsujita, Y. Desalination **2002**, *148*, 359-361.

13. Chen, F.; Chen, C.-L.; Ding, S.; Yue, Y.; Ye, C.; Deng, F. Chemical Physics Letters **2004**, *383*, 309-313.
14. Ito, T.; Fraissard, J. Journal of Chemical Physics **1982**, *76*, 5225-5229.
15. Miller, J. B.; Walton, J. H.; Roland, C. M. Macromolecules **1993**, *26*, 5602-5610.
16. McBrierty, V. J.; Martin, S. J.; Karasz, F. E. Journal of Molecular Liquids **1999**, *80*, 179-205.
17. Ghi, P. Y.; Hill, D. J. T.; Whittaker, A. K. Biomacromolecules **2002**, *3*, 991-997.
18. Refojo, M. F.; Yasuda, H. Journal of Applied Polymer Science **1965**, *9*, 2425-2435.
19. Hodge, R. M.; Simon, G. P.; Whittaker, M. R.; Hill, D. J. T.; Whittaker, A. K. Journal of Polymer Science, Part B: Polymer Physics **1998**, *36*, 463-472.
20. Lou, X.; Dalton, P. D.; Chirila, T. V. Journal of Materials Science: Materials in Medicine **2000**, *11*, 319-325.
21. Chen, Y. C.; Chirila, T. V.; Russo, A. V. Materials Forum **1993**, *17*, 57-65.
22. Chirila, T. V.; Chen, Y. C.; Griffin, B. J.; Constable, I. J. Polymer International **1993**, *32*, 221-232.
23. Chirila, T. V.; Higgins, B.; Dalton, P. D. Cellular Polymers **1998**, *17*, 141-162.
24. Chirila, T. V.; Lou, X.; Vijayasekaran, S.; Ziegelaar, B. W.; Hong, Y.; Clayton, A. B. International. Journal of Polymeric Materials **1998**, *40*, 97-104.
25. Clayton, A. B.; Chirila, T. V.; Lou, X. Polymer International **1997**, *44*, 201-207.
26. Clayton, A. B.; Chirila, T. V.; Dalton, P. D. Polymer International **1997**, *42*, 45-56.
27. Lou, X.; Chirila, T. V.; Clayton, A. B. International. Journal of Polymeric Materials **1997**, *37*, 1-14.
28. Chirila, T. V.; Constable, I. J.; Crawford, G. J.; Vijayasekaran, S.; Thompson, D. E.; Chen, Y. C.; Fletcher, W. A.; Griffin, B. J. Biomaterials **1993**, *14*, 26-38.
29. Refojo, M. F. Journal of Applied Polymer Science **1965**, *9*, 3417-3426.
30. Ferry, J. D. Chemical Reviews **1936**, *18*, 373-455.
31. Haldon, R. A.; Lee, B. E. British Polymer Journal **1972**, *4*, 491-501.
32. Kou, J. H.; Amidon, G. L.; Lee, P. I. Pharmaceutical Research **1988**, *5*, 592-597.
33. Refojo, M. F.; Leong, F.-L. Journal of Polymer Science, Polymer Symposia **1979**, *66*, 227-237.
34. Fischer, E.; Kimmich, R.; Beginn, U.; Moller, M.; Fatkullin, N. Physical Review E: Statistical Physics, Plasmas, Fluids, and Related Interdisciplinary Topics **1999**, *59*, 4079-4084.

35. Fischer, E.; Beginn, U.; Fatkullin, N.; Kimmich, R. Magnetic Resonance Imaging **2005**, *23*, 379-381.
36. Kimmich, R.; Seitter, R.-O.; Beginn, U.; Moller, M.; Fatkullin, N. Chemical Physics Letters **1999**, *307*, 147-152.
37. Provencher, S. W. Journal of Chemical Physics **1976**, 64, 2772.
38. Zimmerman, J. R.; Brittin, W. E. Journal of Physical Chemistry **1957**, *61*, 1328-1333.
39. Brownstein, K. R.; Tarr, C. E. Journal of Magnetic Resonance **1977**, *26*, 17-24.
40. Brownstein, K. R.; Tarr, C. E. Physical Review A **1979**, *19*, 2446-2453.
41. Woessner, D. E. Journal of Magnetic Resonance **1980**, *39*, 297-308.
42. McConville, P.; Pope, J. M. Polymer **2000**, *41*, 9081-9088.
43. Sung, Y. K.; Gregonis, D. E.; John, M. S.; Andrade, J. D. Journal of Applied Polymer Science **1981**, *26*, 3719-3728.
44. Lee, H. B.; Jhon, M. S.; Andrade, J. D. Journal of Colloid and Interface Science **1975**, *51*, 225-231.
45. Araujo, C. D.; MacKay, A. L.; Whittall, K. P.; Hailey, J. R. T. Journal of Magnetic Resonance, B **1993**, *101*, 248-261.
46. Demarquay, J.; Fraissard, J. Chemical Physics Letters **1987**, *136*, 314-318.
47. Clever, H. L. Krypton, Xenon and Radon-Gas Solubilities; Pergamon Press: Oxford, **1979**; Vol. Vol. 2.
48. Kariyo, S.; Kuppers, M.; Badiger Manohan, V.; Prabhakar, A.; Jagadeesh, B.; Stapf, S.; Blumich, B. Magnetic Resonance Imaging **2005**, *23*, 249-253.
49. Schantz, S.; Veeman, W. S. Journal of Polymer Science, Part B: Polymer Physics **1997**, *35*, 2681-2688.
50. Cherubini, A.; Bifone, A. Progress in Nuclear Magnetic Resonance Spectroscopy **2003**, *42*, 1-30.
51. Cheung, T. T. P.; Chu, P. J. Journal of Physical Chemistry **1992**, *96*, 9551-9554.
52. Eldrup, M.; Lightbody, D.; Sherwood, J. N. Chemical Physics **1981**, *63*, 51-58.

Chapter 25

Spectroscopic Investigations into Inactivation of Bacterial Virulence Factors

K. Brandenburg[1], J. Howe[1], M. Rössle[2], and J. Andrä[1]

[1]Forschungszentrum Borstel, Leibniz-Zentrum für M edizin und Biowissenschaften, Biophysik, D-23845 Borstel, Germany
[2]European Molecular Biology Laboratory, c/o DESY, D-22603 Hamburg, Germany

Infectious diseases are still one of the leading causes of death worldwide. For an effective therapeutic strategy, the corresponding bacterial virulence factors must be identified and methods to neutralize them must be developed. One important example of such bacterial factors is lipopolysaccharide (LPS), belonging to the most potent classes of triggers of mammal immune systems. A new and promising therapeutical approach to controlling bacterial virulence factors is the design and synthesis of suitable antimicrobial peptides (AMP), based on the LPS-binding domain of natural defense proteins. These have the potential not only to kill bacteria but to bind to and deactivate the virulence factors as well. A combination of spectroscopic methods (infrared, X-ray diffraction, fluorescence resonance energy transfer) used for the analysis of the inactivation mechanism of bacterial LPS is presented here.

Despite the availability of antibiotics, infectious diseases are an increasing threat for human health worldwide. This is largely due to the development of resistances, partially caused by the misuse of antibiotics in agriculture and animal husbandry. In Gram-negative bacteria, the main virulence (pathogenicity) factor responsible for a severe outcome of the infection is endotoxin (lipopolysaccharide, LPS) a constituent of the bacterial outer membrane (*1*). LPS consists of a lipid moiety, called lipid A, which anchors it in the membrane, and an oligo- or polysaccharide side chain (*2*). Since lipid A itself exerts all the LPS-typical biological effects, it is called the 'endotoxic principle' of LPS. Lipid A consists of a diglucosamine backbone substituted with two phosphate groups and up to seven acyl chains in amide- and ester-linkages. LPS may act beneficially in mammals by inducing cell mediators such as interleukins and tumor-necrosis-factor α, at high concentrations, however, the excessive production of cytokines leads to severe health problems such as the septic shock syndrom, which still cannot be treated effectively, and is a major cause of death in critical care stations (some 200,000 death cases in the U.S. annually) (*3*),. The action of LPS on human cells takes place after its removal from the outer bacterial membrane, and an effective therapeutic treatment would thus afford the neutralization of LPS. One approach is the application of synthetic peptides derived from LPS-binding sequences of defense proteins in humans and animals. For this, we have synthesized peptides derived from the amino acid sequence of *Limulus*-anti-LPS factor (LALF) from the horseshoe crab (*4*). In order to gain a greater understanding of LPS neutralization, biophysical and physico-chemical techniques were applied. We have established various spectroscopic techniques such as Fourier-transfer infrared (FT-IR), synchrotron radiation small-angle X-ray scattering (SAXS), and fluorescence resonance energy transfer (FRET) spectroscopy and other physical and biological (cytokine induction in human mononuclear cells) assays to fully characterize the LPS:peptide interaction. In this way, a profound characterization of the interaction processes (aggregate structure, molecular conformation, secondary structures, membrane interaction) and the molecular requirements for effective neutralization was possible. This should allow - in an iterative process – the establishment of a peptide library, which should eventually lead to highly active anti-septic drugs with negligible side effects.

Materials and Methods

Lipids and peptide

Free lipid A was isolated by acetate buffer treatment of lipopolysaccharide from *Salmonella minnesota* strain R595. After isolation, the resulting lipid A was purified and converted to its triethylamine salt (*5*). Results of all the standard assays performed on lipid A (analysis of glucosamine and total and organic phosphate content, as well as the distribution of the fatty acid residues) were in good agreement with the chemical properties expected for lipid A from LPS R595, the molecular structure of which has already been solved (*6*). The peptide Pep_{LALF} with a sequence of GCKPTFRRLKWKYKGKFWCG was synthesized with an amidated C-terminus by the solid-phase peptide synthesis technique on an automatic peptide synthesizer (model 433 A; Applied Biosystems) on the standard Fmoc-amide resin according to the fastmoc synthesis protocol of the manufacturer. The N-terminal Fmoc-group was removed from the peptide-resin and the peptide was deprotected and cleaved with 90% TFA, 5% anisole, 2% thioanisole, 3% dithiothreitol for 3 h at room temperature. After cleavage the suspension was filtered and the soluble peptide was precipitated with ice-cold diethylether followed by centrifugation and extensive washing with ether. HPLC purification was carried out on a RP-HPLC using an Aqua-C18 column (Phenomenex), and eluted using a gradient of 0-70% acetonitrile in 0.1% trifluoroacetic acid (TFA). Cyclation via cystein residues was achieved by incubating the peptide in 10 % DMSO for 24 h at room temperature. The peptide was further purified by reverse-phase HPLC to a purity greater than 95%. Purity was determined by matrix-assisted laser-desorption-time-of-flight mass spectrometry (MALDI-TOF MS, Bruker).

FTIR spectroscopy

The infrared spectroscopic measurements were performed on an IFS-55 spectrometer (Bruker, Karlsruhe, Germany). For phase transition measurements, the lipid samples were placed between CaF_2 windows with a 12.5 μm Teflon spacer. Temperature-scans were performed automatically between 10 and 70 °C with a heating rate of 0.6 °C/min. Every 3 °C, 50 interferograms were accumulated, apodized, Fourier-transformed, and converted to absorbance spectra. As a sensitive measure of the state of order of the hydrocarbon chains, the peak position of the symmetric stretching vibration $v_s(CH_2)$ of the methylene groups was taken, which is around 2850 cm^{-1} in the gel and 2852.5 to 2853.0 cm^{-1} in the liquid crystalline phase of the acyl chains (*7*).

X-ray diffraction spectroscopy

X-ray diffraction measurements were performed at the European Molecular Biology Laboratory (EMBL) outstation at the Hamburg synchrotron radiation facility HASYLAB using the SAXS camera X33 (*8*). Diffraction patterns in the range of the scattering vector $0.1 < s < 4.5$ nm^{-1} ($s = 2 \sin \theta/\lambda$, 2θ scattering angle and λ the wavelength = 0.15 nm) were recorded at in the range 5-60 °C with exposure times of 1-2 min using an image plate detector with online readout (MAR345, MarResearch, Norderstedt/Germany). The s-axis was calibrated with Ag-Behenate which has a periodicity of 58.4 nm. The diffraction patterns were evaluated as described previously (9), assigning the spacing ratios of the main scattering maxima to defined three-dimensional structures. The lamellar and cubic structures are most relevant here. They are characterized by the following features:

(1) Lamellar: The reflections are grouped in equidistant ratios, i.e., 1, 1/2, 1/3, 1/4, etc. of the lamellar repeat distance d_l

(2) Cubic: The different space groups of these non-lamellar three-dimensional structures differ in the ratio of their spacings. The relation between reciprocal spacing $s_{hkl} = 1/d_{hkl}$ and lattice constant a is

$$s_{hkl} = [(h^2 + k^2 + l^2) / a]^{1/2}$$

(hkl = Miller indices of the corresponding set of plane).

Fluorescence resonance energy transfer spectroscopy (FRET)

The peptide-induced inhibition of the intercalation of lipid A into liposomes made from phosphatidylserine (PS) alone or mediated by lipopolysaccharide-binding protein (LBP), was determined by FRET spectroscopy applied as a probe dilution assay (4). First the peptide, then lipid A, followed by LBP (or vice versa) were added to the liposomes, which were labelled with the donor dye NBD-phosphatidylethanolamine (NBD-PE) and acceptor dye Rhodamine-PE. The final concentrations were: peptide and lipid 1 μM and LBP 0.1 μM. Intercalation was monitored as the increase of the ratio of the donor intensity I_D at 531 nm to that of the acceptor intensity I_A at 593 nm (FRET signal) with on time.

Results and Discussion

The synthetic AMP synthesized here (Pep$_{LALF}$) was derived from the LPS-binding sequence of the *Limulus polyphemus* protein *Limulus*-anti-LPS factor (LALF) (10) (sequence see above), which was connected to a cycle along the two cysteins; since it has been shown that cyclic LALF-peptides are better able to inhibit the LPS-induced cytokine secretion in human mononuclear cells than linear peptides (11). Clearly, the peptide is multiply positively charged (+7 corresponding to the number of arginins and lysins) as a prerequisite for an electrostatic interaction with negatively charged lipid A. We have taken the 'endotoxic principle' lipid A from *Salmonella minnesota* strain R595 as the bacterial virulence factor.

The ability of the peptide to inhibit the lipid A-induced cytokine production of human mononuclear cells (see (*12* in biological experiments)) was tested. It was found that at [lipid A] = 100 ng/ml the cytokine (tumor-necrosis-factor-α) production was 1150 pg/ml, which decreases in the presence of the peptide (10-fold molar excess) to 250 pg/ml, i.e., a drastic neutralization takes place. For an understanding of this ability of Pep$_{LALF}$ to neutralize lipid A, spectroscopic methods were applied to characterize the binding process.

The interaction of lipid A with Pep$_{LALF}$ was studied with (a) FTIR by monitoring the gel to liquid crystalline phase transition of the acyl chains of lipid A, (b) SAXS by analyzing the aggregate structure of lipid A, and (c) FRET by investigating the potential peptide-mediated inhibition of the intercalation of lipid A into model liposomes, induced by lipopolysaccharide binding protein (LBP).

The state of order or fluidity of the acyl chains of amphiphilic molecules is known to be a characteristic property of each membrane lipid, and was analyzed by FTIR using the peak position of $v_s(CH_2)$. In Fig. 1, the phase transition behavior of the acyl chains of lipid A is presented, as a plot of the peak position of $v_s(CH_2)$ versus temperature. It can be seen from the change in the wavenumber values, that in the presence of the peptide only a slight shift to lower wavenumbers, and a concomitant increase in the phase transition temperature T_c takes place. This corresponds to a slight rigidification of the lipid A acyl chains.

The aggregate structure of lipid A has been found to be a determinant of biological activity (*13*). The necessary inactivation of lipid A involving its conversion from a cubic or mixed unilamellar/cubic aggregate structure in pure form into a multilamellar one in the presence of binding structures, which lead to an inhibition of the biological activity, has been described (*14*). The results of the lipid A:Pep$_{LALF}$ system in the temperature range 5 to 60 °C and in Fig. 2b at 40 °C are presented in Fig. 2a. The diffraction patterns are similar at all temperatures (Fig. 2a), and reflect a multilamellar structure. At some temperatures,e.g. at 40 °C a second periodicity becomes obvious, (Fig. 2b); one periodicity lies at 6.49 nm, and another one (in italics) at 5.21 nm. From earlier investigations, it is known that the periodicity at 5.21 nm corresponds to a pure multilamellar lipid A structure, obtained for example at high Mg^{2+} concentration or low water content (*5*). The value at 6.49 nm corresponds to a multilamellar stack composed of lipid A + peptide, i.e., the peptide leads to an increase in the thickness of the water layer between neighboring stacks. This takes place, however, only in the gel phase of lipid A (< 45 °C, Fig. 1), whereas in the fluid

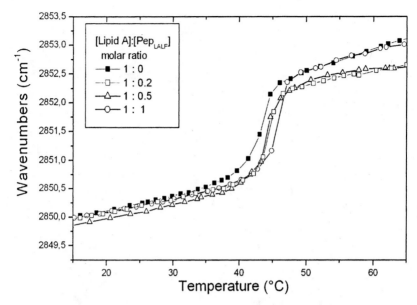

Figure 1. Gel to liquid crystalline phase transition of the acyl chains of lipid A from Salmonella minnesota-LPS in the presence of different amounts of Pep$_{LALF}$, plotted as peak position of the symmetric stretching vibrational band ν$_s$(CH$_2$) versus temperature.

phase (> 45 °C) the peak at 6.49 nm vanishes (Fig. 2a). This may be explained by a dipping of the peptide into lipid A thus reducing the length of the water layer between neighboring stacks.

The incorporation of lipid A, mediated by lipopolysaccharide-binding protein (LBP) into target cell membranes has been described as a prerequisite for endotoxin action, i.e., for cell signalling (*15*). In this study we have tested whether or not the peptide influences this intercalation. For this, FRET spectroscopy was applied by labelling liposomes made from phosphatidylserine (PS) with two fluorophores, and adding the peptide, peptide + lipid A, and LBP in different ways (Fig. 3). The addition of buffer alone at 50 s only leads to a slight decrease in the FRET signal due to dilution, and the addition of LBP at 100 s results in a small increase in the signal corresponding to incorporation of LBP as described previously (*15*). The addition of the peptide alone at 50 s causes a strong increase in the FRET signal, which indicates that the peptide alone can intercalate into PS liposomes. The addition of the peptide at 50 s followed by lipid A at 100 s causes two signal increases, first an intercalation of the peptide and then of lipid A, apparently mediated by the peptide.

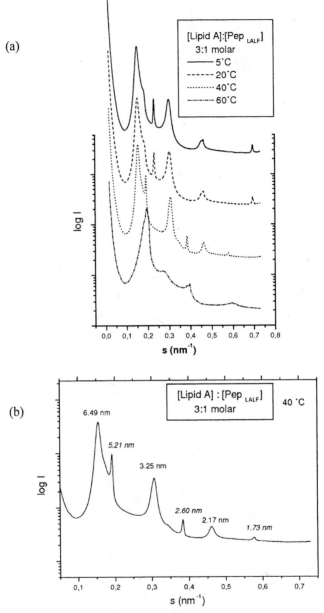

Figure 2. Small-angle X-ray diffraction pattern – logarithm of the scattering intensity versus scattering vector s - of lipid A from Salmonella minnesota-LPS in the presence of PepLALF at 90 % water content and in the temperature range 5-60 °C (a) and at 40 °C (b).

Interestingly, lipid A, without preincubation of the peptide, does not incorporate into the membrane (not shown) in accordance with previous findings (*16*). The addition of LBP at 150 s leads to another signal increase corresponding to a protein-mediated lipid A intercalation, as discussed in earlier reports (*15, 16*). Finally, the preincubation of the peptide with lipid A (Pep$_{LALF}$+ lipid A), added at 50 s, leads to a long lasting increase in the FRET signal with no further effect at 100 s when LBP was added. The same experiments were also performed with liposomes corresponding to the phospholipid mixture of macrophages (*15*). It was found that qualitatively the same effects are observed, but quantitatively to a lower extent (data not shown).

The FRET data clearly suggest that the peptide alone as well as in the presence of lipid A incorporates into liposomal membranes, and does not cause an inhibition of the lipid A incorporation. Thus, a scavenger function role of the peptide can be excluded. Rather, the structural change of the aggregate structure of lipid A seems to be the decisive process. These reoriented, lamellarized aggregates are still able to incorporate into the target membrane (Fig. 3). Therefore , the inactivation process of lipid A does not take place outside the cell. The data presented are strongly in favor of our conformational concept of endotoxicity (*17*). This assumes an incorporation of endotoxin (lipid A, LPS) molecules into the relevant mononuclear cells of the immune system. These can be induced by the binding proteins of the serum or by the membranse such as LBP and CD14. It was also found that in particular membrane-bound LBP is able to cause an intercalation of endotoxins into the membrane by disrupting part of the aggregates (*18*). In the membrane, biologically active endotoxins represent a considerable disturbance of the membrane architecture due to the conical shape of the lipid A moiety (corresponding to a non-lamellar cubic aggregate structure), and thus they may interact with signaling molecules such as TLR4 (*19*) or the MaxiK channel (*20*) leading to their conformational change with subsequent signal transduction. The addition of the peptide to lipid A leads to a lamellarization of lipid A (Fig. 2) and hence to a conformation which does not represent a steric disturbance within the target cell membrane.

Alternate interpretations of the neutralization process of endotoxins, namely that multilamellarization would lead to a strong reduction of the epitopes accessible for binding proteins such as LBP, or that the higher binding energies of the endotoxins within the aggregates would hamper the interaction with these proteins are not supported by our findings (*4*).

It should also be noted that the acyl chain fluidity is not apparently a determinant of biological activity. From Fig. 1 it becomes clear that the overall fluidity changes are only small; in particular, at 37 °C there are only marginal changes. This is in accordance with other findings that the acyl chain fluidity of endotoxins may modulate the bioactivity, but is not a determinant *per se* (*21*).

In order that this peptide or derivatives thereof can ultimately be used in therapeutic applications for infectious diseases, its non-hazardous   effect in

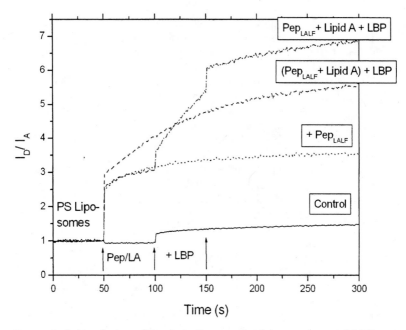

Figure 3. Intercalation of Pep$_{LALF}$, Pep$_{LALF}$-lipid A mixtures, and LBP into phospholipid liposomes made from phosphatidylserine presented as FRET signal (I$_D$/I$_A$)versus time.

human cells must be tested. Preliminary *in vitro* data show that the peptide is not cytotoxic for HeLa cells and only slightly hemolytic. It can therefore be considered a suitable candidate for use in an anti-sepsis strategy. As a next step, we will test this compound in an animal model of sepsis.

Acknowledgments

We would like to thank G. von Busse and C. Hamann by performing the FTIR and FRET measurements, respectively. This work was financially supported by the Deutsche Forschungsgemeinschaft (SFB 617, project A17) and by the Commission of the European Communities under the specific RTD program 'Quality of Life and management of Living Resources', QLK2-CT-2003-01001, 'Antimicrobial endotoxin neutralizing peptides to combat infectious diseases'.

References

1. Nikaido H. *Microbiol. Mol. Biol. Rev.* **2003**, *67*, 593
2. Rietschel E.Th.; Seydel U.; Zähringer U.; Schade U.F.; Brade L.; Loppnow H.; Feist W.; Wang M.-H.; Ulmer A.J.; Flad H.-D.; Brandenburg K.; Kirikae T.; Grimmecke D.; Holst O.; Brade H. *Infectious Disease Clinics of North America* **1991**, *5*, 753
3. Cross A.S. and Opal S.M. *Curr. Opin. Infect. Dis.* **1995**, *8*, 156
4. Andrä J.; Lamata M.; Martinez d.T.; Bartels R.; Koch M.H.; Brandenburg K. *Biochem. Pharmacol.* **2004**, *68*, 1297
5. Brandenburg K.; Koch M.H.J.; Seydel U. *J. Struct. Biol.* **1990**, *105*, 11
6. Rietschel E.T.; Brade H.; Holst O.; Brade L.; Müller-Loennies S.; Mamat U.; Zähringer U.; Beckmann F.; Seydel U.; Brandenburg K.; Ulmer A.J.; Mattern T.; Heine H.; Schletter J.; Hauschildt S.; Loppnow H.; Schönbeck U.; Flad H.-D.; Schade U.F.; Di Padova F.; Kusumoto S.; Schumann R.R. *Curr. Top. Microbiol. Immunol.* **1996**, *216*, 39
7. Casal H.L. and Mantsch H.H. *Biochim. Biophys. Acta* **1984**, *779*, 381
8. Koch M.H.J. and Bordas J. *Nucl. Instr. Meth.* **1983**, *208*, 461
9. Brandenburg K.; Richter W.; Koch M.H.J.; Meyer H.W.; Seydel U. *Chem. Phys. Lipids* **1998**, *91*, 53
10. Hoess A.; Watson S.; Siber G.R.; Liddington R. *EMBO J.* **1993**, *12*, 3351
11. Dankesreiter S.; Hoess A.; Schneider-Mergener J.; Wagner H.; Mietke T. *J. Immunol.* **2000**, *164*, 4804
12. Jürgens G.; Müller M.; Koch M.H.J.; Brandenburg K. *Eur. J. Biochem.* **2001**, *268*, 4233
13. Brandenburg K.; Mayer H.; Koch M.H.J.; Weckesser J.; Rietschel E.Th.; Seydel U. *Eur. J. Biochem.* **1993**, *218*, 555
14. Andrä J.; Koch M.H.J.; Bartels R.; Brandenburg K. *Antimicrob. Agents Chemother.* **2004**, *48*, 1593
15. Gutsmann T.; Schromm A.B.; Koch M.H.J.; Kusumoto S.; Fukase K.; Oikawa M.; Seydel U.; Brandenburg K. *Phys. Chem. Chem. Phys.* **2000**, *2*, 4521
16. Schromm A.B.; Brandenburg K.; Rietschel E.Th.; Flad H.-D.; Carroll S.F.; Seydel U. *FEBS Lett.* **1996**, *399*, 267
17. Brandenburg K.; Andrä J.; Müller M.; Koch M.H.J.; Garidel P. *Carbohydr. Res.* **2003**, *338*, 2477
18. Gutsmann T.; Mueller M.; Carroll S.F.; MacKenzie R.C.; Wiese A.; Seydel U. *Infect. Immun* **2001**, *69*, 6942
19. Beutler B. *Curr. Opin. Immunol.* **2000**, *12*, 20

20. Blunck R.; Scheel O.; Müller M.; Brandenburg K.; Seitzer U.; Seydel U. *J. Immunol.* **2001**, *166*, 1009

21. Brandenburg K.; Schromm A.B.; Koch M.H.J.; Seydel U., in *Bacterial Endotoxins: Lipopolysaccharides from Genes to Therapy*,Levin, J.; Alving, C.R.; Munford, R.; Redl, H. Eds. ; John Wiley & Sons; New York, 1995; p. 167.

Indexes

Author Index

Subject Index

426